# THE SCIENCE OF GYMNASTICS

*The Science of Gymnastics* provides the most comprehensive and accessible introduction available to the fundamental physiological, biomechanical and psychological principles underpinning performance in artistic gymnastics.

The second edition introduces three new sections: applied coaching, motor learning and injury prevention and safety, and features contributions from leading international sport scientists and gymnastics coaches and instructors. With case studies and review questions included in each chapter, the book examines every key aspect of gymnastic training and performance, including:

- physiological assessment
- diet and nutrition
- energetics
- kinetics and kinematics
- spatial orientation and motor control
- career transitions
- mental skills training and perception
- injury assessment and prevention, with clinical cases
- advanced case studies in rotations, vault approach and elastic technologies in gymnastics.

A fully dedicated website provides a complete set of lecture material, including ready-to-use animated slides related to each chapter, and the answers to all review questions in the book.

The book represents an important link between scientific theory and performance. As such, *The Science of Gymnastics* is essential reading for any student, researcher or coach with an interest in gymnastics, and useful applied reading for any student of sport science or sports coaching.

**Monèm Jemni** is Qatar Olympic Committee Chair in Sports Science, based at the College of Arts and Sciences, Qatar University, Qatar. With a history as a gymnast and international coach who contributed to the training of Olympic medal winners, he is now one of the experts within the International Gymnastics Federation's Coaching Academy. He has received outstanding academic awards from universities in France, Tunisia, the USA, the UK and Qatar, and enjoys a world-leading reputation thanks to his cutting-edge research and investigations.

# THE SCIENCE OF GYMNASTICS

## Advanced Concepts

Second Edition

*Edited by Monèm Jemni*

Routledge
Taylor & Francis Group

LONDON AND NEW YORK

Second edition published 2018
by Routledge
2 Park Square, Milton Park, Abingdon, Oxon, OX14 4RN

and by Routledge
711 Third Avenue, New York, NY 10017

*Routledge is an imprint of the Taylor & Francis Group, an informa business*

First edition published by Routledge 2011

*British Library Cataloguing-in-Publication Data*
A catalogue record for this book is available from the British Library

*Library of Congress Cataloging-in-Publication Data*
A catalog record for this book has been requested

ISBN: 978-1-138-70192-2 (hbk)
ISBN: 978-1-138-70193-9 (pbk)
ISBN: 978-1-315-20380-5 (ebk)

Typeset in Bembo
by Keystroke, Neville Lodge, Tettenhall, Wolverhampton

Visit the companion website: www.routledge.com/cw/jemni

This book is dedicated to my dad, who is in a better world. He has educated generations of leaders and inspired his children to be creative and autonomous at their early ages. You won't be forgotten as a father and as a teacher.

Monèm Jemni

For Mr. Peter Schmitz.
Life is not about how fast you run, or how high you climb, but how well you bounce. Peter has allowed me to bounce too many times to count. I am forever grateful.

Bill Sands

To my daughter Angelina, who tirelessly turns to the left and to my son Paolo, who despite his sister's instructions continues turning to the right.

Flavio Bessi

To my wife, Maria Luisa, and my daughters, Leticia and Alicia; to the Brazilian AG coaches and gymnasts who are always overcoming all the difficulties and finally to my research colleagues at GPG/UNICAMP.

Marco Antonio Coelho Bortoleto

My contribution to this book is dedicated to the coaches and gymnasts who have motivated my research and service in gymnastics. Your talent and perseverance in this amazing sport is inspirational. And to my son Alexander whom I love very much.

Liz Bradshaw

My contribution to this book is dedicated to the coaches and to all those who are passionate about gymnastics, who have left the phase of "what do I have to do" behind, and are now focused on "why do I have to do it". Additionally, to my daughter Elisabet for giving me the opportunity to be her coach.

Michel Marina

I dedicate my contributions to my little son Ghassen, hoping that he will be a great gymnast in the future.

Bessem Mkaouer

I dedicate my contribution to this book to my mom, who has inspired and supported me as my gymnastics coach. Together with my own coaching experience this has formed the basis for my research into this interesting area. I also dedicate my work in this book to all other coaches, gymnasts and researchers who continually strive to make this sport the best in the world.

Alexandra Pizzera

My contribution to this book is dedicated to my family. They have always stood beside me and supported me in each endeavor I have sought to take on. And also to my gymnastics "family", the teammates and coaches who defined the joy and fun found in gymnastics, contributing to the love I have for the sport today.

Brooke Lemmen

# CONTENTS

# CONTRIBUTORS

**Flavio Bessi, Institute of Sport and Sport Science, Freiburg, Germany**
Apart from his role as an academic, Dr Flavio Bessi was a federal coach of the male junior team in Germany and after that a national coach of the female juniors (U13). He also fulfils a few roles as expert within the International Gymnastics Federation's Coaching Academy; expert within the German Olympic Committee; and member within the German Gymnastics Federation's Scientific Board.

**Marco Antonio Coelho Bortoleto, University of Campinas, Brazil**
Marco A. C. Bortoleto is a former national-level Brazilian gymnast, coach and judge. He has been a FIG Gymnastics for All (GfA) committee member since 2012 and also Vice-president of the Brazilian Gymnastics' Federation Scientific Committee. Marco has been acting as instructor for FIG Academy Programmes since 2012. He is currently an Associated Professor who coordinates the Gymnastics Research Group (GPG) at the University of Campinas (Brazil) that investigates artistic gymnastics and GfA from the social science point of view. Marco has written and edited four books and 35 articles about gymnastics so far. His research interest is related to the analysis of the "training cultures" from an ethnographic perspective (mainly high-level athletes).

**Elizabeth J. Bradshaw, Australian Catholic University, Melbourne, Australia; Sport Performance Research Institute New Zealand; and Auckland University of Technology, Auckland, New Zealand**
Dr Elizabeth Bradshaw teaches biomechanics and sport technology at Australian Catholic University. Her research interests are in performance enhancement and injury prevention in sport, exercise and dance. She is a consultant to Gymnastics Australia, often attending national training camps at the Australian Institute of Sport in Canberra, and is a Fellow of the International Society of Biomechanics in Sport.

**Patrice Holvoet, Université de Lille 2, France**
Patrice is a former gymnast and coach and is an accredited physical education teacher. His research in the biomechanics of gymnastics has contributed to the development of the sport

in France and around the world. He has written and edited more than 30 articles and manuscripts in French and English.

## Brooke Lemmen

Brooke Lemmen is an osteopathic sports medicine physician in Michigan, USA. A former gymnast, she has worked at all levels of women's artistic gymnastics as well as serving as the team physician for USA Gymnastics Trampoline and Tumbling.

## Michel Marina, INEFC Barcelona, Spain

Former gymnast and member of the national Spanish team, Professor Michel Marina has been teaching sports performance and gymnastics at the INEFC Barcelona since 1997. His main lines of research are related to testing and training athletes, particularly during paediatric ages, strength, power, neuromuscular fatigue both in gymnastics and motorcycling.

## Bessem Mkaouer, Higher Institute of Sport and Physical Education of Ksar Said, Manouba University, Tunisia.

Dr Mkaouer is the head of department of artistic gymnastics at the ISSEP Ksar Said, Tunisis. He is a former gymnast and federal coach. His main lines of research are related to kinetics and kinematics analysis of gymnastics skills and training process.

## Alexandra Pizzera, German Sport University Cologne, Germany

Dr Alexandra Pizzera is one of the FIG experts in coaching education and currently an academic within the Department of Further Education and Institute of Training Science and Sport Informatics of the German Sport University Cologne, Germany. Her main research interests are in judgement and decision making of sports officials, focusing on different factors influencing their performance as well as the selection, training and performance evaluation of sports officials. She is also expanding her interest to visual perception and embodied cognition, examining and transferring the bidirectional link between perception and action in and to sports officiating, athletes and coaches.

## John H. Salmela, University of Ottawa, Canada

John Henry Salmela left this world for a better one on 29 October 2014 in Belo Horizonte, Minas Gerais, Brazil. He took an early retirement after 29 years of working at a 34-year career at the Université Laval, Université de Montréal and University of Ottawa, where he taught motor development and sport psychology courses in both English and French. A former gymnast, he worked as an international coach and sports psychologist, and wrote and edited more than 20 books and 250 articles. He also supervised several graduate students at both Master and Ph.D. levels. He acted as a research chairman and sports psychologist for the Canadian men's national gymnastics team for 19 years.

## William A. Sands, USA Ski and Snowboard Association, Colorado, USA

William A. Sands is a former gymnast and World Championship coach. He has held professional positions with the US Olympic Committee and several universities and has served as Director of Research and Development for USA Gymnastics and Chair of the US Elite Coaches Association for Women's Gymnastics. He has written and edited 18 books and over 300 articles.

# FOREWORD

Dear reader,

Gymnastics is a magical sport; it seeks perfection in movement.

When you see medal contenders at the Olympic Games or at the World Championships, the beauty of their movements makes you think that performing a double layout is as easy as breathing. They make you forget they spend thousands of hours training to obtain this skill. They make you forget that human movements on Earth are governed by the laws of gravity. They make you forget the colossal power needed to generate such dynamic movement. They make you forget everything that happens inside the brain to quickly find one's marks in space while rotating.

In that, gymnastics is considered as a fantastic exploration field for scientists and researchers.

We know that every performance results from a complex interaction of biomechanical, physiological and psychological factors. Thus, it is only science that enables athletes and coaches to better understand the human body's capacity, while respecting its whole integrity. Gymnastics family is indeed grateful to all those who are helping this sport to develop in a positive direction and unveil parts of what makes it so magical.

**Morinari Watanabe**
**President of the International Gymnastics Federation**

# PART I

# Physiology for gymnastics

# PART I
# Physiology for gymnastics

## Learning outcomes (*Monèm Jemni*)

This applied physiology to gymnastics part enables you to:

- Identify the specificities related to the energetics expenditure in gymnastics and the way aerobic and anaerobic metabolisms contribute during performance.
- Find out recent studies reporting gymnasts' maximal oxygen uptake, their metabolic thresholds and power outputs.
- Explore specific biomarkers, such as blood lactate during gymnastic exercises and their energetic significance.
- Elaborate on the particularities of the gymnasts' cardiorespiratory system and its adaptations to performance.
- Explore how to monitor cardiovascular stress and the way to interpret it in gymnastics routines.
- Find out about gymnasts' diet, nutrition habits and supplementation.
- Identify health issues related to high training volume and intensity.

## Introduction and objectives (*Monèm Jemni*)

Physiology is the science that explains how the body systems work and how these systems interact with each other to regulate their functions. The study of the individual organs and their operational biochemistries is also part of the human physiology.

Sports and performances have evolved throughout the last decades thanks to the application of different sciences to exercise. Amongst these sciences, exercise physiology had a major contribution in developing the understanding of "how human systems work under different exercise conditions and regimes". One of the sports that has witnessed significant expansion is gymnastics.

One of the particularities of male or female gymnastics is that it is a sport which encloses different events; each event is different to the other although with some similarities between males' and females' events. Males' competition is a successive rotation between six events, starting in the Olympic order, with the floor exercises, pommel horse, rings, vault, parallel bars and finishing with the high bar. Females compete in four events only: vault, uneven bars, balance beam and floor exercises. Such like the modern triathlon or pentathlon, the duration, the effort, the intensity, the power, the strength, the flexibility, the speed of the stretch, the coordination and the endurance as well as the energy required to perform each of these events differ from one to another.

The overall objective of this part is to provide the reader with a wide overview on several physiological aspects related to artistic gymnastics practice.

This part is divided into many chapters providing the most up to date physiological theories on male and female gymnasts. Starting by highlighting the metabolic energy supply during gymnastic exercises, then the cardiovascular and respiratory markers for training and performance. Physiological adaptations will be highlighted as well as key variables related to the aerobic and the anaerobic metabolisms. Moreover, issues such as diet and supplementation and the everlasting dilemma of growth, sexual developmental and hormonal regulations are also discussed in the following chapters; all supported by proof from the current literature.

# 1

# ENERGETICS OF GYMNASTICS

*Monèm Jemni*

## 1.1 Learning outcomes

- Identify the specificities related to the energetics expenditure in gymnastics and the way aerobic and anaerobic metabolisms contribute during performance.
- Find out recent studies reporting gymnasts' maximal oxygen uptake, their metabolic thresholds and power outputs.
- Explore specific biomarkers, such as blood lactate during gymnastic exercises and their energetic significance.

## 1.2 Introduction

Although gymnastics as a sport is becoming increasingly more publicised, attracting global media coverage for competitions such as the World Championships and the Olympics, it has not attracted the same scientific interest compared to the main stream sports. As early as the 1960s, authors reported that gymnasts were characterised by a low maximal aerobic power but a high level of strength (Horak, 1969; Montpetit, 1976; Saltin & Astrand, 1967; Szogy & Cherebetiu, 1971). This has been confirmed in a more recent review of the literature which has incorporated the modern artistic gymnastics (Jemni et al., 2001). Very few studies have explored the differences between the physiological responses in male and female gymnasts nor the energetics of the different gymnastic events. Facts which may have contributed to the paucity of the scientific database in this field include the complexity of the sport, the lack of adequate equipment and specific physiological testing protocols. During the 1970s it was suggested that male gymnasts, irrespective of their events, were considered as having an energy expenditure similar to that of running at 13 km/h on a treadmill (Montpetit, 1976). More recently, studies have shown significant different physiological, biomechanical and psychological requirements between running and gymnastics.

The aim of this chapter is to provide a current overview of the physiological and energetic requirements of the modern gymnastics, taking into consideration each artistic event separately. A current physiological profile of the modern artistic gymnast will be identified.

## 1.3 Aerobic metabolism

Aerobic metabolism, also known as aerobic respiration, cell respiration and oxidative metabolism, is a series of chemical processes by which individuals produce energy through the oxidation of different substrates in the presence of oxygen. The main metabolic substrates for this process include: carbohydrates (sugars), lipids (fat) and, very rarely, proteins which are used to produce the energy source, adenosine triphosphate (ATP). Other by-products from this process include water, carbon dioxide and heat. Aerobic metabolism is the main supplier of energy for endurance exercise which may last for hours. It is known that athletes who compete in long duration/distance events have a highly developed aerobic metabolism.

### Maximal oxygen uptake ($VO_2$ max)

One of the indicators of the aerobic metabolism is $VO_2$ max, or the maximal amount of oxygen that an individual can utilise during intense or maximal exercise. This factor is generally considered to be the best indicator of an athlete's cardiovascular fitness and aerobic endurance. It can be measured by a variety of methods, mainly in the laboratory environment using different types of exercise protocol and collecting the gas exhaled while progressively increasing the workload.

Although $VO_2$ max has a genetic component (Bouchard et al., 1992), it can be increased through appropriate training. It is generally known that sedentary individuals have a lower $VO_2$ max compared to that of more active subjects. Meanwhile, amongst many factors, three have a major influence on $VO_2$ max, namely age, gender and altitude (Jackson et al., 1995; Jackson et al., 1996; McArdle, Katch, & Katch, 2005; Trappe et al., 1996).

Most elite endurance athletes have $VO_2$ max values in excess of 60 ml/kg/min. Although such high values may indicate an athlete's potential for excellent aerobic endurance, other factors such as metabolic threshold, exercise economy and energetic cost are better predictors and correlate better with endurance performance (Wilmore & Costill, 2005).

### Gymnasts' $VO_2$ max

The $VO_2$ max of international level gymnasts reported over the last 40 years (Table 1.1.) is around 50 ml/kg/min, although a higher value (around 60 ml/kg/min) was reported for the American elite female gymnasts (Noble, 1975). Table 1.1 reports some of these investigations.

Barantsev (1985) reported that gymnasts' $VO_2$ max decreases between adolescence and adulthood with average values decreasing from 53.2 + 6.3 at age 12 to 47.2 + 6.7 ml/kg/min at age 25. This regression is associated with an increase in both the volume and intensity of training for strength and power required for higher technical skills. Other studies have reported that increasing anaerobic power via specific training programmes leads to a decrease in aerobic capacity. However, it appears that this effect is not evident before puberty. In fact, muscular fibre specification and selective fibre type recruitment are less evident in children compared to adults (Bar-Or, 1984; Inbar & Bar-Or, 1977). It is well known that, during puberty, individuals experience significant morphological and hormonal transformations associated with an increase of the maximal anaerobic power (Bedu et al., 1991; Falgairette et al., 1991). It is accepted by coaches that this peri-pubertal stage constitutes a crucial period for maximum potential in technical learning, strength and power gain. Taking into consideration all the above, very clear and progressive periodisation of the training seasons is necessary to avoid burnout.

**TABLE 1.1** Average VO₂ max and (SD) of different level gymnasts, from the 1970s up to 2006

|  |  | n | Level | Age (yrs) | VO₂ max (ml/kg/min) |
|---|---|---|---|---|---|
| Females | Sprynarova & Parizkova (1969) | – | non-elite | – | 42.5 (3.7) |
|  | Noble (1975) | 3 | elite | 16–22 | 61.8 (8.0) |
|  | Montgomery & Beaudin (1982) | 29 | non-elite | 11–13 | 50.0 (0.9) |
|  | Elbæk & Froberg (1992) | 19 | elite | 20.1 (1.7) | 50.4 (2.9) |
| Males | Bergh (1980) | – | elite | – | 51.0 |
|  | Barantsev (1985) | – | non-elite | 12–13 | 53.2 (6.3) |
|  |  |  |  | 14–15 | 50.9 (6.2) |
|  |  |  |  | 17–25 | 47.2 (6.7) |
|  | Goswami & Gupta (1998) | 5 | non-elite | 24.2 (3.1) | 49.6 (4.9) |
|  | Lechevalier et al. (1999) | 9 | elite | 17–21 | 53.1 (3.2) |
|  | Jemni et al. (2006) | 12 | elite | 18.5 (1) | 49.5 (5.5) |
|  |  |  |  |  | 33.4 (4.8)★ |
|  |  | 9 | non-elite | 22.7 (2) | 48.6 (4.6) |
|  |  |  |  |  | 34.4 (4.6)★ |

★: Upper body VO₂ peak measured with an arm-cranking ergometer.

Table 1.1 shows recent updated upper and lower body VO₂ max/peak values of male national and international gymnasts (Jemni et al., 2006). This update was warranted by the important evolution/transformation of the modern artistic gymnastics since the 1980s. Gymnasts' upper body VO₂ peak represents two-thirds of that measured on the treadmill (around 35 ml/kg/min). This value is quite high; the reason might be the fact that four out of six events in male artistic gymnastics involve exclusively the upper body (pommels, rings, parallel and high bars).

The rules of the competitions (code of points) as well as the compulsory elements and the free routines are changed every Olympic cycle with greater emphasis given to strength and power based skills. Consequently, both training and preparation for competitions have changed in order to match the new requirements/rules. Nowadays, gymnasts spend increasingly more time training compared to the past. Training intensity and volume have increased considerably, especially at the highest levels where training often exceeds 34 hours per week (Richards, Ackland, & Elliott, 1999). However, the average VO₂ max values of both elite and non-elite gymnasts has not changed appreciably for the last five decades, remaining around 50 ml/kg/min (Jemni et al., 2006) (Table 1.1). Moreover, comparing elite and non-elite VO₂ max has not reached any statistical significant difference, even with the large disparity in training volumes between the two levels (Jemni et al., 2006).

Conventional wisdom would indicate that the elite group should have a greater aerobic capacity due to higher practice volume, although, unlike running, swimming and cycling, VO₂ max in gymnasts is not directly related to performance parameters. Gymnasts' VO₂ max values have been reported to be significantly lower to those of athletes participating in short intense activities, such as high level male sprinters (Barantsev, 1985; Bergh, 1980; Goswami & Gupta, 1998; Marcinik et al., 1991) and some reports suggested that they are comparable to values for a sedentary population (Crielaard & Pirnay, 1981; Willmore & Costill, 1999).

Finally, coaches' opinions differ towards gymnasts' $VO_2$ max. Many coaches are still encouraging gymnasts to develop their $VO_2$ max by planning jogging sessions and/or long activities such as cycling on gym-ergometers. They justify this training by the fact that "having an important oxidative foundation enhances recovery during high intensity sessions". Other coaches still believe that long "endurance type sessions" help weight control, especially in female gymnasts (Sands et al., 2000). There is evidence that aerobic endurance training may interfere with power, the main physical quality of a gymnast. This section clearly demonstrates that gymnasts' $VO_2$ max has not changed appreciably for the last five decades (remaining at around 50 ml/kg/min) while gymnastics' performance has increased significantly. The findings provide strong evidence that it is unnecessary to enhance gymnasts' $VO_2$ max.

## Gymnasts' metabolic thresholds

Several metabolic thresholds commonly assessed during maximal incremental exercise tests are considered to be better predictors of endurance performance than the $VO_2$ max (Willmore & Costill, 2005). These include the anaerobic threshold (AT), lactate threshold (LT), lactic turn point (LTP), onset of blood lactate accumulation (OBLA) and the ventilatory threshold (VT). It is to be expected that high level endurance athletes would have delayed metabolic threshold as the consequence of an improved oxidative system. Few studies have investigated these metabolic thresholds in gymnasts. Table 1.2 shows the LT assessed in elite and sub-elite male artistic gymnasts during a maximal incremental test performed on the treadmill (Jemni et al., 2006). Both categories achieved their LT at a very high percentage of their $VO_2$ max (mean of 79%) and at a very similar heart rate (mean of 169 bpm).

A low $VO_2$ max as demonstrated in the previous section associated with a high metabolic threshold is not a common feature of a conventional physiological profile. Analysis of this feature is to be described progressively throughout this chapter.

The increasing training volume in gymnastics would be expected to result in physiological adaptations that reflect an improvement in the removal of blood lactate through increased buffering capacity, increased uptake through neighbouring slow twitch fibres, or increased removal through the Cori cycle. Alternatively, there might be less production, as a greater training volume may enhance the capacity for a higher intensity to be sustained by relying on limited anaerobic energy sources. This is suggested by the low blood lactate values measured in gymnasts at the end of the maximal exercise test and the high percentage of $VO_2$ achieved at the lactic turn point (around 10 mmol/l and 79% respectively) (Jemni

**TABLE 1.2** Average male gymnasts' lactic thresholds and (SD) assessed during maximal incremental test performed on the treadmill (Jemni et al., 2006)

|  | HR (bpm) | VO$_2$ max (ml/kg/min) | % VO$_2$ max |
|---|---|---|---|
| Elite level (n = 12, 18.5 yrs) | 169.8 (10.1) | 44.5 (5.4) | 82.1 (6.5) |
| Non-elite (n = 9, 22.7 yrs) | 169.3 (4.0) | 37.0 (5.1)[S] | 76.3 (9.9) |

[S] (p < .05): significant difference.

et al., 2006). Interestingly, the lactic turn point of gymnasts is comparable to that achieved by elite endurance athletes (Willmore & Costill, 1999, p. 137). Long distance runners and cyclists reach their lactic turn point at similar levels (77% of $VO_2$ max in a group of endurance cyclists measured by Denadai et al. (2004) for example). Several authors have noted a relationship between increased strength and increased anaerobic power and measures of endurance as a result of strength training (Hickson et al., 1988; Marcinik et al., 1991). It has been demonstrated that lactate threshold can be delayed markedly through resistance training (Marcinik et al., 1991). Jemni et al. (2006) have proved that the amount of strength and conditioning performed during gymnastic training not only increases strength and power but also enhances fatigue tolerance.

## 1.4 Energy cost of gymnastics exercises

Energy cost of gymnastics was first measured in the early 1950s in the old Eastern Bloc (Blochin, 1965; Krestovnikov, 1951). Different procedures have been used to collect exhaled gas but mainly used the Douglas bags. Seliger et al. (1970) have supposed that such heavy equipment effects might have underestimated the energetic cost by 10%. However, technological evolution in lab equipment has allowed higher precision in measurement. Investigation methods and techniques have differed between the studies, but in all cases it is agreed that energy cost differs between events (males and/or females) (Table 1.3).

Hoeger and Fisher (1981) measured the energy cost in male gymnasts while performing their six compulsory routines. Gymnasts performed their six routines while equipped with a mouthpiece linked to a Douglas bag. At the end of each routine, gymnasts had to hold their breaths for a few seconds until the investigators attached another bag to collect the expired air during recovery. The bags were analysed using a pneumograph MTG and a gas meter "Parkinson Cowan CD 4" (Hoeger & Fisher, 1981). Results showed that the most costly event was the floor exercises (37 kcal), followed in decreasing order by the pommels, rings, high bar, parallel bars and finally the vault. Following this, Rodríguez, Marina and Boucharin (1999) measured the excess of post-oxygen consumption (EPOC) during the

**TABLE 1.3** Average energy cost and (SD) of male and female gymnastics events

| Females | Vault | Uneven bars | | Balance beam | Floor |
|---|---|---|---|---|---|
| Seliger et al. (1970) ml/kg/min of $O_2$ | 16.9 (3.5) | 16.5 (3.6) | | 15.1 (4.9) | – |
| Rodríguez et al. (1999) ml/kg/min of $O_2$ | 34.3 (7.7) | 36.6 (4.6) | | 31.3 (6.1) | 40.8 (4.0) |

| Males | Floor | Pommels | Rings | Vault | Parallel bars | Horizontal bar |
|---|---|---|---|---|---|---|
| Ogawa et al. (1956) kcal/min | – | 7–11 | 11–16 | 5–10 | 8–10 | 12–15 |
| Seliger et al. (1970) ml/ kg/min of $O_2$ | 20.5 (6.3) | 16.4 (2.5) | 17.3 (2.7) | – | 17.1 (3.6) | 18.5 (3.4) |
| Hoeger & Fisher (1981) kcal/routine | 37.01 | 36.6 | 32.5 | 25.8 | 32.3 | 32.5 |
| Sward (1985) cal/kg/min | 0.13 | 0.16 | 0.08 | 0.13 | 0.09 | 0.14 |

first 30 sec following female gymnastic routines, using online gas analysis which allowed them to estimate the real oxygen consumption. Their results were similar to the male routines: the floor exercises induced the highest value (40.8 + 4.0 ml/kg/min) and the vault exercise induced limited oxygen consumption (34.3 + 7.7 ml/kg/min). The findings of these two investigations are in agreement with the blood lactate values of male and female routines measured in other studies (Jemni et al., 2000; Montgomery & Beaudin, 1982; Montpetit, 1976) (see section on "Blood lactate measurement during gymnastic exercises").

Some studies have indirectly estimated the energy cost of gymnastic exercises by using the regression relation between heart rate and oxygen consumption measured during a maximal graded test and heart rate values measured during gymnastics (Noble, 1975; Montpetit, 1976). The findings suggest that aerobic metabolism contribution was 20% and from anaerobic metabolism, 80%. It was further suggested that gymnasts would use only 35% of their aerobic capacity as measured during the $VO_2$ max test to perform gymnastic routines. However, this relation is valid only when measurements are taken in a steady-state (heart rate and oxygen uptake). It is almost impossible to achieve such a steady-state while performing any gymnastic routines due not only to the short and different durations of the routines (Table 1.4) but also to the different intensity/pace and rhythm change. Chapter 2, Figure 2.1, shows examples of heart rate fluctuation during male routines. The steady-state is difficult to achieve because of the variety of muscle contractions that gymnasts use while performing. In addition, gymnasts often perform some elements while maintaining an apnoea for a few seconds. It has indeed been shown that apnoea has an effect on the cardiovascular system (Shaghlil, 1978).

It is possible to achieve a steady-state quite quickly while running or cycling at a steady pace (around 3 min), whereas it is impossible in gymnastics because of all the above. Therefore, extrapolating gymnastics energy expenditure from the regression relation between heart rate and oxygen consumption measured during a maximal graded test and heart rate values measured during gymnastics, as some authors have previously reported, is totally invalid.

Figure 1.1 shows the energy supply continuum. Whether aerobic or anaerobic activity, the three processes responsible for ATP production, i.e., the anaerobic ATP-Phosphocreatine, the anaerobic glycolysis system and the aerobic oxidation system, work together rather than in sequence. It is well known that at a certain time during the exercise, one of these three processes would be the main source of ATP, depending on the duration

**TABLE 1.4** Averages and standard deviations of the male and female routines duration measured in different international competitions (in seconds) (Jemni et al., 2000)

|                | Males      | Females    |
|----------------|------------|------------|
| Floor exercise | 60.9 (3.5) | 82.9 (3.2) |
| Pommel horse   | 30.5 (4.5) | –          |
| Rings          | 40.7 (5.1) | –          |
| Vault          | 5.2 (0.5)  | 4.8 (0.9)  |
| Parallel bars  | 31.2 (6.2) | –          |
| High bars      | 36.5 (6.6) | –          |
| Uneven bars    | –          | 46.5 (3.5) |
| Balance beam   | –          | 81.8 (4.5) |

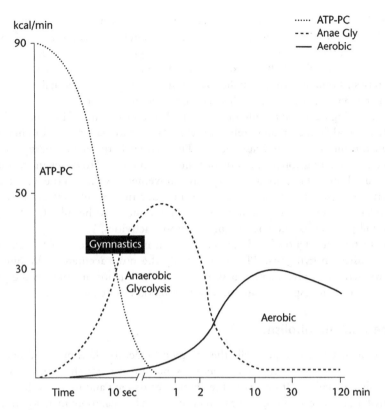

**FIGURE 1.1**  The energy continuum.

(Adapted from McArdle, Katch, & Katch, 2005)

and the intensity. Short and very intense burst exercises rely predominantly on the anaerobic ATP-PC system, which provides an immediate but very limited source of ATP for explosive contractions and power development as required for the vault. Restoration of the ATP-PC system is rapid during the following rest period post-exercise, and requires oxygen supply and colossal reduction of the exercise intensity.

If the exercise is meant to be continued after the very powerful start of the first few seconds (such as the case for a 400m sprinter), then it is generally expected that the intensity of the exercise would drop slightly. In the meantime the energy supply would progressively rely on the anaerobic glycolysis which can be fully functioning around 20 sec. At this stage, the high intensity of the exercise can only be maintained for a short period because of the high acidity of the muscle environment as a consequence of the lack of oxygen and the considerable production of lactic acid via anaerobic glycolysis. This hypoxic environment prohibits the efficacy of the muscle contractions and therefore muscular fatigue and power output decline are the main outcomes. Powerful athletes are able to push the boundaries of the high acidity with little drop in the efficacy of their exercise. Anaerobic and resistance training would enhance this specific performance. The anaerobic system is considered as the main supplier of most of gymnastics' competitive events. Blood lactate measurements might reflect the contribution of this metabolism as shown in the section on "Blood lactate measurement during gymnastic exercises". There is no doubt that the intensity of the competitive events is very high. This intensity is even pushed to the extreme in high level gymnasts. Nowadays, it is common to see gymnasts performing double front or back somersaults including twists within the vault, floor exercises or as dismounts from several apparatus. In addition, the longest events are the floor exercises and the balance beam (90 sec maximum for female floor exercises and balance beam; 70 sec maximum for male floor exercises). Pommels, rings, parallel bars, uneven bars and horizontal bar routines last approximately 35 sec, whereas a vault lasts only 6 sec on average.

As shown in Figure 1.1, if the physical exercise has to be continued beyond the anaerobic glycolysis, the body would mainly rely on the oxidative pathway system for energy supply. A drop in the intensity would also occur. This system is the main energy supplier for endurance sports lasting longer periods of time (section 1.3). However in gymnastics, all the events are short and require quick explosive movement. Although the oxidative system may contribute slightly in several events, their short durations would not allow a full energy supply. Indeed the oxidative system requires a few minutes to be fully functional and to act as the full provider of the energy during aerobic activities.

Unfortunately, until now, and for mainly ethical reasons, no one has attempted to perform biopsies on gymnasts. This is essentially the most accurate technique to assess energy production/supply. Table 1.5 shows energy estimation in each of the gymnastic events based on their respective duration and on blood lactate assessment.

## 1.5 Anaerobic metabolism

The anaerobic metabolism is generally the energy provider for short bursts of high intensity exercise lasting a few minutes. The energy is produced through a series of chemical reactions called glycolysis. It occurs in the absence of oxygen and involves the breakdown of carbohydrates (glucose, glycogen) as a fuel source. The anaerobic metabolism is limited in time by the accumulation of protons ($H^+$) and the increase of the acidity in the muscle cells within a few minutes because the rate of production of lactic acid exceeds its removal

**TABLE 1.5** Estimation of energy supply in each gymnastic event

|  | ATP-PCr | Anaerobic glycolysis | Oxidative | Blood lactate (mmol/l) |
|---|---|---|---|---|
| *Females* | | | | |
| Vault (6 sec) | 100% | 5–10% | 1–2% | 2.5★ |
| Uneven bars (45 sec) | 100% | 80–90% | 3–5% | 7.4★ |
| Balance beam (90 sec) | 90% | 50–60% | 20–30% | 4.3★ |
| Floor exercises (90 sec) | 100% | 80–90% | 20–30% | 7.0★ |
| *Males* | | | | |
| Floor exercises (70 sec) | 100% | 60–70% | 20–30% | 6.2★★ |
| Pommel horse (35 sec) | 100% | 80–90% | 3–8% | 5.8★★ |
| Rings (35 sec) | 100% | 80–90% | 3–8% | 5.8★★ |
| Vault (6 sec) | 100% | 5–10% | 1–2% | 3.8★★ |
| Parallel bars (35 sec) | 100% | 70–80% | 5–10% | 4.0★★ |
| High bars (35 sec) | 100% | 70–80% | 3–8% | 5.0★★ |

★: (Rodríguez et al., 1999); ★★: (Jemni et al., 2000).

rate. Acid lactic is in fact a natural by-product of the anaerobic glycolysis. The acidic environment impairs muscle cell contraction and causes an increasing level of fatigue which leads to a reduction in physical performance and to exhaustion. The impairment of the muscle contractions might affect technical performance, especially in artistic sports such as gymnastics. Gymnasts are indeed marked on "how good they perform" not on how many skills they are able to present. Any imperfection is penalised, even one single step at a landing of an acrobatic element. It is therefore fundamental that gymnasts are able to perform under a high level of "metabolic stress" giving the short duration of their routines and the high intensity levels of their exercises.

Many authors have reported that modern artistic gymnastics requires greater strength and power because of the ever increasing technical difficulty required through revision of the Code of Points (F.I.G., 2009) (Brooks, 2003; French et al., 2004). The Federation of International Gymnastics (FIG), the international governing body, reviews and updates the code of judging every four years. Hence, performance demands on gymnasts are continually changing to meet the new code requirements. In the 1970s, the Code of Points showed only three levels of difficulties: A, B and C. In 2009, the Code of Points shows not only an increase in the number of technical skills but also seven levels of difficulty: A to G. Routines which include E, F and G skills have higher start values than routines constructed with B, C and D skills only. Gymnasts are encouraged from an early age to learn more difficult skills to ensure a higher start value that leads to a higher score in order to reach the highest levels of competition. This suggests that modern artistic gymnasts have to develop their anaerobic metabolism at an early age in order to perform at the highest technical levels. Nowadays, it is very common to see two-year-old children in the gym (initiation programme) because it typically takes about ten years for a gymnast to reach the elite level.

This section highlights several components of the gymnasts' anaerobic metabolism.

## Power output of the gymnasts

The power of the anaerobic metabolism can be assessed through a series of biochemical analyses following a muscle biopsy. This technique is indeed the best approach to evaluate the physiological mechanisms that underpin a physical performance. Unfortunately, this technique is not only painful but also very costly and requires highly qualified technicians and a medical facility. Indirect, non-invasive techniques can give an idea about the power of the anaerobic metabolism. Ergometry results have been strongly correlated with the outcomes of the invasive techniques and therefore are widely accepted as means to estimate the anaerobic metabolism.

Currently, there is no specific ergometer to assess gymnasts. Sport scientists commonly apply standardised ergometry tests in the laboratory or in the gym. The "Force-Velocity" (Vandewalle et al., 1987) and the Wingate tests (Bar-Or, 1987) are considered as gold standards to estimate the mechanical power output and are used most frequently. These tests might be performed by the upper and/or the lower body limbs. The Force-Velocity test consists of measuring peak cycling velocity during short maximal sprints (about 6 sec) at different resistances. The Wingate test consists of measuring not only the peak power generated during the first 10 sec of a full out 30-sec cycling exercise but also the mean power generated during the 30-sec period. Regrettably, gymnasts are not used to cycling or arm cranking and therefore they are somehow penalised compared to other athletes. However, performing such recognised standard tests allows for comparison with other athletes.

Tables 1.6 and 1.7 show the results of Force-Velocity and Wingate tests respectively, performed by elite and non-elite gymnasts. Note there is a significant difference between the upper and lower body performance within the same level, whereas there are no significant differences between the levels of practice. Maximal power output relative to body mass developed by the upper body of the gymnasts represents two-thirds of that developed by the lower body in both levels.

These high peak power values (~15 W/kg in the Force-Velocity, ~13.5 W/kg in the Wingate for males and ~10.5 W/kg in the Wingate for the females) place the gymnasts near the top levels of power athletes. For example, upper and lower body powers of the male gymnasts are higher than those measured in elite wrestlers (7.8 + 1 W/kg and 10.9 + 1.2 W/kg, respectively for upper and lower body) (Horswill et al., 1992).

When comparing the Force-Velocity results with those from similar investigations using the same protocol with other sport groups, male gymnasts have similar upper body values to swimmers (~10.5 W/kg by Vandewalle et al., 1989), similar lower body values to volley-ball players (~15.8 W/kg (Driss, Vandewalle, & Monod, 1998)) and just below the lower body values scored by sprinters (~17.0 W/kg (Garnier et al., 1995)).

It is evident that "peak power" is a key component in gymnasts' singular physiological profile. In fact, the contribution of strength and power to gymnastic performance has increased during the last four decades (Jemni et al., 2001). The total effort time of a gymnast during a three-hour competition is only between 12 and 15 min (warm-up included) (Jemni et al., 2000). During these minutes of effort time, gymnasts must perform a limited number of skills in a limited time (at each event). To achieve powerful skills, speed and strength are indeed key components, especially in modern artistic gymnastics. The specificity of gymnastic training has certainly had an effect on shaping the energetic requirements of the practitioners. The literature provides evidence that specific training has an influence on physical aptitudes, fibre type characteristics and, indirectly, on aerobic and/or anaerobic metabolism (Jansson, Sjodin, & Tesch, 1978). Clearly, repeating powerful gymnastic skills and carrying out and/or supporting the body weight on the apparatus would enhance the above qualities. In addition, Table 1.7 shows high blood lactate values measured after Wingate tests performed by upper and lower bodies in male and female gymnasts (around 10.5 mmol/l). These high values are indirectly symptomatic of an established anaerobic metabolism. More detail about this metabolism is highlighted in the following section through investigations in blood lactate during gymnastic routines.

**TABLE 1.6** Average upper- and lower-body peak power output and (SD) measured in male gymnasts during Force-Velocity test (Jemni et al., 2001; Jemni et al., 2006)

| | | [Peak power]6 sec (W) | [Peak power]6 sec (W/kg) |
|---|---|---|---|
| Elite level (n = 12, 18.5 yrs) | Upper body | 688.3 (87.7) | 10.6 (0.9) |
| | Lower body | 1028.0 (111.5)[S] | 15.9 (1.3)[S] |
| Non-elite level (n = 9, 22.7 yrs) | Upper body | 652.4 (79.9) | 9.8 (1.1) |
| | Lower body | 980.7 (266.4)[S] | 15.1 (4.3)[S] |

[S] ($p < .05$): significant difference between upper and lower body results but no difference between the levels.

**TABLE 1.7** Average upper- and lower-body power outputs and (SD) measured in male and female gymnasts during Wingate test (Jemni, 2001; Jemni et al., 2006)

| Females | | MPO (W/kg) | PPO (W/kg) | BL max (mmol/l) |
|---|---|---|---|---|
| | | | Lower body | |
| Heller et al. (1998) | Elite level (n = 6, ~15.5 yrs) | 8.6 (0.1) | 10.4 (0.4) | 11.6 (1.7) |
| Sands et al. (1987) | (n = 25, ~14 yrs) | 7.1 (1.3) | 7.9 (2.0) | – |
| | | | Upper body | |
| Sands et al. (1987) | (n = 25, ~14 yrs) | 3.1 (0.7) | 3.6 (1.0) | – |
| Males | | | Lower body | |
| Jemni et al. (2006) | Elite level (n = 12, 18.5 yrs) | 9.7 (1.00)[s] | 13.5 (1.34) | 11.7 (2.03) |
| | Non-elite (n = 9, 22.7 yrs) | 10.1 (1.3)[s] | 14.1 (3.0) | 11.0 (3.1) |
| Savchin & Biskup (2003) | Mixed (n = 24, 11.6 yrs) | – | 9.6 (1.9) | 12.1 (0.9) |
| Heller et al. (1998) | Elite level (n = 5, >18 yrs) | 10.7 | 13.2 (1.0) | 11.2 (1.5) |
| | | | Upper body | |
| Jemni et al. (2006) | Elite level (n = 12, 18.5 yrs) | 7.1 (0.5)[s] | 9.6 (0.6) | 12.2 (1.5) |
| | Non-elite (n = 9, 22.7 yrs) | 6.6 (0.6)[s] | 9.2 (1.1) | 10.4 (0.7) |

MPO: Mean Power Output; PPO: Peak Power Output; BL: Blood Lactate post Wingate test; n: number;

[s] ($p < .05$): significant difference between upper and lower body results but no difference between the levels.

## Blood lactate measurement during gymnastic exercises

Blood lactate (BL) analysis allows an indirect estimation of the anaerobic glycolysis contribution. Sadly, there have been few attempts to measure BL during gymnastic routines, in particular in male gymnasts. Back in the 1970s, some authors supposed that lactate production was negligible (Montpetit, 1976), suggesting that anaerobic glycolysis is not the main supplier of energy and that the main part of the energy production is assured by the ATP-PC. Beaudin (1978) reported an average BL of 2.8 mmol/l following the four female routines, with higher values measured post the floor and the uneven bars (~4.4 mmol/l). These low values might be attributed to the low level of the gymnasts, who were unable to perform high difficulty exercises. Higher BL values were reported in the 1980s and 1990s in low and higher levels, with Montgomery and Beaudin (1982) reporting an average of 4.0 mmol/l and Rodríguez et al. (1999) an average of 5.3 mmol/l (Table 1.8). Also, Goswami and Gupta (1998) and Lechevalier et al. (1999) found similar

**TABLE 1.8** Blood lactate for each event in male and female gymnasts in mmol/l

*Females*

|  | Montgomery & Beaudin (1982) (n = 29) non-elite | Bunc & Petrizilkova (1994) (n = 7) elite | Rodríguez et al. (1999) (n = 8) mixed |
|---|---|---|---|
| Vault | 3.1 | 2.4 | 2.5 |
| Uneven bars | 2.2 | 9.5 | 7.4 |
| Balance beam | 3.0 | 10.2 | 4.3 |
| Floor exercises | 8.5 | – | 7.0 |

*Males*

|  | Goswami & Gupta (1998) (n = 5) non-elite | Jemni et al. (2000) (n = 7) non-elite | Groussard & Delamarche (2000) (n = 5) mixed | Jemni et al. (2003) (n = 12) mixed |
|---|---|---|---|---|
| Floor exercises | 7.11 | 6.2 | 11 | 6.3 |
| Pommel horse | 5.18 | 5.8 | 6.5 | 6.0 |
| Rings | 6.77 | 5.8 | 6.6 | 6.7 |
| Vault | – | 3.8 | 5.0 | 3.8 |
| Parallel bars | 6.23 | 4.0 | 5.8 | 5.1 |
| High bars | 5.97 | 5.0 | 6.0 | 5.2 |

averages in male gymnasts despite the difference of their research methods (6.2 + 0.7 and 6.2 + 1.6 mmol/l respectively in five apparatus except the vault).

Figure 1.2 shows the maximum and the minimum BL values assessed during a male gymnastics competition. The competition was composed of the six Olympic events and began with the floor exercises, followed by pommel horse, rings, vaulting, parallel bars and horizontal bar. The routines were separated from each other by 10 min of recovery. The programmes performed were free but technical composition did not vary between gymnasts.

The average maximum BL of all the six events was 4.8 + 1.1 mmol/l. This average would be even higher if the vault's value did not count (5.1 + 0.9 mmol/l).

As shown in Figure 1.2, BL varied from one apparatus to another. The highest value was observed at the floor exercise (6 to 11 mmol/l); this was significantly higher than those of vault, parallel bars and the horizontal bar. BL measured at the vault (3 to 4 mmol/1) was significantly less than those of the other exercises. These two findings confirm similar investigations performed on female gymnasts (Rodríguez et al., 1999). Figure 1.2 also indicates that even BL decreased after the 10-min recovery period; the values were always higher than the expected rest values (less than 2 mmol/l).

This average could not be considered as negligible as it is above the onset of blood lactate accumulation (OBLA) of 4 mmol/l, indicating the moment when the rate of lactate production becomes more important than its removal. This also indirectly indicates the moment when anaerobic metabolism is becoming the main energy supplier.

Coaches pursue varying training objectives based on the particular period of training. Gymnasts, much like sprinters, perform in both aerobic and anaerobic conditions (Jemni et al., 2000; Sands, 1998). During intense sessions, gymnasts are asked to perform routines

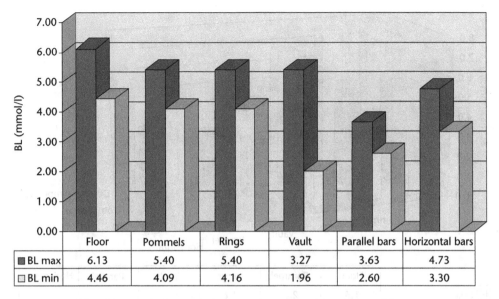

**FIGURE 1.2**   Blood lactate during males' gymnastics competition in mmol/l.

(Jemni et al. 2000)

while fatigued. They are often asked to find the best compromise between technical effectiveness, safety and high intensity effort. During the competitive phase of the season, gymnasts usually repeat their six events several times per practice session (Arkaev & Suchilin, 2004). Figure 1.3 is a typical example of different practice sessions. BL was measured during 1) "an easy session" which was mainly rehearsing skills at three apparatus (floor, pommels and rings); 2) a competition where the gymnasts performed their six routines; 3) double competition, where the gymnasts performed their six routines twice (also called back to back). This figure proves that the anaerobic metabolism becomes a major contributor when the gymnasts have to repeat their routines more than once. The average of BL in the double competition was 5.8 + 1.9 mmol/l (6.3 + 1.7 mmol/l in five apparatus except the vault). This value might increase considerably if the gymnasts are requested to repeat their routines more than twice. These types of practices are similar to interval training due to the intermittent and intense activities that are involved. Therefore one of the questions that should be addressed is how should gymnasts recover in these types of practices? (See Chapter 5, section 5.7 on "Regeneration and the new concept of recovery in gymnastics"). Figure 1.3 also shows that BL was very low during the "skill learning session" (2.0 + 0.7 mmol/l). In addition, average BL was similar to the one shown in Figure 1.2 (4.2 + 1.0 mmol/l in six events or 4.5 + 0.8 mmol/l in five events).

Interestingly, by rehearsing the same skills/routines, the gymnast becomes increasingly "economic". Progressively, the gymnast finds the most efficient movements and indirectly the suitable contractions allowing him/her to perform the exercises without undue fatigue. Gymnastics practice is indeed based on "repetition". High level gymnasts often repeat more than 1,700 elements per micro-cycle (seven to eight days), not counting the strength and conditioning exercises (Arkaev & Suchilin, 2004). They have to master each element separately before including it with a mini routine or a combination of elements. Learning these skills separately is indeed the way to refine the technique, but also another way to

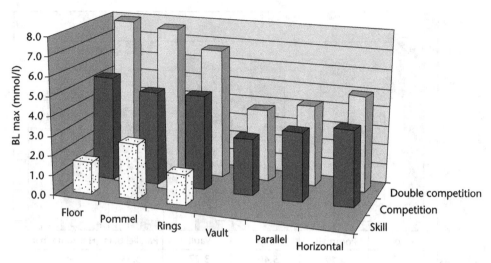

| | Floor | Pommel | Rings | Vault | Parallel | Horizontal |
|---|---|---|---|---|---|---|
| ⊡ Skill | 1.6 | 2.6 | 1.6 | | | |
| ■ Competition | 5.4 | 4.8 | 4.9 | 2.9 | 3.5 | 3.9 |
| ▨ Double competition | 7.9 | 7.7 | 6.7 | 3.7 | 4.1 | 4.9 |

**FIGURE 1.3**   Blood lactate in mmol/l measured in three different males' gymnastic sessions: Skills learning, Competition, Double competition.

(Adapted from Lechevalier et al., 1999)

find the best hemodynamics allowing the minimum energy expenditure. One would expect to see the energy expenditure of a gymnast reduce between the start and the end of a learning period. It is also suggested that lactate production would be reduced. In addition, a high level gymnast would produce less lactate compared to a lower level gymnast if they are asked to perform the same task. The high level gymnasts would use less force and would rely on his/her most "economic" muscle fibres and therefore would produce less lactate. The lower level gymnast would use a higher percentage of his/her strength capacity, and would rely more on type $II_B$ fibres and thus would produce more lactate. Moreover, one of the specific effects of training is the enhancement of "lactate tolerance". Sands (2003) suggested that because of the risk of injuries associated with practice, maximum levels of lactate production are not likely to be achieved due to the increasing acidity, contractile failure and fatigue. The author concluded that the maximum "safe" level of lactate for competent skill performance is still unknown.

## Variation of blood lactate production according to the competition rotation (Bessem Mkaouer)

According to the earlier-mentioned and also to others (Dallas & Kirialanis, 2010; Grossfeld, 2014; Hernández, Balón, & Galarraga, 2009; Irurtia et al., 2007; Jemni, 2010, 2011; Jemni et al., 1998; Jemni & Sands, 2000, 2003; Jemni et al., 2000, 2001; Jemni, Friemel, & Sands, 2002; Jemni et al., 2003, 2006; Kirkendall, 1985; Lange, Halkin, & Bury, 2005; Lechevalier et al., 1999; Marina & Rodríguez, 2014; Papadopoulos et al., 2014; Prados, 2005; Sands,

Caine, & Borms, 2003; Viana & Lebre, 2005), it has been confirmed that the cardiovascular and metabolic responses are quite different from an apparatus to another in artistic gymnastics. In other words, some gymnastics apparatus are more or less "taxing" than others. The highest cardiorespiratory stress associated with higher anaerobic contributions was noticed during the floor exercise routines for males' artistic gymnastics (MAG) with maximal heart rates reaching 186 ± 11 b·min-1 and BL values ranging from 6 to 11 mmol·l-1 (Jemni & Sands, 2003; Jemni et al., 2000, 2003). High cardiorespiratory and metabolic stresses were also found at the pommel horse followed by the rings, the high bars and the parallel bars in MAG and at the uneven bars followed by balance beam in women's artistic gymnastics (WAG). The less taxing apparatus was found to be the vault, which lasts only 4–6 sec.

The majority of the previously mentioned authors have assessed BL responses while the routines order performed according to the instructions imposed by the FIG (2013), that is to say competitors start with the floor exercise, followed by the pommel horse, rings, vault table, parallel bar and horizontal bar in MAG (Figure 1.4a) and vault, uneven bars, balance beam and floor exercise in WAG (Figure 1.4b). However, these instructions vary from one gymnast to another according to the competition draw, except for competition type II (general individual) in which the six chosen gymnasts are obliged to begin with floor exercise in MAG and by the vault in WAG.

In this particular context, it is supposed that gymnasts could get physical advantages if they start their competition with the floor routine, however, those who finish at this apparatus could be the most penalised. Is the metabolic stress of a gymnast who begins with the floor exercise similar to the one who ends with it? Can a gymnast who begins with the floor exercise have an advantage physically and technically over the other candidates?

The following paragraphs highlight the metabolic differences whereas the cardiovascular outcomes are presented in Chapter 2, section 2.4 on "Cardiac response during gymnastic exercises".

## Variation of blood lactate production according to the competition rotation in MAG

Six elite-level male gymnasts (age 20.20 ± 3.61 years; weight 63.56 ± 8.51 kg; height 1.64 ± 0.05 m) performed two different simulations of two gymnastics competitions in random

(a) MAG Olympic rotation                    (b) WAG Olympic rotation

**FIGURE 1.4**    Males and females' competitions' Olympic rotation orders.

order (Zar, 1984). One simulation (R1) was performed following the FIG's standards rotation order, which began with the floor exercises (FX), followed by the pommel horse (PH), the rings (SR), the vault (VT), the parallel bars (PB) and ended by the horizontal bar (HB). The other competition (R2) was also performed according to the FIG's standards, but began with the PH followed by the SR, the VT, the PB, the HB and ended with the FX (Mkaouer et al., 2017).

Analysis of the blood lactate concentration (BLa) between the apparatus rotation orders has shown a significant variation with the FX, SR and VT levels [$\Delta = 15.38\%$, $\Delta = 11.43\%$ and $\Delta = 14.87\%$ respectively] (Mkaouer et al., 2017). However, for the other three apparatuses (i.e., PH, PB and HB), results showed similar BLa values (Mkaouer et al., 2017) (see Figure 1.5).

Results of BLa at the FX and the PH were similar to those reported by Groussard and Delamarche (2000) [11 mmol·l-1 and 5 mmol·l-1, respectively] and superior to the values reported by Jemni et al. (2000, 2003) [6.2 and 3.8 mmol·l-1; 6.3 and 3.8 mmol·l-1, respectively]. As for PB values, they are similar to Groussard and Delamarche (2000) and Jemni and Sands (2003) [5.8 and 5.1 mmol·l-1 respectively], inferior to Goswami and Gupta (1998) [6.23 mmol·l-1] and superior to Jemni et al. (2000) [4 mmol·l-1]. The BLa values of PH and SR are still similar to Goswami and Gupta (1998), Groussard and Delamarche (2000), Jemni and Sands (2003) and Jemni et al. (2000), but contrary to the previous authors mentioned earlier. HB values are superior to all of them.

## Variation of blood lactate production according to the competition rotation in WAG

Modern WAG seeks to exhibit power at the floor and the vault exercises, elegance and technicity at the uneven bars and balance with choreography at the beam. Each of these four apparatuses requires the highest degree of perfection, adjustment and coordination of movements, all with different physical and energetic demands (Jemni, 2011).

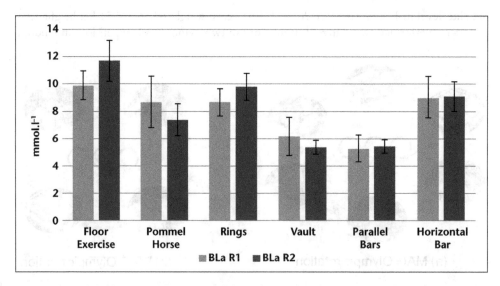

**FIGURE 1.5** Blood lactate concentration during the MAG competition rotation orders.

Eight elite-level female gymnasts (age 17.26 ± 2.19 years; height 1.56 ± 0.09 m; weight 48.37 ± 5.86 kg) performed a similar study as the ones described in the previous section. One of the competitions was performed following the FIG's standards rotation order (R1), which began with the vault (VT) and followed by uneven bars (UB), balance beam (BB) and floor exercise (FX). The other competition (R2) was also performed according to the FIG's standards but began with the FX, followed by VT, UB and BB (Marina & Rodríguez, 2014; Viana & Lebre, 2005).

Results showed a significant variation of BLa between the two rotations' orders. BLa was significantly higher at the VT and BB during R2 but was considerably lower at the FX [$\Delta = 47.62\%$, $\Delta = 9.67\%$ and $\Delta = 16.78\%$ respectively]. However, for the UB, results showed similar BLa values (Marina & Rodríguez, 2014) (see Figure 1.6). The results of BLa are comparable to those reported by Viana and Lebre (2005) and Rodríguez et al. (1999).

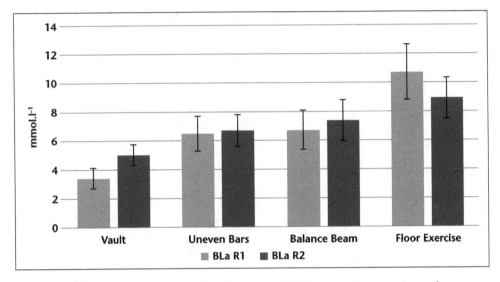

**FIGURE 1.6**   Blood lactate concentration during the WAG competition rotation orders.

## 1.6 CONCLUSION

The following conclusions are drawn from the previous two sections:

- There is evidence that gymnasts' aerobic metabolism is classified amongst the lowest compared to other athletes and is similar to non-active persons.
- Gymnasts' VO$_2$ max has not increased for the last five decades in spite of an increased training volume and intensity in conjunction with "tougher" rules (~ 50 ml/kg/min).
- There is evidence that gymnastic practice is not a sufficiently significant stimulus to enhance the oxidative system.
- Gymnasts are demonstrating an increased maximal power output as measured by standardised tests such as Force-Velocity and Wingate tests (up to 16 W/kg in males and up to 14 W/kg in females). This classifies them amongst the most powerful athletes.

- Gymnasts' metabolic thresholds are achieved at a late stage during maximal incremental tests (~80% of their $VO_2$ max). This delay is the result of the important volume of strength and conditioning.
- There is a paucity of literature resources regarding energy cost of male and female gymnastic routines. Meanwhile, there is evidence showing the increasing contribution of the anaerobic metabolism (blood lactate up to 11 mmol/l).
- Blood lactate concentrations vary between apparatus in MAG and WAG with the lowest values found at the vault and the highest at the floor exercises.
- Changing the rotation order of an Olympic gymnastics competition would lead to different lactate production and fatigue levels, hence starting or finishing the competition with the floor exercises could be an advantage or a disadvantage.

# 2

# CARDIOVASCULAR AND RESPIRATORY SYSTEMS OF THE GYMNASTS

*Monèm Jemni*

## 2.1 Learning outcomes

- Elaborate on the particularities of the gymnasts' cardiorespiratory system.
- Find out how the cardiovascular system adapts during gymnastics performance.
- Explore how to monitor cardiovascular stress and the way to interpret it in male and female gymnasts.

## 2.2 Respiratory and ventilation system

There is an evident paucity of literature regarding the respiratory system in male and female gymnasts. This lack of information is mainly due to the difficulties of operating medical instruments while performing gymnastics. It is indeed very tricky, even with advanced technology, to collect the gas exchange during aerial catch and release skills. This is the case, for example, in the horizontal and parallel bars. Investigators have tried to use portable telemetric gas analysers; unfortunately, the weight of the equipment obliged the gymnasts to make some adjustments to the skills. Aerial figures, for example, were harder to perform. The gymnasts were not only afraid to miss their landings/dismounts and therefore damage the equipment, but also they were prohibited from performing certain elements where they had to go on their backs and/or their fronts. In addition, the face mask and the turbine of the system prohibit head flexion and tucked or piked positions in all apparatus (as in the case of the somersaults).

The only two studies reported in the literature within this field were carried out in the late 1970s. The authors gave mainly details of the ventilatory system of the gymnasts at rest: Shaghlil (1978) showed that breathing frequency in gymnasts at rest was lower than the one measured in a group of sedentary subjects of a similar age (12 to 14 vs. 16 to 18 breaths per minute respectively). Gymnasts' ventilation varies, particularly in certain situations, such as handstand when respiration becomes more difficult due to viscera putting more weight on the diaphragm. This would increase the intra-thorax pressure, slow down the ventilation and cause congestion of the neck and the face. However, this is not due to any modification of the characteristics or the functioning of the respiratory muscles.

Barlett, Mance, and Buskirk (1984) have assessed the expiratory reserve volume (ERV), expressed as a percentage of the vital capacity (VC), in a group of sub-elite female gymnasts and compared it to a group of runners of similar age (29.7 + 7.1% vs. 43.1 + 6.4% respectively). This significant lower ERV was associated with a significant larger upper-body composition in the female gymnasts. The authors have suggested the greater upper-body mass in female gymnasts (the extra muscle on the chest) impinges upon the thorax to reduce its resting end-expiratory dimensions and, hence, the ERV of the lungs. Nevertheless, this decrease in lung efficiency is not a limiting factor in gymnasts' performance because they do not use their full lung capacity as in endurance sports. (Note, information about maximal oxygen uptake in gymnasts is given in Chapter 1, under section 1.3.)

## 2.3 Cardiovascular adaptation to gymnastic exercises

Practising gymnastics like any other regular physical activity would induce some cardio-vascular adaptations; amongst them, a normal hypertrophy of the myocardial, a decreased resting heart rate and an increased systolic ejection volume. However, only very few studies have investigated the cardiovascular adaptations to gymnastics training and, therefore, an extensive conclusion cannot be drawn at this stage.

According to Potiron-Josse and Bourdon (1989), regular exercise increases the cavity dimensions of the heart by 30% in active adults compared to sedentary people. Nonetheless, Roskamm (1980) has compared the volume of the heart relative to body mass of different athletes from the German teams with similar matching age groups of inactive persons. He concluded that weightlifters have the lowest volume followed by the inactive individuals and gymnasts (10.8 ml/kg and 11.7 ml/kg respectively). Similar findings have been confirmed by a later study conducted by Obert et al. (1997); their comparisons of pre-pubertal gymnasts with similar age group children did not show any significant differences for the following variables: myocardial mass, systolic diameter, systolic and diastolic ejection fractions, cardiac output, heart rate, and systolic volume.

According to Shaghlil (1978), the blood pressure of gymnasts does not differ to that of the sedentary people of the same age. However, a slight increase might occur at the approach of competitions. It is indeed during this critical period when the physical, physiological and psychological profile of the gymnast would take the "competitive shape". The increasing competitive anxiety might be one of the reasons behind the slight increase in blood pressure. The same author confirmed that during handstands, acrobatic elements and full swings at the horizontal bar, some modifications of the local blood flow occur due to the centrifugal and/or centripetal force. However, the vascular system conserves a fair distribution of the blood volume. In addition, local blood flow is quickly re-established back to normal post exercise.

## 2.4 Cardiac response during gymnastic exercises

Amongst the most recent investigations of cardiac response, Montpetit and Matte (1969) have shown a significant decrease of the heart rate (HR) when holding a handstand for 30 sec. HR drops from 120 bpm and stabilises around 95 bpm after 5 sec. It increases slightly after the cessation of the exercise to finally drop back to the original level after 10 sec.

This decrease is explained by an increased stroke volume following a sudden increase in venous return and the reverse occurs when returning to the standing position.

Seliger et al. (1970) and Faria and Pillips (1970) were also amongst the first who studied cardiac responses in varieties of gymnastic routines. Thanks to the electrocardiogram, Seliger et al. (1970) were able to assess the cardiac stress in male and female gymnasts. HRs reached 148 bpm in the balance beam, 135 at the uneven bars and 133 in vault. Meanwhile, it was slightly higher in males; it ranged between 139 bpm at the parallel bars and 151 bpm at the floor exercises. These values seem to be quite low and can be explained by the nature of the "easy" exercises of the 1970s (Table 2.1). During the same period, Noble (1975) was able to measure the cardiac response among female gymnasts using electrocardiographs emitting signals every 5 sec. The values found in the floor exercise, balance beam and uneven bars are higher than those found by Seliger et al. They lie between 162 and 189 bpm. The average values were respectively 169 + 6, 159 + 6 and 167 + 2 bpm.

In 1976, Montpetit found a fairly high cardiac response among male gymnasts performing simple exercises. A telemetric electrocardiograph was used to record the HRs. The values found ranged between 130 and 170 bpm. They are higher than those found by Seliger et al. in 1970 (139 and 151 bpm). The highest value was obtained on high bar (170 + 2 bpm) then, in descending order, parallel bars (158 + 4 bpm), rings (149 + 5 bpm), pommel horse (145 + 7 bpm) and vault (130 + 4 bpm). It was not until 1982 that Montgomery and Beaudin assessed the HR along the four female routines using telemetric recording. The peak HR was much higher than the previous investigations 178 + 11 bpm. The average HR was 166 + 10 bpm. Later in the 1990s, Goswami and Gupta (1998)

**TABLE 2.1** Heart rates in bpm during gymnastic exercises

| Females | Seliger et al. (1970) | Noble (1975) | Montgomery & Beaudin (1982) | Viana & Lebre (2005) | |
|---|---|---|---|---|---|
| Vault | 133 ± 13 | – | 162 | – | |
| Un bars | 133 ± 10 | 167 ± 2 | 187 | 195 ± 10 | |
| B beam | 130 ± 13 | 159 ± 6 | 177 | 179 ± 8 | |
| Floor exe | – | 169 ± 6 | 185 | 193 ± 2 | |
| Max | 148 | 189 | 187 | 205 | |
| Average | 132 | 165 | 178 | 189 | |
| Males | Goswami & Gupta (1998) | Lechevalier et al. (1999) | Groussard & Delamarche (2000) | Jemni et al. (2000, 2002, 2003) | Viana & Lebre (2005) |
| Floor exe | 183 ± 11 | 186 ± 5 | 160–179 | 186 ± 11 | 182 + 2 |
| Pommels | 173 ± 9 | 188 ± 7 | 158–170 | 185 ± 11 | 174 + 8 |
| Rings | 175 ± 10 | 188 ± 6 | 156–165 | – | 171 ± 15 |
| Vault | – | 160 ± 9 | 160–179 | 162 ± 14 | – |
| Parallel b | 175 ± 15 | 183 ± 7 | 154–162 | 181 ± 11 | 176 + 5 |
| High b | 182 ± 12 | 187 ± 7 | 164–180 | 185 ± 9 | 183 |
| Max | 195 | 186 | 180 | 180 | 190 |
| Average | 178 | 181 | 160 | 167 | 177 |

studied the cardiac responses during full male routines in separate sessions using HR monitors (Sport Tester PE-3000). Peak HR of five events (floor, pommels, vault, parallel bars and horizontal bar) was 180 + 5 bpm and the average was 161 + 9 bpm. These values are much higher than those found in the 1970s although the recording was performed in gymnasts practising at a similar level (non-elite). This confirms the increasing difficulties of the gymnastic elements.

Lechevalier et al. (1999) recorded the cardiac response during three sessions of different intensities using a pulse by pulse HR monitor (BHL Bauman 6000): 1) "an easy session" based on skills' rehearsing on the floor, pommels and rings; 2) a competition where the gymnasts performed their six routines; 3) a double competition, where the gymnasts performed their six routines twice (also called back to back). During the competition and the back to back sessions, gymnasts' maximal heart rate ($HR_{max}$) was close to their $HR_{max}$ measured in a maximal incremental test (average HRs 122 + 7 bpm and 124 + 8 bpm respectively). However, their HRs were around ~60% of the $HR_{max}$ during the skill rehearsing session and therefore significantly lower than the two other sessions (average HR 114 + 10 bpm).

Figure 2.1 shows different recordings of the cardiac response during a high level male competition. HR was recorded continuously with HR monitors (BHL Bauman 6000). The recording shows the maximal values reached, the average and the HR range that the gymnasts have mostly used during each event. The highest HRs were recorded in the floor exercises and the lowest were obtained at the vault; these were significantly lower than all

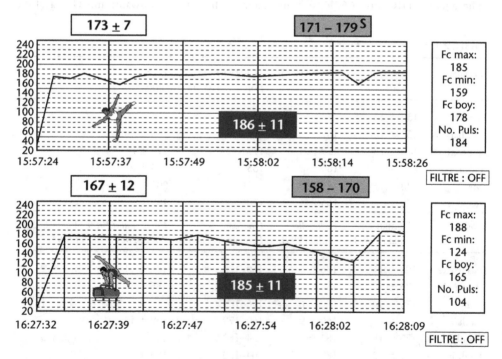

**FIGURE 2.1** Cardiac response during high level male gymnastics competition.

(Jemni, 2001)

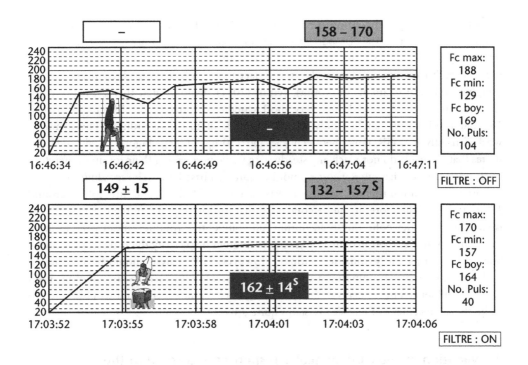

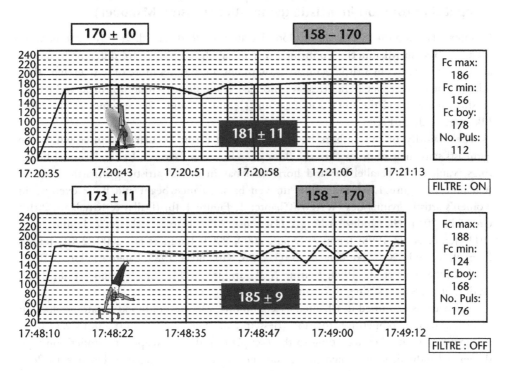

the other recordings and confirm the low blood lactate values found in the same apparatus (see Chapter 1, section on "Blood lactate measurement during gymnastic exercises"). It is interesting to notice that peak HR was reached at the end of the floor, pommels, parallel bars and high bar routines. This is due to the fact that gymnasts finish their routines with a higher technical element (dismount). Jemni (2001) demonstrated that gymnasts work closer to their $HR_{max}$ only during very short sequences. Indeed, the HR range between 180 and 190 bpm was applied only during 16 + 9% of all the recordings. The gymnasts worked mostly in a HR range between 158 and 170 bpm (28%). Figure 2.1 also demonstrates that gymnasts do not reach any steady state while performing their routines, therefore confirming the fact that energy cost cannot be estimated based on such recordings. It is also important to mention that the same gymnasts have reached their lactic threshold at around 170 + 10 bpm. Therefore, it can be concluded that cardiac response of high level gymnasts is close to their metabolic threshold during full competitive routines, meanwhile peaking from time to time at their maximal HRs. These findings help clarify the nature of the metabolic stress during full gymnastic routines. However, HR analysis should be interpreted with caution because running on a treadmill is different to gymnastics. In addition, HRs could be influenced by several variables, such as catecholamine levels, and would therefore make the achievement of any steady states quite a hard task.

## 2.5 Variation of the cardiovascular responses according to the competition rotation in artistic gymnastics (Bessem Mkaouer)

Chapter 1 (section on "Variation of blood lactate produced according to the competition rotation (Bessem Mkaouer)") confirmed that metabolic responses are different from one apparatus to another in artistic gymnastics, with few apparatuses more or less "taxing" than others. The highest metabolic stress was noticed at the floor routines for male and female artistic gymnasts and the lowest at the vault (Jemni & Sands, 2003; Jemni et al., 2003; Mkaouer et al., 2017).

The majority of the authors have also assessed cardiovascular responses whilst the gymnasts start their routines with the floor exercise, then followed by the pommel horse, rings, vault table, parallel bar and horizontal bar in males' artistic gymnastics (MAG) (Chapter 1, Figure 1.4a) and vault, uneven bars, balance beam and floor exercise in women's artistic gymnastics (WAG) (Chapter 1, Figure 1.4b) (Dallas & Kirialanis, 2010; Grossfeld, 2014; Hernández, Balón, & Galarraga, 2009; Irurtia et al., 2007; Jemni, 2010, 2011; Jemni et al., 1998; Jemni & Sands, 2000, 2003; Jemni et al., 2000, 2001; Jemni, Friemel, & Sands, 2002; Jemni et al., 2003, 2006; Kirkendall, 1985; Lange, Halkin, & Bury, 2005; Lechevalier et al., 1999; Marina & Rodríguez, 2014; Papadopoulos et al., 2014; Prados, 2005; Sands, Caine, & Borms, 2003; Viana & Lebre, 2005).

However, as explained in Chapter 1 (section on "Variation of blood lactate production according to the competition rotation (Bessem Mkaouer)"), these instructions vary from one gymnast to another according to the competition draw, except for competition type II (general individual) in which the six chosen gymnasts are obliged to begin with floor exercises in MAG and by the vault in WAG.

Similarly to the metabolic advantage that gymnasts could get if they start their competition with the floor routine, they are supposed they get the same advantage from the cardiovascular stress. From the other side, those who finish at this apparatus could be the most penalised. The following paragraphs highlight the cardiovascular differences when swapping the rotation orders in MAG and WAG.

### Variation of the cardiovascular responses in males' artistic gymnastics

The same experimental protocol described in Chapter 1 (section on "Variation of blood lactate production according to the competition rotation in MAG") has enabled the assessment of the cardiovascular stress (Mkaouer et al., 2017) during two competitions with different rotation orders.

The results showed significant variations of the cardiovascular responses between the two rotations' orders (Table 2.2).

Significant variation of the maximal HR ($HR_{max}$) at the floor routines between both rotation sequences with R2 implicating higher values [$\Delta = 2.83\%$]. Similar findings were observed at the vault and parallel bars exercises, with R2 higher than R1 [$\Delta = 2.72\%$ and $\Delta = 1.72\%$ respectively]. HR records did not statistically differ at the pommels, rings and horizontal bar. It is worth noting that $HR_{max}$ values reported in the present study were similar to those reported by Hernández et al. (2009), Jemni (2011), Prados (2005) and Viana and Lebre (2005).

Table 2.2 shows that HRs differed during recovery; this later was 4.69% higher at the first minute following the floor exercises during R2 when compared to R1. Similarly, they were 5.12% and only 1.72% higher at the parallel bars and the high bars respectively during

**TABLE 2.2** Cardiovascular response during two different competitions under two different Olympic rotation orders in MAG.

| Apparatuses | | $HR_{max}$ (b min$^{-1}$) | $HR_{Rtm}$ (b min$^{-1}$) |
|---|---|---|---|
| | | M ± SD | M ± SD |
| Floor exercise | R1 | 185.20 ± 11.67 | 142.40 ± 6.91 |
| | R2 | 190.60 ± 9.15 ↗ | 149.40 ± 4.61 ↗ |
| Pommel horse | R1 | 157.80 ± 6.90 | 132.40 ± 6.80 |
| | R2 | 148.80 ± 5.24 | 130.60 ± 2.40 |
| Rings | R1 | 181.60 ± 12.74 | 147.40 ± 8.87 |
| | R2 | 184.40 ± 8.64 | 147.6 ± 5.03 |
| Vault | R1 | 171.40 ± 10.06 | 124.00 ± 11.46 |
| | R2 | 176.20 ± 7.95 ↗ | 119.80 ± 2.49 |
| Parallel bars | R1 | 171.40 ± 6.18 | 129.80 ± 7.69 |
| | R2 | 174.40 ± 7.02 ↗ | 136.80 ± 8.78 ↗ |
| Horizontal bar | R1 | 183.60 ± 9.65 | 160.40 ± 6.50 |
| | R2 | 185.40 ± 8.20 | 163.20 ± 4.08 ↗ |

(R1) Rotation order (floor exercise, pommel horse, rings, vault table, parallel bars and horizontal bar); (R2) Rotation order (pommel horse, rings, vault table, parallel bars, horizontal bar and floor exercise); ($HR_{max}$) maximal heart rate; ($HR_{R1m}$) recovery heart rate after 1 min; (↗) increase significantly at p < 0.05; (↘) decrease significantly at p < 0.05.

the same rotation. Recovery HRs ($HR_{R1m}$) were similar to those reported by Jemni (2010), Jemni and Sands (2003) and Jemni et al. (2003).

### *Variation of the cardiovascular responses in womens' artistic gymnastics*

The same experimental protocol described in Chapter 1 (section on "Variation of blood lactate production according to the competition rotation in WAG") has enabled the assessment of the cardiovascular stress during two competitions with different rotation orders (Mkaouer et al., 2017).

The results showed significant variations of the cardiovascular variables between the two rotations' orders (Table 2.3).

Similar to MAG, the results showed that R2 provoked statistically higher cardiovascular stress (with $\Delta$ = 10.18%; 2.37%; 2.22% respectively at the vault, uneven bars and floor exercise).

Table 2.3 also shows that HRs differed during recovery; this later was 12.98%, 6.37% and 3.53% higher at the first minute following the vault, uneven bars and the beam during R2 when compared to R1. However, it seems like R2 has provoked 2.72% less stress than R1 following the recovery at the floor exercise. The above results are comparable to those reported by Viana and Lebre (2005) and Rodríguez et al. (1999).

In summary, cardiovascular responses have noticeably changed according to the sequence order of the apparatus during the two different Olympic rotations. $HR_{max}$ has noticeably varied at the floor exercise, vault, parallel bars and horizontal bar in MAG and also at the vault, uneven bars, balance beam and floor exercise in WAG. They have noticeably increased during the second rotation. Thus, $HR_{R1m}$ have remained lower during the first rotation, as gymnasts began with the floor exercise.

**TABLE 2.3** Cardiovascular response during two different competitions under two different Olympic rotation orders in WAG

| Apparatuses | | $HR_{max}$ (b min$^{-1}$) | $HR_{Rtm}$ (b min$^{-1}$) |
|---|---|---|---|
| | | M ± SD | M ± SD |
| Vault | R1 | 180.50 ± 7.35 | 132.88 ± 6.20 |
| | R2 | 198.88 ± 9.15 ↗ | 150.13 ± 4.36 ↗ |
| Uneven bars | R1 | 179.00 ± 5.21 | 131.38 ± 11.21 |
| | R2 | 183.25 ± 4.23 ↗ | 139.75 ± 5.70 ↗ |
| Balance beam | R1 | 188.00 ± 16.39 | 127.50 ± 8.59 |
| | R2 | 186.00 ± 16.40 | 132.00 ± 7.11 ↗ |
| Floor exercise | R1 | 196.75 ± 9.90 | 142.50 ± 4.57 |
| | R2 | 201.13 ± 10.53 ↗ | 138.63 ± 4.31 ↘ |

(R1) Rotation order (vault, uneven bars, balance beam, floor exercise); (R2) Rotation order (floor exercise, vault, uneven bars, balance beam); ($HR_{max}$) maximal heart rate; ($HR_{R1m}$) recovery heart rate after 1 min; ( ↗ ) increase significantly at $p < 0.05$; ( ↘ ) decrease significantly at $p < 0.05$.

## 2.6 CONCLUSION

- More details still need to be investigated in order to draw a full picture of the respiratory and cardiovascular responses to gymnastics.
- Several gymnastic elements, particularly those involving upside-down positions, slow down ventilation.
- It has been suggested that the greater upper-body composition of the gymnasts, when compared to similar age groups, may impinge upon the thorax and reduce its resting end-expiratory dimensions and, hence, the expiratory reserve volume of the lungs.
- There is evidence that gymnastic practice is not a sufficiently significant stimulus to enhance several heart variables including myocardial mass, cardiac output, HR, systolic volume and blood pressure.
- Some modifications of the local blood flow occur due to the centrifugal and/or centripetal force while performing several elements. However, it is quickly re-established to normal post exercise.
- The development of the measurement tools from the 1960s to nowadays allowed more reliable results and shows an increased cardiovascular stress in parallel to the increasing technical requirements.
- HRs of high level gymnasts are close to their metabolic threshold values during full competitive routines. Maximal HRs are only reached for short peak periods.
- Gymnasts' HR recordings did not show any steady states. This makes them difficult to be interpreted for energy cost.
- Cardiovascular responses differ according to the sequence order of the apparatus during Olympic competitions.
- Maximal HR is noticeably higher at the floor exercise, vault, parallel bars and horizontal bar in males' artistic gymnastics and also at the vault, uneven bars, balance beam and floor exercise in women's artistic gymnastics, when finishing the competition with the floor exercises.
- Recovery HRs have remained lower when gymnasts began with the floor exercise.

# 3

# DIET, NUTRITION, SUPPLEMENTATION AND RELATED HEALTH ISSUES IN GYMNASTICS

*Monèm Jemni*

## 3.1 Learning outcomes

• Find out about gymnasts' diet, nutrition habits and supplementation.
• Identify health issues related to high training volume and intensity.

## 3.2 Diet in male and female gymnasts

In gymnastics, likewise all artistic sports, a special consideration is given to the external appearance. Female gymnasts and particularly rhythmic gymnasts suffer from several pressures: the most direct one is the pressure of the coach who in most of the cases is indirectly concerned by the weight of the gymnasts; the pressure of the judges/referees who attribute a mark on their aesthetic abilities/appearance and the pressure of putting on weight which may reduce their acrobatic abilities. These pressures may push the gymnasts and the coaches towards severe dietary habits which may lead to a negative energy balance (Rosen & Hough, 1988). It has been demonstrated that the risks of eating disorders such as anorexia nervosa and bulimia are increased in athletes performing sports which emphasise leanness (Brotherhood, 1984; Jankauskienė & Kardelis, 2005). Almost 100% of the reviewed articles agreed on the issue of eating disorder amongst gymnasts. Evidence showed that such diets led to a decreased performance and an increased risk of injuries which might be associated with some severe health issues. Benardot (1999), for example, has attributed the incidence of stress fracture in rhythmic gymnastics to their negative energy balance and demonstrated that 86% of the high level gymnasts have at least one serious injury per year.

In both genders, total energy consumption and nutritional intake are insufficient in spite of high energy expenditure, although to a lesser extent in male gymnasts (Filaire & Lac, 2002; Fogelholm et al., 1995; Lindholm, Hagenfeldt, & Hagman, 1995). These athletes may practise more than six hours per day in certain periods, such as during camps (Sands, 1990b; Stroescu et al., 2001). Interviews with former rhythmic gymnasts showed that it is very common to practise for more than ten hours per day. The energy intakes reported in Chinese sub-elite and elite female gymnasts were 1,637 kcal/day and 2,298 kcal/day respectively (Chen et al., 1989). The authors demonstrated that both groups consume a high percentage of fat compared to proteins (~43% vs. 13% respectively). However, this might be the result of a lack of knowledge about dietary and healthy eating. Filaire and Lac (2002), for example, found different diet portions in French female gymnasts (14% protein, 48% carbohydrate and 37% fat). Nevertheless, both studies, as well as Lindholm et al. (1995), have demonstrated that female gymnasts have deficiency in several minerals in particular: calcium, thiamine and riboflavin, iron, fibre and E and B6 vitamins when compared to similar age females.

In spite of the unbalanced diet and the above deficiencies, it has been shown that protein intake in pre- and early pubertal female gymnasts was comparable to the control group (~7 g/kg/day) (Boisseau et al., 2005). The authors have also shown that protein synthesis and degradation were also similar (~6 g/kg/day and ~5 g/kg/day respectively).

Male gymnasts, on the other hand, have a more balanced diet. Some authors have even suggested that protein intake is quite high in order to enhance lean tissue formation (Brotherhood, 1984). Nutritionists recommend that gymnasts who train for more than eight hours per week should always consume nutrient rich foods such as vitamin enriched and wholegrain cereals, fruits, vegetables and lean meats in order to ensure an adequate B-vitamins status. Total energy intake should also be adequate to maintain weight. Arkaev and Suchilin (2004) recommend a daily caloric intake between 4,500 and 5,000 Kcal during intensive training periods for high level male gymnasts. Table 3.1 gives some generic indications of the most recommended nutrients

As a final point, there is an evident paucity of the literature regarding male gymnasts' diet. However, female gymnasts' diet is more problematic and has been quite well investigated. Most of the investigations are in accordance and conclude that the unbalanced diet associated with the high energy expenditure could partly explain their low body mass index (BMI), small percentage of body fat, late pubertal development and the irregular menstrual patterns as explained later in this chapter under the heading "Hormonal regulation, growth and sexual development".

## 3.3 Supplementation in gymnastics

Coaches always ask: do gymnasts need any supplementation to perform the high level of strength and power required, especially at high level?

Several investigations have confirmed that a significant percentage of elite gymnasts are familiar with the use of nutritional aids. The most frequently used supplements are vitamins,

**TABLE 3.1** Food sources for B complex vitamins according to the South African Gymnastics Federation (Humphy, 2010)

| Vitamin | Best food sources |
| --- | --- |
| Thiamine (B1) | Dried brewer's yeast, yeast extract, brown rice |
| Riboflavin (B2) | Yeast extract, dried brewer's yeast, liver, wheatgerm, cheese and eggs |
| Pyridoxine (B6) | Enriched cereals, potatoes with skin, bananas, legumes, chicken, pork, beef, fish, sunflower seeds, spinach |
| Folate | Dried brewer's yeast, soya flour, wheatgerm, wheat bran, nuts, liver and green leafy vegetables |
| Vitamin B12 | Liver, kidney, fish, red meat, pork, eggs and cheese |

proteins and calcium. A questionnaire study amongst the Greek elite gymnasts showed that 58% have taken nutritional supplements following their doctor's recommendation (Zaggelidis et al., 2005). Likewise any other sports, a significant number of gymnasts are not aware of the prohibited substances list and neither of the possible detrimental effects of excessive supplementation.

A four-month period of intense training and supplementation with soy protein in several members of the Rumanian Olympic female team has demonstrated an increase in lean body mass, serum levels of prolactin and T4 and also a decrease in serum alkaline phosphatises (Stroescu et al., 2001). However, a 12-month randomised control trial comparing the effect of 500 mg calcium supplementation between gymnasts and a similar age group did not show any significant difference in volumetric bone mineral density (BMD) at the radius, tibia, spine and whole body (Warda et al., 2007). The authors have concluded that there is no beneficial effect of additional calcium in gymnasts who intake their recommended calcium (555–800 mg/day for eight to 11 year olds). They also suggested that the lack of significant difference in volumetric BMD might be related to the fact that gymnasts adopt their bones under the effect of the repetitive loading and therefore do not benefit from additional calcium supplementation.

To conclude, most authors confirmed several nutrient deficiencies; however, investigations on supplementation were not conclusive. In fact, most of the supplementation studies did not check if there were any deficiencies at baseline and thus makes interpretation quite devious. While balanced nutrition might be sufficient for some gymnasts, common sense would suggest supplementation for those who present some deficiencies. Nonetheless, these supplementations should be prescribed by a sport dietician in order to avoid any risk of taking any potential banned substances. The remaining issue and the most important thing to consider/monitor is: if they eat enough, what do they eat?!

## 3.4 Effect of high volume and intensity of training on body composition, hormonal regulation, growth and sexual development

Several health issues related to gymnastics (artistic and rhythmic) have been the centre of ongoing debate for the last 15 years. Amongst these issues: body composition and its relation to a high volume of training combined with inadequate diet and also hormonal regulation, growth and sexual development. Online search engines showed up to 229 articles based on the following key words: gymnastics, growth, energy intake, amenorrhea and bone mineral density. This section highlights some of the main health problems that gymnasts would encounter at the highest level of performance.

### Body composition

It has become evident that gymnasts have a very low body fat percentage and a very low BMI when compare to normal age and gender matched groups, particularly noticeable in females (rhythmic and artistic) (Cassell, Benedict, & Specker, 1996; Claessens et al., 1992; Courteix et al., 1999; Malina, 1994; Soric, Misigoj-Durakovic, & Pedisic, 2008). Ninety-eight per cent of the reviewed articles confirmed a cross sectional decrease in body composition parameters in particular weight and height, meanwhile 43% of them confirmed a longitudinal effect. Table 3.2 shows BMI and body fat percentage reported in male and female gymnasts. It has also been demonstrated that these decreased body composition variables are sustained over the years. Benardot and Czerwinski (1991) reported a

**TABLE 3.2** Body mass index and fat percentage in male and female gymnasts

| Females | Courteix et al. (2007) 13.4 yrs n = 36 RG | Douda et al. (2006) 13 yrs n = 39 RG | Jemni et al. (2006) 20.6 yrs n = 21 AG | Georgopoulos et al. (2004) 16 yrs n = 169 AG | Markou et al. (2004) 16 yrs n = 120 AG | Muñoz et al. (2004) 16.2 yrs n = 9 RG | Bale & Goodway (1990) 13.3 yrs n = 20 AG |
|---|---|---|---|---|---|---|---|
| BMI (kg/m²) | 16.5 | 15.9 | | 19.0 | 18.6 | 18.6 | 18.2 |
| % BF | 14.4 | 13.9 | | 19.5 | 18.4 | – | 10.9 |
| *Males* | | | | *17 yrs n = 93 AG* | *18 yrs n = 68 AG* | | |
| BMI (kg/m²) | – | | 23.2 | 21.5 | 21.1 | – | – |
| % BF | – | | 9.7 | 10.6 | 10.3 | – | – |

AG: artistic gymnastics; RG: rhythmic gymnastics.

cross-sectional BMI range of 12.9 to 20.8 kg/m² in elite junior gymnasts aged between seven and ten years and 14.6 to 20 kg/m² between 11 and 14 years. The percentage of body fat reported in the same group was as follows: 5.1 to 16.7% in seven to ten year olds, which does not change at 11 to 14 years old and remains around 6 to 15.1%.

## Bone development and mineral density

It has been shown that low energy intake associated to high energy expenditure might be considered as two potential complicating factors which may affect bone development and later osteoporosis (Benardot, Schwarz, & Weitzenfeld, 1989; Drinkwater et al., 1984). Around 80% of the studies confirmed the skeletal maturation delay in artistic gymnasts whereas this agreement reaches almost 98% in rhythmic gymnastics. Theodoropoulou et al. (2005) undertook one of the wider scale investigations by assessing 400 rhythmic gymnasts and 400 female artistic gymnasts from 32 countries, all of them female world class aged between 11 to 23 years. The authors confirmed a delayed bone development of 1.3 years in rhythmic gymnasts and of 2.2 years in artistic gymnasts. A similar finding was confirmed by Muñoz et al. (2004): around two years' delay in rhythmic gymnasts compared to controls. However, it seems that male gymnasts are less affected; Markou et al. (2004) proved that bone age was delayed by two years in 169 females compared to only one year in 93 males aged between 13 to 23 years, all elite level. Similarly, Jemni et al. (2000) have also shown a six-month developmental delay in post-pubertal high level French male gymnasts (18.5 years old).

It appears that artistic gymnasts endure more intense mechanical load for upper and lower limbs and trunk because of the higher acrobatic skills including mounts and dismounts. Authors have confirmed that gymnastic exercises are associated with an increase in bone mineral content and density (BMC – BMD) particularly in the most exposed joints (wrist and ankle) and of the lumbar spine (Courteix et al., 2007). It was suggested that the higher levels of osteocalcin might be the main contributor to higher BMD. Furthermore, the authors suggested that preserved bone health might be a counterbalanced effect of hormonally disturbed athletes. In addition, it appears that gymnastics has long lasting benefit effects on bones. It has been shown that total or partial BMD increases after retirement in active and/or inactive gymnasts (Dowthwaite & Scerpella, 2009). Ducher et al. (2009) gave evidence of an enhanced mineral content of the radius and the ulna in retired artistic gymnasts (18–36 years). These findings confirm previous studies: Markou et al. (2004) proved that long term high intensity exercise is negatively correlated with BMD, whereas it is positively correlated with chronological age. In their review of the literature, Dowthwaite and Scerpella (2009) recommended further research in order to elucidate skeletal loading dose-response curves and the sex- and maturity-dependence of skeletal adaptation which varies up on the anatomical sites and tissue compositions.

As a final point, it can be concluded that bone mineral density excess is exercise related.

## Hormonal regulation, growth and sexual development

Markou et al. (2004) have reported around 32% and 12% of female amenorrhoeic rhythmic gymnasts (absence of menstruation by the age of 16 years) in Canada and in Greece respectively. Meanwhile the percentage of the oligoamenorrhoeic rhythmic gymnasts (menstrual cycle duration greater than 36 days) in both countries was more dramatic (around 65%). A similar finding was also confirmed in Spanish rhythmic gymnasts with 45% oligoamenorrhoeic (Muñoz et al., 2004). Pubertal delays have been confirmed in both artistic and rhythmic gymnastics, with a more pronounced effect in artistic gymnastics due to higher stress imposed by training intensity and the higher number of competitions. Muñoz et al. (2004) revealed that rhythmic gymnasts reached their menarche at 15 + 0.9 years compared to controls (12.5 + 1 years). Similarly, Courteix et al. (2007) found a two-year delay in French rhythmic gymnasts, whereas Theodoropoulou et al. (2005) have shown not only a delayed puberty in 400 rhythmic and 400 artistic gymnasts but also 17% and 20% of them respectively had no menarche also with comparison to their mothers' menarche and to their untrained sisters. Likewise, Sands, Hofman, and Nattiv (2002) confirmed the difference in menarche age and menstruation irregularities between high level American gymnasts and their mothers via a questionnaire study. These findings in reality demonstrate that growth and development issues are not related to any genetics variables but actually the consequence of training stress.

Several authors had a particular interest investigating leptin in gymnasts. Leptin is in fact an ob-gene protein secreted by fat cells; it has a role in the regulation of body weight and in the stimulation of the reproductive axis (through fat). Evidence shows that mechanisms leading to a deregulation of the reproductive-axis in patients with anorexia nervosa are comparable with those leading to delayed puberty in elite female gymnasts with very low body fat and low diet. This may lead to the "Female Triad" which is a combination between eating disorder, amenorrhea and osteoporosis.

Of the studies, 65% confirmed a decreased level of leptin in both rhythmic and artistic gymnasts (Courteix et al., 2007; Jemni, Keiller, & Sands, 2008; Markou et al., 2004; Muñoz et al., 2004; Weimann, 2002; Weimann et al., 1999). Noticeably, it has been shown that leptin levels of gymnasts with particularly low fat stores are less than those measured in a group of patients with anorexia nervosa (1.2 ± 0.8 μg/l v 2.9 ± 2.7 μg/l respectively) (Matejek et al., 1999). The reason for this hypoleptinemia was attributed to the insufficient caloric intake. Weimann et al. (2000) have suggested that hypoleptinemia, in turn, causes delayed puberty and growth. In fact, several authors have confirmed decreased oestradiol and oestrogen levels in pubertal gymnasts with no significant rise, which occurs with normal sexual maturation. Whereas, IGF-1, IGF BP-3, TSH, T3 and

T4 showed normal age-dependent serum levels (Courteix et al., 2007; Weimann, 2002; Weimann et al., 1999).

Leptin and oestrogen production by fat tissue plays a crucial role in triggering menarche, reflecting a natural adaptation of the body to high energy demands (Moschos, Chen, & Mantzoros, 2002). Female fat tissue is a significant source of oestrogens (converting androgens to oestrogens). A decreased conversion of androgens to oestrogens because of decreased fat tissue in athletes may contribute to a delayed breast development and menarche (Perel & Killinger, 1979; Schindler, Ebert, & Friedrich, 1972). However, recent experiments suggested that menstrual disorders are part of a metabolic response to face an energy deficit relationship: stress of high energy output + inappropriate low energy intake + energy cost + energy availability (Loucks & Redman, 2004).

## 3.5 CONCLUSION

- Considerable pressure upon gymnasts (especially females, rhythmic and artistic) to maintain leanness.
- Evidence of eating disorders which have serious health implications.
- In both genders, total energy consumption and nutritional intake are insufficient in spite of high energy expenditure; although to a lesser extent in male gymnasts as they have a more balanced diet.
- Evidence that female gymnasts suffer from several minerals deficiencies, in particular: calcium, thiamine and riboflavin, iron, fibre and E and B6 vitamins.
- Significant percentages of elite gymnasts are familiar with the use of nutritional aids. The most frequently used supplements are vitamins, proteins and calcium.
- Only a few studies have investigated the effect of supplementation and therefore results are inconclusive. However, protein supplementation has shown some effects on body composition whereas calcium did not show any benefits on bone health.
- Gymnastics is a significant stimulus to maintain a very low body fat percentage and a very low BMI, particularly noticeable in female rhythmic and artistic gymnasts.
- Strong agreement on skeletal maturation delay which affects more artistic than rhythmic gymnasts because of the increased stress of training and competitions.
- Gymnastics increases bone mineral content and density, particularly in the most exposed joints and the lumbar spine. There is accumulating evidence of benefit persistence after activity cessation, into adulthood and retirement in active and/or inactive gymnasts.
- Pubertal delays have been confirmed in both artistic and rhythmic gymnastics, with a more pronounced effect in artistic gymnastics.
- The literature shows a very high number of female amenorrhoeic gymnasts and alarming figures of oligoamenorrhoeic.
- It is premature to conclude that amenorrhoeic gymnasts do not experience premature osteoporosis.
- Confirmed decrease of leptin level in both rhythmic and artistic gymnasts with particularly low fat stores, with some values less than those measured in a group of patients with anorexia nervosa.
- It is suggested that hypoleptinemia causes delayed puberty and growth because of the decreased oestradiol and oestrogen indirectly affected by the reduced fat store.
- Evidence that growth and development issues are not genetics but exercise induced.

# PART I REVIEW QUESTIONS

Q1. Describe maximal oxygen uptake of the gymnasts and comment about its importance/relevance to gymnastics performance.

Q2. Analyse the evolution of the $VO_2$ max of the gymnasts throughout the last few decades versus the evolution of the sport.

Q3. Do you think it is important to enhance gymnasts' $VO_2$ max and why?

Q4. Explain the physiological and training concepts affecting the lactic threshold's delay of the gymnasts.

Q5. Describe a few methods and techniques used to estimate and/or measure the energy cost of male and female gymnastic exercises/routines.

Q6. Using relevant data and illustrations, discuss the energetic cost of gymnastics exercises. *(Essay type question.)*

Q7. Analyse the peak power output of the gymnasts as assessed by standardised tests.

Q8. Analyse the metabolic responses to gymnastics routines/exercise.

Q9. Do you think metabolic responses to gymnastics routines vary if the competition rotation's order changes? Justify your answer.

Q10. Compare the aerobic and the anaerobic metabolisms of the gymnasts. *(Very long essay type question.)*

Q11. Analyse the respiratory system in gymnasts.

Q12. Analyse the cardiovascular adaptations to gymnastic exercises.

Q13. Analyse heart rate responses during gymnastic exercises.

Q14. Do you think cardiovascular responses to gymnastics routines vary if the competition rotation's order changes? Justify your answer.

Q15. Discuss the relation between the nutritional intake and the energy expenditure in male and female gymnasts. *(Essay type question.)*

Q16. Discuss the ergogenic supplementation in gymnastics.

Q17. Discuss the effect of high volume and intensity of training on the gymnast's body composition, hormonal regulation, growth and sexual development. *(Long essay type question or average length answer if taken individually.)*

# PART II

# Applied coaching sciences in gymnastics

# PART II

# Applied coaching sciences in gymnastics

## Learning outcomes (*Monèm Jemni*)

By the end of this part you will be able to:

- Appreciate the importance of strength, speed, power, flexibility, stamina and skills, as an integrative inter-related fitness model.
- Identify the training principles that one could apply in gymnastics.
- Explore techniques, tools and procedures to assess the above fitness components in the lab and in the field.
- Appreciate the importance of the jumping skills from the early ages to high levels of practice and the way to assess and to develop them.

## Introduction and objectives (*Monèm Jemni*)

The first three chapters of this section were part of the physiology for gymnastics section in the book's first edition. We decided to complete them with another important chapter related to jumping skills and put them together to stand alone as an entire new section entitled "Applied coaching sciences in gymnastics". Surely this provides some crucial information to students, young and confirmed coaches about real training situations and the science underpinning the actions. As an academic, myself, I regularly witness how my students transfer their learning within my course to different sports they practise and/or coach. We hope, through the content of this section, to provide you with some basic, but also some advanced, knowledge and practices to enhance your field experiences and to open your minds to further research ideas.

The objective of this part is to give the reader holistic information about gymnasts' fitness make-off and the way to improve them via scientific principles of training and coaching. It starts by redefining the physical abilities required to perform this sport at a high level of fitness and their interactions, followed by the principles of training that could be applied to enhancing them. These first two sections would have provided enough background to embark on a review of the most specific physical and physiological assessments of these abilities and specifically for gymnasts. The last chapter is dedicated to the very important basic gymnastics skill: the jump. This will be decorticated into the smallest details showing the underpinning neuro-physiology and biomechanics principles behind them, followed by ways not only to assess them but also to improve them.

# 4

# FITNESS MODEL OF HIGH LEVEL GYMNASTS

*William A. Sands*

## 4.1 Learning outcomes

• Appreciate the importance of strength, speed, power, flexibility, stamina and skills, as an integrative inter-related fitness model.

## 4.2 The fitness model

Gymnastics physical fitness implies that there are some specific biological/physiological/ physical regulatory processes that are at least somewhat unique to the gymnast. Indeed, there are specific physiological functions that are unique in terms of magnitude, timing and dominance in all sporting activities. In general, we describe a gymnast's physiological condition as his/her "fitness". This term implies that the body must be prepared, enhanced and maintained in order to perform gymnastics with competence. In short, fitness is "readiness" for some type of activity.

Interestingly, no athlete can be 100% fit in all characteristics of physiology at the same time (Sands, 1994a; Sands, McNeal, & Jemni, 2001b). The body naturally economizes its adaptations to the demands that are imposed on it by training and performance. In fact, this economic adaptation has an acronym that accompanies it and a principle – SAID – specific adaptations to imposed demands. SAID means that the body will adapt to the demands made on it, no more and no less. A corollary to the SAID principle is that training must provide the athlete's body with an "unambiguous message" of what you want the body to become.

> There are limits to the capacity of the athlete which are determined by his stage of development. During the competitive season (competitive period), the load tolerance and adaptability of the athlete is taxed by intensive competition and maximum workload at competition levels which take him to his limits. This leads to a particularly rapid development of the standard of performance; the previously developed bases are transferred to athletic performance, thus establishing optimum relationships between the performance factors, which lead to a definitive performance structure.

Practical experience shows that this process cannot continue steadily in a linear fashion. It must be assumed that the load tolerance and the adaptation processes are so highly strained by threshold sustained activity that restrictions in activity in certain biological systems appear and adaptations which have not yet fully stabilized are temporarily lost.

*Harre, 1982, p. 78*

The unambiguous message means that training must proceed in certain ways and that training for gymnastics will not be like training for other sports (Jemni et al., 2000; Jemni et al., 2006; Sands, 1991b; Sands, McNeal, & Urbanek, 2003).

The varied tasks of training cannot all be worked on at the same time. Care should be taken that a specific and systematic arrangement of immediate, intermediate, and long-term goals is made. Thus, observing the complexity of the main tasks and the continuous influence on all performance determining abilities, skills, and qualities of the athlete, certain areas of the standard of performance have to be emphasized for limited periods while others are simultaneously only stabilized or maintained.

*Harre, 1982, p. 79*

For example, gymnastics training and performance requires considerable strength fitness. Strength fitness is more effectively and efficiently acquired over the long term by beginning with strength-endurance, followed by maximal strength, then a focusing of the maximal strength to gymnastics-specific strength, and completed by a maintenance period that follows the duration of competitions. The aforementioned progression and development of sport-specific strength is a "model" of training.

Models are simplifications of complex things such that the universe of things and ideas that could be taken into account is reduced to a more manageable number so that we can gain an understanding more rapidly and easily (Banister, 1991; Sands, 1993a, 1995). However, it is important to remember that models are not the "real thing" and models are only as good as the assumptions and constraints that are used to build them. With these qualifications let's look at a fitness model for gymnastics.

Fitness models are also sometimes called "Profiles". A number of investigations have looked specifically at determining a fitness model or profile for gymnasts (Sands, 1994a; Sands, McNeal, & Jemni, 2001a; Sands, McNeal, & Jemni, 2001b). These models attempt to determine a sort of "recipe" to determine the relative contributions of various fitness components for gymnastics performance. The model of Siff is particularly helpful as a starting point for fitness models for all sports (Siff, 2000). A modified model of gymnastics fitness consists of the following categories of fitness (Figure 4.1):

- Strength
- Speed
- Flexibility
- Skill
- Stamina (commonly used to mean "muscular endurance")

Body composition can also be considered as the result of the interaction between the five components of the model mentioned because gymnasts must lift their body weight against gravity. Body composition demands for gymnasts include a premium on leanness,

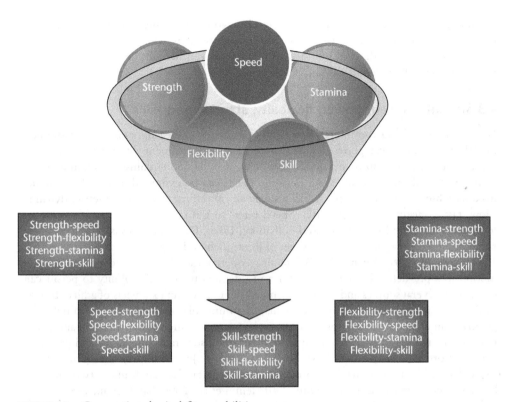

**FIGURE 4.1** Gymnastics physical fitness abilities.

(Adapted from Sands, 2003)

Note that strength, speed, skills, stamina and flexibility are considered as being equal to each other. Stamina is used to mean "muscular endurance".

but the idea probably doesn't rise to a level that merits sixth fitness category. The gymnast's job is made considerably easier by being light as well as strong (Malina, 1996; Sands, 2003; Sands, Hofman, & Nattiv, 2002; Sands, Irvin, & Major, 1995). Moreover, gymnastics is an "appearance sport" which means that the attractiveness of the athlete's body and his/her performance is relevant. Anthrompetric characteristics of gymnasts have been shown to be related to score, particularly among female gymnasts (Claessens et al., 1999; Pool, Binkhorst, & Vos, 1969).

Each of these physical fitness categories is connected to every other category. We simply don't have the unique words in English to describe the blending of physical fitness categories that often spell the qualities of sport-specific fitness. For example, a fitness characteristic composed of strength and speed refers to the application of high levels of force but rapidly. Speed-strength is a similar idea but the emphasis is on the speed component. Thus, after connecting all of the possible there should be 20 pairs of fitness components.

The combinations of fitness characteristics show how complex a fitness model might become; and we're not going to delve deeper into the three-way types of fitness, such as speed-strength-flexibility, and so on. The combinatorial explosion evident in the information outlined is why training the athlete is so complex. Moreover, within this model are all of the inherited characteristics of the athlete such as his/her genetic makeup, psychological characteristics, and so forth. In other words, while the model helps to narrow our universe of things that merit consideration, we must acknowledge that actually getting the model to work in the real world with real people can be considerably more difficult (Sands, 1994a, 2000c; Sands et al., 2001b).

## 4.3 Strength, speed, power, flexibility, stamina and skills

While acknowledging the interactions described earlier, the components of the simplest model of five fitness categories deserves more definitive treatment. Strength refers to the amount of force that can be exerted, under some pre-defined circumstances (Knuttgen & Komi, 1992; Sale & Norman, 1982; Wilk, 1990). It can be expressed in a different mode: maximal/absolute, usually expressed in a single all-out effort; static or isometric; dynamic (slow, fast, concentric and eccentric). Speed refers to the rate at which motions are performed (Joch, 1990; Mero, 1998; Verkhoshansky, 1996). Flexibility is usually defined as the range of motion in a joint or related series of joints such as the spine (Alter, 2004; Cureton, 1941; Holt, Holt, & Pelham, 1995). Stamina is defined as the ability to persist in some motion in a pre-defined circumstance for a pre-defined period. The ability to persist can come from several sources and is thereby one of the more complex aspects of athlete fitness. In the gymnastics context, stamina refers mainly to muscular endurance. For example, the gymnast can enhance his/her ability to persist by increasing metabolic efficiency and effectiveness via training of specific energy system enzymes, substrates, mediators and pathways (Sale & Norman, 1982; Sands, 1985; Sands, McNeal, & Jemni, 2001a). The gymnast can also enhance his/her stamina for short-term endurance (task demands under two minutes) by increasing maximal strength (Jemni, 2001; Jemni et al., 2006; Sands et al., 1987; Sands, McNeal, & Jemni, 2001a; Sands et al., 2004a; Stone et al., 1984). Skill is the coordinated application of forces, positions and movements to accomplish a pre-defined task (Abernethy, Wann, & Parks, 1998; Schmidt & Young, 1991). In addition to the model outlined earlier is the concept of body composition. A gymnast must be on the "lean side of lean" because he/she must lift his/her body mass against gravity through complex and difficult positions while maintaining exquisite control.

## 4.4 CONCLUSION

- Strength, speed, flexibility, skill and stamina (or muscular endurance) are the main physical abilities of the gymnasts. Body composition and coordination are considered as prerequisite at high level.
- Modern artistic gymnastics stresses the importance of "strength" and "speed" or in another word: power.
- All these physical abilities interact between each other showing 20 pairs of fitness components essential to build up a physical fitness model for gymnasts.

# 5

# TRAINING PRINCIPLES IN GYMNASTICS

*Monèm Jemni and William A. Sands*

## 5.1 Learning outcomes

- Identify the training principles that one could apply in gymnastics.

Different characteristics of fitness are combined like a recipe to result in optimal fitness for gymnastics. However, the recipe for optimal fitness also demands several other concepts. These are the principles of training.

## 5.2 Specificity

Although specificity was briefly described earlier, it is important to appreciate just how specific training and testing are in the development and monitoring of a gymnast. Early training of young athletes or the early stages of a long preparatory period may involve more "general" training, also called "multilateral development" (Bompa & Haff, 2009). However, specificity remains one of the most important principles involved in athlete training. A gymnast trains by performing on the apparatus not by swimming or running. Although the gymnasts may get some training benefits from swimming and running, specificity garners the greatest return on training investment.

Training and testing are specific to the position of the body (Behm, 1995; Oda & Moritani, 1994; Sale & MacDougall, 1981; Sale, 1986), to the speed of movement (Moffroid & Whipple, 1970; Sale & MacDougall, 1981), to the range of motion of the movement (Sale & MacDougall, 1981; Sale, 1992; Siff, 2000), to sport (Müller, Raschner, & Schwameder, 1999; Yoshida et al., 1990), type of tension (Jurimae & Abernethy, 1997; Oda & Moritani, 1994), gender (Drabik, 1996; Mayhew & Salm, 1990), and limb or side (Hellebrandt, Parrish, & Houtz, 1947; Sale, 1986). As such, the coaching and training of gymnasts should always consider the specificity of the exercises so that maximum transfer of training to gymnastics performance is more likely to occur.

## 5.3 Readiness

The principle of readiness refers to the idea that the road to competent performance in gymnastics is long and the path is seldom simple. Predictability of training exercises and loads is not guaranteed and the transfer of training approaches to actual performance is poorly understood (Bondarchuk, 2007; Christina & Davis, 1990; Sands, 2003). However, readiness for performance has been addressed from a practical standpoint with the basic idea involving the questions a coach should ask and answer prior to teaching a skill or allowing a gymnast to perform a skill unaided (Sands, 1990a, 1990b). Clearly, ignoring the simple-to-complex approach to training and the physical and psychological competence of the gymnast leads to poor adaptation at best and injury at worst. The gymnast's readiness will develop with age and maturation. A gymnast at six years of age will benefit very little from anaerobic training. His/her body is not ready for strength and conditioning at high intensities. However, an older gymnast is more "receptive" to strength and conditioning/anaerobic training and can develop them quicker.

## 5.4 Individualization

This training principle is commonly listed among the most important in that each athlete has individual strengths and weaknesses and thereby needs or deserves an individualized training program (Bompa & Haff, 2009; Sands, 1984; Stone, Stone, & Sands, 2007). It is obvious that gymnasts interpret and respond differently to the same training. This might be the result of different foundations based on heredity, maturity, nutrition, rest, sleep, level of fitness, illness/injury, motivation, and environmental influence. A wise coach should detect individual responses and formulates appropriate reactions for each athlete.

However, there are also circumstances when all athletes can and should do the same things. In many cases, coaching a team makes individualizing every training session a practical impossibility. Moreover, there are aspects of both development and enjoyment that profit from a team-based approach (Gould et al., 1998; Hanin & Hanina, 2009; Loehr, 1983; Martin, 2002; Ravizza, 2002). The need for individualization of training increases as the athlete progresses in development. In the first few years of training, the young athletes all need to learn a large but complete list of basic skills (Sands, 1981a). As the athlete reaches a higher standard, he/she is often using different skills and has clearly expressed his/her training and performance strengths and weaknesses. Fortunately, the size of the group of athletes or team is usually smaller in the latter stages of the athletes' careers due to simple attrition (Sands, 1995; Stone, Stone, & Sands, 2007).

## 5.5 Variation

Variation applies largely to the conditioning aspects of preparation, but also refers to the simple problem of boredom with training tasks and loads that remain fixed for periods beyond their effectiveness in causing adaptation (Arce et al., 1990; Stone, Stone, & Sands, 2007). Training theory has placed a high premium on training variation due to the observation that unidimensional approaches to training do not appear to result in continued adaptations (Verkhoshansky, 1981, 1985, 1998, 2006). The general rule of thumb for training adaptation is that something about training such as the tasks, loads, durations, frequencies, etc. should change approximately every two to four weeks (Bompa & Haff,

2009; Verkhoshansky, 1981, 1998). Moreover, variation applies to training on a variety of levels. Athletes should receive varying training demands, varying skill levels, varying competition demands, varying yearly plans, and varying levels of opponents. However, more recent information questions the idea of changing training loads so quickly, advocating training load changes only after approximately six weeks. Indeed, coaches often ask the following questions: Why wait six weeks before introducing a new training load? Why not change the training plan after two weeks when we notice that adaptations occurred? Olbrecht, (2000, p. 7) has answered these questions: "The reason is that weeks 3 to 6 are necessary to stabilize the adaptations brought about in weeks 1 and 2."

## 5.6 Diminishing returns

Related to the idea of training variation is the observation that a training program and training stimuli suffer "wear" (Harre, 1982). A training program applied this year will not produce the same results if applied next year. Moreover, the early adaptations to training tasks are typically neural and occur rapidly while later changes (weeks or months later) involve structural changes (Moritani & DeVries, 1979; Sands & Stone, 2006; Stone, Stone, & Sands, 2007). From an evolutionary perspective, this approach to organism adaptation is smart, using neural adaptation first (i.e., learning) is not as demanding on available calories as structural changes. A novice gymnast would expect to learn new skills and gain some fitness at a much faster rate than those who have already been training for a number of years. However, most of the changes sought in athlete development require months or years to achieve and thereby require an enormous caloric/structural investment by the body. Wolff's law [Julius Wolff (1836–1902), a German anatomist and surgeon] states that "function determines structure" and there is no better example of this than in athletic training (Alter, 2004). The demands of training alter the functional demands placed on the body which in turn results in modifications to structure. One author has observed that to achieve Olympic podium performance, the athlete must improve through approximately 20 preparatory and competitive periods (Bondarchuk, 2007). Fifteen periods of improvement usually lead to a high national and low international performance level. Ten periods of improvement lead to a national level of performance. Five periods of improvement usually result in competitive prowess at only the regional level. If an athlete cannot continue to improve for at least ten preparatory and competitive periods then he/she is likely to be untalented or chose the wrong sport (Bondarchuk, 2007). A related observation regarding diminishing returns is that improvement can be represented by a decelerating curve. Improvements are rapid in the beginning and slow dramatically or plateau as the athlete reaches his/her genetic ceiling (Sale, 1992).

## 5.7 Regeneration and the new concept of recovery in gymnastics

Training should be considered a unity of both training load and recovery-adaptation; both allow adaptations. Recovery does not start after the training session, it actually starts within the session by imposing the right "duration and method." However, gymnastics coaches seldom question the effectiveness of their recovery "procedure." Gymnasts rarely adopt "active types of recovery" compared to other sports where it has shown substantial benefits (Dodd et al., 1984; Stamford et al., 1981). It is now understood that "overtraining could be avoided if recovery/regeneration is planned effectively."

Establishing the unit of work and recovery is essential for the effectiveness of athletic training. Training causes fatigue which occasions a temporary lowering of performance. Hard work is followed by two processes which have already been introduced during the work itself and which may temporarily run along parallel lines:

- The recovery process leading to the re-establishment of the full ability to function, and
- The adjustment processes leading to the functional improvement of performance and the morphological reorganization of the functional systems under stress.

*Harre, 1982, pp. 65–66*

Once recovered the body will enter a period of overcompensation where the body's systems have adapted beyond the original threshold. Optimal post training recovery allows the athletes part in their other daily activities, such as studying, working or socializing. Harre (1982) recognized that overloading is not actually possible unless the athletes have recovered. Quicker is the recovery; faster is the shift to the new training stimulus. Moreover, if the recovery is repeatedly insufficient, fatigue builds up and performance deteriorates and therefore adds an increased risk of injuries.

Gymnastics competition is a series of performances on several apparatus interspersed with rest periods. It is in fact similar to circuit training due to the intermittent and intense activities that are involved. It is therefore important to design the optimal means of recovery so that the gymnast can begin each event without undue fatigue. Fatigue is indeed more than a simple physiological problem; it may lead to falls and injuries in gymnastics (Sands, 2000a; Sands, 2000b; Tesch, 1980).

During intense sessions, gymnasts are asked to perform routines while fatigued (Jemni et al., 2000). They are usually asked to repeat their events several times per practice session which leads to a high level of lactate production and accumulation (Lechevalier et al., 1999). They are therefore required to find the best compromise among technical effectiveness, safety and high intensity effort. Jemni et al. (2003) have set new recovery guidelines that could assist gymnasts in reducing their blood lactate between competitive events and therefore enhance regeneration. They have shown that a combined period of rest and active recovery, where heart rate is kept below the anaerobic threshold, incorporated between the events enhances not only lactate clearance but also helps the subsequent performance. Blood lactate clearance was indeed significantly higher by using the combined passive/active recovery when compared to the "classical passive only" (40.51% vs. 28.76% respectively).

To conclude, training plants the seeds of performance, careful nurturing of the seeds allows them to sprout (periodization), and recovery-adaptation makes the plants grow and bear fruit. Gymnastics training should include periods of regeneration such that the athlete is fully prepared and not fatigued when a new training challenge is posed.

## 5.8 Overload and progression

The gymnast must attempt tasks that are initially beyond his/her capability in order to force the organism to fatigue and later recover and adapt. Overload is the term used to describe the task(s) that the gymnast performs that exceed his/her current performance limits. Progression in gymnastics is a key principle; it refers to the idea that skills should be learned in a systematic progression. Starting with the most basic set of skills and evolving to the more complex. This allows not only building up the "technical repertoire" of the gymnasts step by step, but in particular it avoids systematic errors which are difficult to correct at a late stage. A current example of this is a gymnast who struggles to perform a correct "round-off"; in most of the cases it would be revealed that this gymnast has "jumped" a learning stage, obviously the "cartwheel." In addition, there are limits to the amount of overload an athlete can withstand without injury, breakdown and excessive fatigue. The ideal training program provides optimal challenges that the athlete can barely handle and then the athlete rests (reduces training demands) to promote recovery-adaptation. Too high and/or too challenging loads may affect the motivation of the athlete and may lead to lost interest. Meanwhile, too low load/challenge does not allow any benefit. The athlete is rewarded during recovery-adaptation with new skills and abilities that were caused by the previous overload. Thus, training is an undulating series of training challenges given in a planned successive increasing loads (i.e., overload) taking into consideration the initial state (i.e., progression) and allowing a shift to a higher level, allowing supercompensation in order to achieve training goals (Sands, 1984, 1987, 1991a, 1994b; Stone, Stone, & Sands, 2007). Overload is multi-dimensional, including all aspects of psychology and fitness. Maintaining order and a systems approach requires a specialized methodology. That methodology is called "periodization."

## 5.9 Periodization

All of the previous training principles are often included or embodied in the concept of periodization. "Periodisation is the division of a training year into manageable phases with the objective of improving performance for a peak(s) at a predetermined time(s)" (Smith, 2003, p. 1114). Periodization involves two simultaneous concepts: cyclic variation in training load and the division of training demands into separate phases (i.e., periods) (Bompa & Haff, 2009; Harre, 1982; Matveyev, 1977; Verkhoshansky, 1985). Periodization is currently a very confused term with a variety of investigators, coaches and administrators weighing in on exactly what periodization means, what concepts are most important and how to implement the ideas.

Periodization begins by developing an annual plan that is composed of one or more macrocycles that include a preparatory period, competitive period and a transition period. The macrocycle is broken down into mesocycles that are usually four to six weeks in duration and compartmentalize the total number of training demands in a way that only a few are emphasized during each mesocycle. Each mesocycle has its own goals and each mesocycle builds on previous mesocycles. Within each mesocycle are microcycles. Microcycles are approximately one week in duration and involve a more focused group of tasks and goals that function to achieve an accumulation of training stimuli that will force the athlete to experience the unity of an overload and recovery-adaptation. Interestingly, it takes about a week of continuous load demands to cause a sufficient training stimulus that the athlete's body will show the effects of overload (i.e., fatigue) and unify with recovery and adaptation following the overload.

Linking microcycles and mesocycles together in a systematic way rewards the coach and athlete with improved training and performance ability. The entire "map" of this development is called periodization. There are many periodization models involving the controlled application of volume (how much the athlete does, might be expressed in time or number) with intensity (how hard the challenges are, might be expressed in number per unit time) to achieve a balanced training experience that leads to progress without injury or the threat of overtraining. Of course, periodization tenets also attempt to ensure that the athlete maximizes his/her abilities such that the athlete is not defeated by opponents who simply work harder.

Planning a training cycle in high level gymnastics should take into consideration the two peaks of the season: one for the continent Championships (such as European Championships), usually between April/June and the second for the World Championships or Cup, around October/November. Therefore, there are two macrocycles, each one lasts approximately six months and is composed of the three periods, i.e., preparation, competition and transition. Each period is composed of mesocycles with specific objectives, and these in turn are divided into weekly microcycles. Members of the national squads are always gathered into regular training camps and also pursue one or two training sessions per week in the regional center in order to achieve the objectives set.

Arkaev and Suchilin (2004) showed that the structure of an Olympic cycle has to take into consideration the four annual training cycles. The whole cycle would be composed of eight six-month macrocycles. A specific objective is set for each macrocycle starting by raising the level of specific physical training and ending by the final selection of the Olympic team a few months before the games. Coaches and technical staff have to plan ahead and consider all the following preparations with their plan: physical/functional preparation, monitoring, tactical preparation, psychological preparation, jump preparation, acrobatic preparation, choreographic preparation, camps preparation, friendly competitions and medical preparation.

## 5.10 CONCLUSION

- As with all other sports, gymnastics coaching is based on the main generic training principles which are: specificity, readiness, individualization, variation, diminishing returns, overload and progression, periodization and regeneration. Applying science to gymnastics has allowed this sport to evolve by adopting new concepts to training and coaching such as the new concept of recovery in gymnastics.
- Long-term planning facilitates the achievement of the objectives via a structured periodization which takes into account all the above principles.

# 6

# SPECIFIC PHYSICAL AND PHYSIOLOGICAL ASSESSMENTS OF GYMNASTS

*Monèm Jemni*

## 6.1 Learning outcomes

• Explore techniques, tools and procedures to assess the above fitness components in the lab and in the field.

The complexity of gymnastics' events (six for males and four for females) requires not only different training approaches, but also a wide range of physical and physiological testing in order to monitor the progress of each gymnast. However, the measurements are made difficult by having a number of complex parameters, such as a wide variety of technical skills, muscular contractions and speed of the stretch but only a limited number of standardised specific tests.

The first section in this chapter highlights some of the most used tests in assessing the physical ability groups.

## 6.2 Strength and power tests for upper and lower body

### Standardised laboratory tests

The standardised laboratory tests that have been applied in gymnastics include: maximal oxygen uptake protocols for upper and lower body, force velocity tests and Wingate tests. These are widely accepted valid and reliable tests in sport science, often considered as gold standards for assessing aerobic and anaerobic metabolisms. Results of such tests performed by different level male and female gymnasts are presented in Chapter 1, section 1.3 on "Aerobic metabolism".

### Specific jumping and plyometric tests

The literature provides a variety of jumping tests which might be used in different sports/contexts (Fry et al., 2006; Gabbett, 2006; Lidor et al., 2007; Melrose et al., 2007). They all assess the height of the jump (and/or the displacement of the centre of mass) and indirectly the power output via different equations using mainly body mass. Some authors have estimated the power output by using contact mats and therefore have highlighted other indices, such as flight and contact times (Bosco, Luhtanen, & Komi, 1983; Loko et al., 2000; Markovic et al., 2004; Sipila et al., 2004). The force plate, however, gives better measurement of the power output developed during the jump (Carlock et al., 2004). Nevertheless, the most accurate jumping profile might be set, including kinetic and cinematic analysis, especially if the force plate is synchronised with video capture/analysis. All authors agreed on the strong relation between the height of the jump and the power output (Van Praagh & Dore, 2002). Other components, such as the body's vertical velocity at the take off, the peak jumping velocity and the impulse (force by time) are also very important to consider in studying jumping ability (Bobbert & Van Ingen Schenau, 1988; Haguenauer, Legreneur, & Monteil, 2005; Winter, 2005). However, to maximise all of these factors, segmental coordination and applying a proper technique are essential in order to obtain the best performance (Bobbert, 1990; Carlock et al., 2004). As a matter of fact, it is widely accepted that analysing jumping performance should take into consideration more than one factor in particular when comparing subjects (Marina, Jemni, & Rodríguez, 2012).

In all cases, the calculated power output does not reflect the physiological mechanisms underpinning the athletes' performance. Numbers are in fact different to metabolic reactions. Jumping tests have been widely accepted and are applied in gymnastics (Table 6.1). It has indeed been shown that the use of the jumping mat is an effective assessment tool for

**TABLE 6.1** Jumping test results for male and female gymnasts

| | *Jumping tests* | | | |
| --- | --- | --- | --- | --- |
| | *Females* | | *Males* | |
| | *Results (cm)* | *Norms for high level\* (cm)* | *Results (cm)* | *Norms for high level\* (cm)* |
| Vertical jump with arms swing | 47.8 (Sands, 1993b) 45.2 (Marina, Jemni, & Rodríguez, 2012) 49.2 (Heller et al., 1998) 42.5 (Bale & Goodway, 1987) | 50–60 | – 52.8 (Marina, Jemni, & Rodríguez, 2012) 53.9 ± 10.9 (Jankarik & Salmela, 1987) | 60–70 |
| Vertical jump without arms swing | 38.2 (Marina, Jemni, & Rodríguez, 2012) | 40–45 | 41.5 (Marina, Jemni, & Rodríguez, 2012) 41.3 ± 2.3 (León-Prados, 2006) | 50–55 |
| Standing long jump | | 210–230 | 50.9 (from 60 cm) | 225–245 |
| Drop jump | 45.6 (from 60 cm) (Marina, Jemni, & Rodríguez, 2012) | | (Marina, Jemni, & Rodríguez, 2012) 26.1 (Faria & Faria, 1989) | |
| Drop jump with jump off | | 60–65 | | 70–80 |
| Squat jump | | | 36.6 ± 2.3 cm (León-Prados, 2006) | |

\*: Arkaev and Suchilin (2004).

gymnasts, which simulates training and allows monitoring (Sands et al., 2004a). Bouncing and jumping indeed make up important parts of the floor, balance beam and vault routines. Gymnasts learn this skill at the very early age of specialisation as part of their daily training. It allows take off and further high aerial acrobatics skills. Table 6.1 shows the results of some jumping tests' investigations, as well as norms for high level males and females.

Plyometric work with upper and lower body is extensively used in all gymnastics' apparatus. Gymnasts are very often required to combine an aerial element to another as soon as they land. In addition, all catches and releases of the apparatus require plyometric contractions, while maintaining a straight body line. Gymnasts learn and develop this quality at a very early age in order to be able to maintain the basic handstand. Marina et al. (2012) have compared plyometric performances of elite male and female gymnasts with similar age and gender-matching groups (see Figure 6.1). Elite gymnasts obtained significantly better performances at drop heights from 20 to 100 cm. Most of the male and female gymnasts obtained their best performance between 40 and 60 cm, with exception of the best elite gymnasts of both genders who achieved their best at 80 cm.

## Muscular endurance tests

Muscular endurance is a significant discriminator of high versus low level gymnasts (Sands, 2003). It is the ability to perform for extended sequences, such as long routines, without undue fatigue. Gymnasts develop this quality progressively with age and practice. It is habitually assessed through specific tests performed on the apparatus engaging upper and/ or lower limbs or several muscular groups/joints at the same time. Table 6.2 shows examples of muscular endurance tests for male and female gymnasts. For example, a leg lift test requires the gymnast to fully lift his/her straight legs together or up to 90 degrees (horizontal) while hanging at the espalier or the bar. This test has been used in different contexts including talent identification and selection. The repeated jumping test is also commonly used for lower body muscular endurance. One of the most used is the 60 second Bosco test, which involves repeated use of the stretch-shortening cycle of the lower extremity. Sands (2000b) and Sands, McNeal, and Jemni (2001a, 2001b) have used this test to select the women's Olympic gymnastics members during their seven-month trials leading to Sydney's Olympic Games. The average power outputs of these trials were as follows:

- Female US senior National (n = 34, 17.2 yrs): 23.7 + 5 W/kg
- Female US senior National Team (n = 6, 17.3 yrs): 23.0 + 4.8 W/kg
- Female US Junior National Team (n = 15, 13.9 yrs): 21.6 + 2.8 W/kg

## Agility, speed, strength and power tests

In order to acquire strength, agility, speed and power, gymnasts vary their training according to the specificities of the skills, as well as the specificity of the apparatus. While some apparatus require relatively slow moving or held positions (i.e., isometric) and extraordinary strength such as the cross and the maltese at the rings, others events need explosive power development in very short sequences such as tumbling. Gymnasts also need exceptional agility together with remarkable spatial awareness in order to achieve very complex elements, including catches and releases of the apparatus after somersaulting and twisting. Agility, speed of movements and coordination between hands and the rest of the body are also required to perform at the pommels.

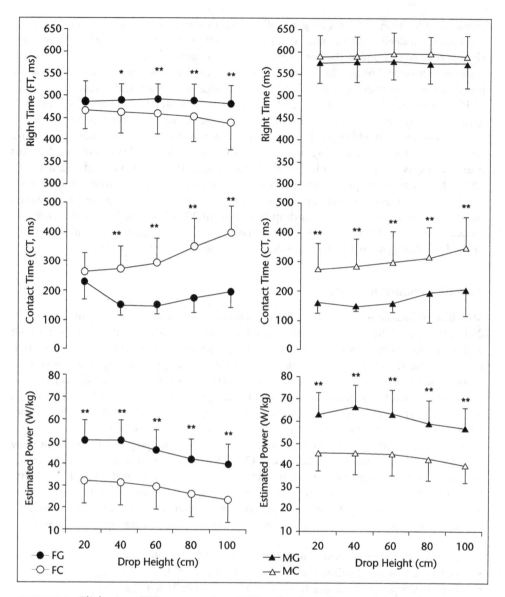

**FIGURE 6.1** Flight time (FT), contact time (CT) and estimated power (W/kg) during plyometric jumps from 20, 40, 60, 80 and 100 cm.

(Adapted from Marina et al. 2012)
Note: FG: Female Gymnasts; FC: Female Control; MG: Male Gymnasts; MC: Male Control. *: p ≤ 0.01. **: p ≤ 0.001.

Although, speed could be assessed with several sprinting distances, it is rare that applied gymnastics tests exceed 20 m which corresponds to the maximum distance allowed as a run up for the vault. Otherwise, agility, strength and power are often assessed by specific tests on the apparatus involving hanged or support positions. One of the classical tests recognised worldwide is the rope climbing. Different climbing distances are set for males and females; in most cases the athletes are not allowed to use their feet, therefore the test

**TABLE 6.2** Muscular endurance tests for gymnasts

*Muscular endurance tests*

| | | Females | Males |
|---|---|---|---|
| Bosco 60-sec jumps | Sands et al. (2001b) | 23.7 W/kg | – |
| Max leg lifts | Lindner et al. (1991) | 13.2 ± 3.8 | – |
| Max pull-ups | Seck et al. (1995) | – | 16.6 ± 4 |
| Push-ups in 60-sec | Grabiner & McKelvain (1987) | – | 292.5 ± 45.8 W/kg |
| Sit-ups in 60-sec | Grabiner & McKelvain (1987) | – | 201.1 ± 28.5 W/kg |
| Dips in 60-sec | Grabiner & McKelvain (1987) | – | 163.4 ± 41.1 W/kg |
| Pull-ups in 60-sec | Grabiner & McKelvain (1987) | – | 97.6 ± 31.4 W/kg |
| Support half-lever | Arkaev & Suchilin (2004) | 28–30 sec | – |

**TABLE 6.3** Specific tests for strength, agility, speed and power in gymnastics

*Strength, agility, speed and power tests*

| | | Females | Males |
|---|---|---|---|
| Rope climb without leg help | Arkaev & Suchilin (2004) | (4m) 5.5 to 6 sec | (3m) 5 to 5.5 sec |
| | León-Prados (2006) | | (5m) 5.7 ± 1.3 sec |
| Full leg lifts in 10 sec from hanging position at espalier | Sands (1993b) | 6.2 reps | – |
| Push-ups in 10 sec | Sands (1993b) | 14.1 reps | – |
| Pull-ups in 10 sec | Sands (1993b) | 7.2 reps | – |
| 20-m dash | Sands (1993b) | 3.1 sec | – |
| | Jankarik & Salmela (1987) | | 3.6 ± 0.5 sec |
| 30-sec jumps | León-Prados (2006) | | 30.4 ± 10.9 W/kg |

is performed with bent knees or in half lever. Table 6.3 shows some data from examples of strength, agility, speed and power tests.

## 6.3 Flexibility tests

Gymnastics is a sport which requires a great range of motion in all the joints. Flexibility plays a considerable part in the success of a routine. In many cases, the score is directly influenced by the possibilities of a gymnast's "body motion". Gymnasts are indeed marked on "how perfect they perform" and not on "how many skills they present". Full amplitudes of the movements are continuously required. A slightly bent ankle, for example, is penalised; therefore, gymnasts learn to be "over flexible" in order to achieve the required range of motion in a natural fashion. It is very common to observe an over-split, for example, while a gymnast is stretching or performing. The gymnast is actually stretched beyond the horizontal position, i.e., more than 180 degrees. In addition, lack of flexibility in one or more joints may slow down the learning process and/or make it quite difficult. Upper and lower limbs, and also neck and spine, should demonstrate a variety of positions: forward, backward, sideward and sometimes in a longitudinal plane, such as the case for dislocation elements in the high bar and rings. It has in fact been demonstrated that gymnasts are virtually the most flexible athletes (Kirby et al., 1981). However, it has also been shown that rhythmic gymnasts are more flexible than artistic gymnasts (Douda & Toktnakidis, 1997).

Flexibility assessment in gymnastics is a daily routine at the very beginning of the specialisation stage when athletes are developing their basic physical skills. Once an optimal range of motion is gained, it will be maintained by daily stretching. Flexibility tests include passive and active forms, most commonly in different planes: forward, backward and sideward. Table 6.4 shows some of the tests performed in gymnastics. It is indeed important to maintain a symmetrical range of motion from both sides in order to perform the skills at their full amplitudes. It has been shown that artistic female gymnasts have better symmetry than rhythmic gymnasts (Douda & Toktnakidis, 1997). This in fact might be explained by more active flexibility being performed in artistic gymnastics which is indirectly influenced by the level of strength of the agonist muscles. Personal interviews with the coaches of high level rhythmic gymnasts revealed that gymnasts tend to put more emphasis on stretched skills with their leading leg because they have to handle the apparatus at the same time.

Finally, an increasing interest has been given to the enhancement of the range of motion by the new "vibration therapy". Several studies have shown that this technology might be a promising means of increasing flexibility beyond that obtained with static stretching. Sands et al. (2006b) have not only increased the forward split amplitudes of a high level group of male gymnasts when compared to a control group (both are already flexible) but have also succeeded in saving time.

**TABLE 6.4** Flexibility tests in gymnastics

| Flexibility tests | | | Females | Males |
|---|---|---|---|---|
| Passive flexibility | Sands (1993b) | Right forward split | 35.5 cm | – |
| | Sands (1993b) | Left forward split | 31.4 cm | – |
| | Jankarik & Salmela (1987) | Hip flexion | – | 149,0 ± 76,9 degrees |
| | | Back extension | – | 97,3 ± 1,82 degrees |
| | | Shoulder extension | – | 47,2 ± 1,35 degrees |
| Active flexibility | Sands (1993b) | Shoulder flexion | 48 cm | – |
| | | Left forward leg lift | 7.3 pts★ | – |
| | | Right forward leg lift | 8.6 pts★ | – |
| | | Left sideward leg lift | 7.9 pts★ | – |
| | | Right sideward leg lift | 8.9 pts★ | – |
| | Jankarik & Salmela (1987) | Hip flexion | – | 111.7 ± 52.4 degrees |
| | | Back extension | – | 71.1 ± 46.9 degrees |
| | | Shoulder extension | – | 27.4 ± 12.7 degrees |

★: ankle above chin = 10 pts; above shoulder = 9 pts; above chest = 8 pts; above hip = 6 pts; below hip = 3 pts.

**TABLE 6.5** Technical tests in gymnastics

*Specific tests*

|  |  | *Females* | *Males* |
|---|---|---|---|
| Handstand push-ups in 10 sec | Sands (1993b) | 7.5 reps | – |
| Held handstand | Arkaev & Suchilin (2004) | 9 sec | – |
| Handstand push-ups in 60 sec | Grabiner & McKelvain (1987) | – | 38.8 ± 11.8 reps |
| Cross on rings | Arkaev & Suchilin (2004) | – | 5–6 sec |
|  | León–Prados (2006) | – | 6.7 ± 4.3 sec |
| Maltese on ring | León–Prados (2006) | – | 3.3 ± 1.8 sec |
| Front horizontal hang | Arkaev & Suchilin (2004) | 20–23 sec | 5–6 sec |
| Back horizontal hang |  | 28–32 sec | – |
| Horizontal support (planche) | Arkaev & Suchilin (2004) | – | 5–6 sec |
| Inverted cross | Arkaev & Suchilin (2004) | – | 5–6 sec |
| From support half-lever, lift to | Arkaev & Suchilin (2004) | 8–10 reps | – |
| handstand with straight arms and bent body | León–Prados (2006) | – | 10.9 ± 2.9 reps |
| From swing on low bar, upstart | Arkaev & Suchilin (2004) | 10–12 reps | – |
| to handstand and repeat | Sands (2000b) | 10–15 rep | – |

## 6.4 Technical tests

Technical tests have been applied in gymnastics since the very early age of the discipline. They allow not only checking the readiness of the gymnasts but also selection. Becoming a member of the national squad is mainly based on performance and ranking in competitions; however, several countries have set up a parallel stream for young talents who engage in a selection process based on physical and technical abilities, such as the "Talent Opportunity Programme" (T.O.P) in the USA (Sands, 1993b). In addition, monitoring basic technical skills is regular procedure at a higher level of performance. However, it appears that at sub-elite level, coaches often wait for competitions to check their athletes' preparation. Table 6.5 shows some of the common technical tests in artistic gymnastics.

## 6.5 CONCLUSION

A wide range of physical and physiological tests are applied in order to monitor the progress of the gymnasts. However, most of these tests are non-invasive, field based and indirectly reflect the metabolism. The lack of specific laboratory tests for these athletes makes authors "speculate" about the energy requirement, force and power. Investigators still have to come up with "the gold standard tests" for gymnastics. Meanwhile, the tests currently being applied are widely accepted, such as jumping tests and flexibility. The accuracy, reliability and validity of the specific technical tests are quite low because of the "learning effect", and also because different coaches and scientists perform the tests differently.

# 7

# JUMPING SKILLS

## Importance, assessment and training

*Michel Marina*

## 7.1 Learning outcomes

- Learn the most used techniques and procedures to assess jumping performance on a contact mat.
- To identify the most relevant variables obtained from the jump tests.
- To understand the relationship between the variables that best characterize the specific jumping profile in artistic gymnastics.
- Having simple criteria to select the adequate drop heights in plyometrics as well as the optimum overloads when training young gymnasts.

## 7.2 Introduction and objectives

In gymnastics, jumps are usually characterized by short run-ups followed by dynamic take-offs that require several factors, such as: 1) maximal force (Peterson, Alvar, & Rhea, 2006; Schmidtbleicher, 1992), 2) muscle power (Kawamori & Haff, 2004), 3) stretch-shortening cycle (SSC) of muscles (Bosco, Komi, & Ito, 1981), and 4) an appropriate technique with segmental coordination (Bobbert & Van Ingen Schenau, 1988). Bouncing jumps (also called plyometrics) are indeed extensively practised during daily work-outs. Gymnasts acquire this skill from a very early age as part of their daily training routines (Marina & Torrado, 2013). Inappropriate planning of these sessions and/or wrong drop height selection vs. age could have serious implications on gymnasts' health and safety.

The objective of this chapter is to provide the reader with the scientific basis underpinning jumping and plyometrics testing and training in gymnastics.

## 7.3 How to assess jumping capacity

The use of a contact mat has been proven as an effective assessment tool for gymnasts, which allows real training conditions' monitoring (Sands et al., 2004).

## Instruments

Since the arrival of force platforms, the use of contact mats has been reduced, but the latter are still the easiest way to run field tests. Their use is widely accepted even though force platforms are the favourite tools in a lab setting because of their high level of accuracy. Unfortunately, the cost of the force platforms reduces their accessibility. The strong correlation between both systems (r > 0.99) has been highlighted by previous studies (Carlock et al., 2004; Christou et al., 2006). This chapter will be focused on the utilization of the contact mat.

## Jump tests

Bosco and his colleagues/team are considered amongst the leading authorities in jumping assessments (Bosco, 1985; Bosco & Komi, 1979; Bosco, Komi, & Ito, 1981; Bosco, Luhtanen, & Komi, 1983; Bosco et al., 1982). They have described several assessments of the vertical jumps with or without SSC. Some of these tests are shown in Figure 7.1:

1. Squat jump (SJ) with progressive loads of 0, 25, 50, 75 and 100% of body mass (SJ0-25-50-75-100). From a maintained semi-squat position with knee flexed 90° approximately, and arms akimbo.
2. Counter-movement jump (CMJ). From a straight position a fast knee flexo-extension before jumping; arms akimbo.
3. CMJ with arm swing (CMJA), also known as Abalakov (Bosco, 1985; Lidor et al., 2007; Loko et al., 2000). Same as CMJ, but with free swing of the arms.
4. Drop jump (DJ) from stairway steps of 20-40-60-80 and 100 cm height. Rebound jump from a drop height, can be done with arms akimbo or free arm swing.
5. Repeated jumps test for 5 s (R5), 10 s (R10) and 60 s (R60) (Marina & Rodríguez, 2013). Repeated rebound jumps on the floor, with or without arms akimbo.

It is of utmost importance to emphasise that DJs and Rs should be performed with the instruction of "performing the maximum flight time (FT) combined with the shortest contact time (CT)" in order to get a "quick jump" as described by previous studies (Bobbert, Huijing, & van Ingen Schenau, 1987; Eloranta, 1997; Young, Prior, & Wilson, 1995; Young, Wilson, & Byrne, 1999b).

## 7.4 Variables which could be measured and/or calculated

The contact time (CT; ms) and flight time (FT, ms) are the only variables collected directly by a contact mat. Time is adequate to assess exercise performance (Winter & Fowler, 2009). The FT could also be normalized to the body mass ($FT_{bm}$) as suggested by Benefice (1992) (cited in Benefice & Malina, 1996; Markovic & Jaric, 2005). The subject's ability to jump high with a low body mass is particularly relevant in gymnasts.

In this context, the flight-contact ratio (FC) gives an idea about the direct relation between FT and CT. A high value of FT divided by a reduced CT (FT/CT) could result in a high ratio and must be interpreted as a good performance.

The mechanical power could be calculated using FT and CT, according to formulae suggested by Bosco, Luhtanen, and Komi (1983). In spite of certain known biases (Hatze, 1998),

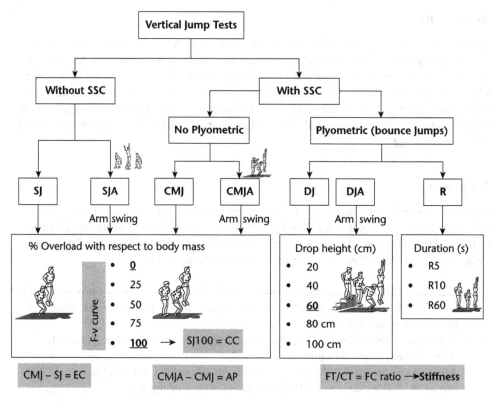

**FIGURE 7.1** Diagram of the jumping conditions used to characterize the multifactorial nature of the jumping profile in gymnasts.

Note: SSC (Stretch Shortening Cycle), SJ (Squat Jump), SJA (Squat Jump with Arm Swing) CMJ (Counter Movement Jump), CMJA (Counter Movement Jump with Arm Swing), DJ (Drop Jump), DJA (Drop Jump with Arm Swing), R (Repeated Reactive Jumps).

the Bosco expression "BE" (Equation 1) is still considered as a useful and valid variable if used in strict conditions (Arampatzis et al., 2004).

$$BE = \frac{g^2 \times FT \times Tt}{4 \times CT}$$

Equation 1: Estimated average mechanical power of one DJ according to Bosco et al. (1983). *[FT: Flight Time; CT: Contact Time; Tt: Total time of the jump (Tt = FT + CT); g: gravity force; BE: Bosco Expression].*

Other variables could be calculated as follow:

*Estimated elastic component (EC) = CMJ – SJ.*
The increase in FT during the CMJ is due to the muscle pre-stretch (the descendent phase) when compared to the SJ that does not contain a pre-stretch phase. This difference (EC) has been attributed to the release of an *elastic energy* (Asmussen & Bonde-Petersen, 1974; Bosco & Komi, 1979; Komi & Bosco, 1978). A short transition between the eccentric and

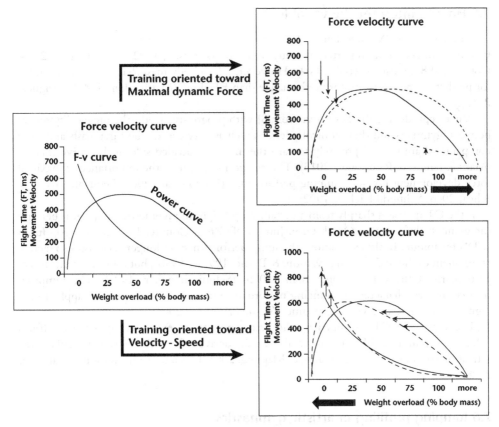

**FIGURE 7.2** Estimation of the force-velocity curve and power curve using SJ from 0 to 100% overload with respect to body weight. Explanation about how the training orientation can modify these two curves.

concentric phase of the jump is indeed associated with high reutilization of elastic energy (Bosco, 1985; Wilson & Flanagan, 2008).

*Estimated contractile component (CC) = SJ100*
CC is measured by performing a SJ with an added weight equal to his/her body mass; we call it SJ100.

*Force-velocity curve (F-v)*
This curve is plotted while measuring successive FTs obtained from the SJ with added weight relative to their body mass, respectively (SJ0-SJ25-SJ50-SJ75-SJ100) (Marina, Jemni, & Rodriguez, 2013; Viitasalo, 1985a, 1985b). (See Figure 7.2.)

*Estimated arms participation (AP) = CMJA − CMJ*
The arms swinging provides a significant amount of lifting energy for many gymnastics skills. If segmental timing and coordination are properly done (Luhtanen & Komi, 1978) it is assumed that the FT obtained with the CMJA is longer than that obtained with the CMJ (Marina, Jemni, & Rodriguez, 2013).

## 7.5 How to interpret the variables?

Our investigations (Marina, Jemni, & Rodriguez, 2013) showed FT lower in gymnastics than most of the values reported in many other sports (Bosco, 1992; Carlock et al., 2004; Cometti, 1988; Chamari et al., 2004). However, FT to body mass ($FT_{bm}$) is a better index for predicting gymnasts' motor performance than FT alone (Marina, Jemni, & Rodriguez, 2013).

When reproducing the F-v curve from SJ with progressive loads, the bigger differences between gymnasts and their control peers, as well as between the best gymnasts and their companions, are located precisely toward the more overloaded side of the curve (SJ100) (Marina, Jemni, & Rodriguez, 2013). This relation strengthens the importance of maximal force development for best jumping performance (Bosco et al., 2000; Peterson, Alvar, & Rhea, 2006; Schmidtbleicher, 1992).

If the CT increases sharply from a certain drop height in plyometric jumps, it is because the gymnast shifts from a Quick Drop Jump (QDJ) to a Counter Movement Drop Jump (CDJ) technique. In these situations the jump performance no longer relies on its reactive component of the SSC (Young, Wilson, & Byrne, 1999a, 1999b), but on its maximal force, particularly if the CT is longer than 250 m (Schmidtbleicher, 1992). The temporal restrictions that characterize bouncing performance in gymnastics compel to apply a very high level of force in a very short time (Arampatzis et al., 2004).

The very short CTs obtained by the best gymnasts are explained by a great stiffness (Marina & Jemni, 2014; Marina et al., 2012), derived from the neuro-muscular pre-activation (Arampatzis, Bruggemann, & Klapsing, 2001; Eloranta, 1997; Wilson & Flanagan, 2008).

## 7.6 Jumping profiling in artistic gymnastics

$FT_{bm}$, FC and BE must be considered as an indivisible set of variables when profiling the jumping performances in gymnastics. During a DJ, a subject could score a great height (long FT) if he/she increases the CT. As a result he/she would obtain a modest BE and a low FC. It is, however, very important to mention that FT is the least discriminating factor when it comes to distinguishing gymnasts' DJ performances compared to control groups and possibly to other sports (Marina et al., 2012).

A low SJ and high CMJ performance suggests a considerable EC in elite gymnasts, in particular when it comes to distinguishing the best ones in floor exercises and vault from their competitors (Viitasalo, 1988).

Gymnasts could perform quite good arms participation (AP), possibly because of their greater segmental coordination and arm strength.

## 7.7 How to improve jumping capacity

Time constraints are too short for gymnasts to express their maximum strength/force seeing that tumbling take-offs are less than 150 m. Therefore gymnasts, more than other athletes, need to reach a high level of their rate of force development in an extremely short time when performing their daily routines (Marina & Jemni, 2014). To increase maximal force components, apart from using strength conditioning machines we recommend jump and plyometric exercises with overload from 50% body mass onward (Marina, Jemni, & Rodriguez, 2014). This, of course, has to be performed at the right time of the year and preceded by a substantial force development. During the sessions of conditioning routines with machines, we suggest maximal loads that can be moved eight to 12 times in a series at the maximal possible speed, and until reaching muscle failure during the concentric phase of the movement (Marina, Jemni, & Rodriguez, 2014; Sale, 1989).

Another important question emerges when the coach must choose the appropriate drop height for plyometric routines. There is a considerable inter-subjects variability concerning the best drop height that produces the longest FT (Radcliffe & Osterning, 1995). While the longest FT should only be considered for CDJ, the higher FC is more appropriate to decide the best drop height in QDJ (Young, Wilson, & Byrne, 1999b).

We recommend the maximal BE as the best criterion to choose the suitable drop height for training purposes, because it also depends on the FT and CT relationship and is more sensitive to FT than the FC (Figure 7.3). Generally, the drop heights should range from 40 cm to 80 cm depending on the level of fitness.

As general recommendations, we suggest:

- To set apart two workouts per week of ~30 min duration each and separated by 72 h. If possible, insert the plyometric session in the middle of the week and the maximal strength routine at the end.
- Select five to eight different exercises within each workout, two to four series each. A global volume of 150 to 250 repetitions or jumps should be enough. Optimum intensity is the key to achieve power development in each repetition without inducing injuries (Table 7.1).

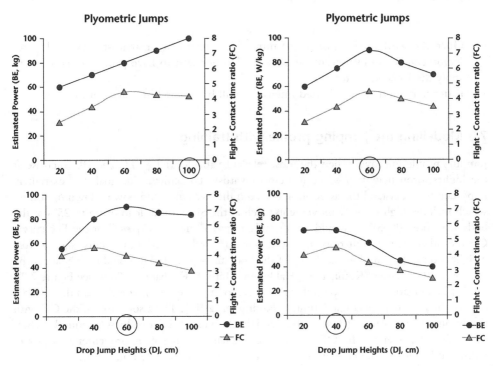

**FIGURE 7.3**  FT, CT, FC, and estimated Power (W/kg) during DJs at 20-40-60-80-100 cm height.

**TABLE 7.1** Proposal of a plyometric workout. Particularly in DJ and R jumps we suggest free swing of arms instead of arms akimbo, as this coordination is very "gymnastics oriented".

| Types of jumps | Series x Rep | Clarifications |
| --- | --- | --- |
| SJ/SJA (90° knee flex) | 3 × (6–8) | • Reception stabilized and maintained 1–2 s |
| SJ/SJA (full knee flex) | 3 × (6–8) | • Foam obstacles of different heights on the floor |
| CMJ/CMJA | 3 × (6–8) | • Slight forward displacement |
| DJs (40-60-80) | (2–4) × (6–8) | • Combine different drop heights according to the estimated power (BE) and flight-contact time ratio (FC) |
| Drop height depending on the level | | • At least three different drop heights in the same session<br>• Avoid contact times superior than 200 m |
| R | $R_5$: 2x<br>$R_{15}$: 3x<br><br>$R_{30}$: 1x<br>$R_{60}$: 1x | • Duration ranging from 5 to 60 s<br>• Maximal possible short range elastic stiffness<br>• Performed with one or both legs<br>• Optional Swedish bench, foam bench, etc… during lateral or zig-zag displacements |

SJ with arms akimbo (SJ), SJ with arm swing (SJA), CMJ with arms akimbo (CMJ), SJ with arm swing (CMJA), Drop Jump (DJ), Repeated reactive jumps (R).

- Once the workout is finished, 10 min of ice treatment or at least very cold water (cryotherapy) would prevent joints issues. Do not wait to have pain or discomfort to apply the ice/cold water.
- Patience is another key: desired intensities will be reached in years, not months.

## 7.8 Modelling the jumping profile with training

Jumping profile improves when the most relevant variables ($FT_{bm}$, FC and BE) increase at the highest drop heights possible (60 cm onwards). To achieve this goal it is extremely important to understand the association between the F-v curve and power. Training stimuli which includes high velocity movements, such as the SJs without load or up to 25% of the body mass (i.e. SJ0–SJ25), would significantly improve the "*velocity end*" of the F-v curve. On the other side, a high resistance stimuli with slow velocity movements, such as free weight and machine mobilizing 100% of the body mass (i.e. SJ100) would mainly affect the "*force end*" of the curve (Komi, 1986; Hakkinen, Mero, & Kauhanen, 1989) (see Figure 7.2).

Our investigations have shown that it is easier to improve the "*force end*" than the "*velocity end*" of the F-v curve (Marina, Jemni, & Rodriguez, 2014). The association of the CC with the dynamic maximal force (Bosco, 1985), and the short range elastic stiffness (Schmidtbleicher, 1992) that characterizes the QDJ technique, explains the bigger improvements of power (BE), and FC at the higher drop heights in gymnasts (Marina & Jemni, 2014).

Finally, it is noteworthy to state that the EC is very difficult to improve, possibly due to the strong dependence of this component on genetics (Marina, Jemni, & Rodriguez, 2014).

## 7.9 CONCLUSION

### From the assessment perspective

- Coaches should not take the flight time (FT) as the "golden" variable to evaluate the jumping performance of their gymnasts. FT normalized to body mass ($FT_{bm}$) seems to better distinguish them.
- The elastic component (EC) is useful to distinguish the best gymnasts from their team mates/competitors.
- We encourage gymnasts to specifically develop their maximal force and the contractile component (CC) using squat jumps (SJ100 – CC).
- Excessive drop heights should be avoided in order to preserve plyometric technique as this may increase the contact time (CT) and therefore the jump would not be considered as "Quick Drop Jump".
- We recommend the estimated power "BE" as the main criteria to individually assess the most appropriate drop height for plyometric training purposes (see Figure 7.3).

### Training recommendations

- We recommend the inclusion of two to three intense physical conditioning workouts per week aiming to optimize gymnasts' jumping skills.
- If the elastic component fails to increase, we recommend focusing on the squat jumps with overloads since they are easy to achieve. We also recommend that training should be focused on components of the vertical jump that are related to the maximal force.
- A combination of heavy resistance training with high impact plyometric jumps is more effective than repeatedly performing gymnastic movements alone. This last does not optimize the level of strength or power required for elite gymnasts.

# PART II REVIEW QUESTIONS

Q1. Analyse the strength, speed, power, flexibility and the muscular endurance of the gymnasts.

Q2. Analyse each of the following training principles in relation to gymnastics: *(long essay type question or average length answer if taken individually)*

- Specificity
- Readiness
- Individualization
- Variation
- Diminishing returns
- Regeneration
- Overload and progression
- Periodization

Q3. Give examples of specific tests used to assess jumping abilities in gymnastics.

Q4. Give examples of specific tests used to assess muscular endurance in gymnastics.

Q5. Give examples of specific tests used to assess agility, speed, strength and power in gymnastics.

Q6. Give examples of specific tests used to assess flexibility in gymnastics.

Q7. Give examples of specific technical tests used in gymnastics.

Q8. A back somersault on the spot is a:

a. flexibility related skill
b. agility related skill
c. strength–speed related skill
d. strength related skill

Q9. Maximal strength training induces:

    a.   CSA (cross-sectional) muscle hypertrophy

    b.   intramuscular coordination

    c.   increased recruitment of type IIB Fibers

    d.   all the above

Q10. Gymnasts must learn to:

    a.   apply less than maximal force as rapidly as possible

    b.   apply maximal force as rapidly as possible

    c.   apply less than maximal force for as long as possible

    d.   apply minimal force as rapidly as possible

Q11. Plyometric jumps develop:

    a.   muscles cross-sectional area

    b.   speed

    c.   strength

    d.   the rate of force development

Q12. Bouncing in gymnastics should be:

    a.   Quick Drop Jump (QDJ)

    b.   Bounce Drop Jump (BDJ)

    c.   Drop Jump for Height-Time (DJ-H/t)

    d.   all the above

Q13. Kinetic energy could be:

    a.   potential energy

    b.   elastic energy

    c.   three types of energy

    d.   developed by the entire body or by body part/s

# PART III

# Biomechanics for gymnastics

# PART III

# Biomechanics for gymnastics

## Learning outcomes (*Monèm Jemni*)

After reading Part III, you will be able to:

- Appreciate the role of biomechanics in gymnastics.
- Distinguish the difference between linear kinematics applied to displacement, velocity and acceleration in gymnastics.
- Understand the kinematics of flight and trajectories and the relations between angular and linear motions.
- Differentiate the laws applied to linear kinetics, in particular, Newton's laws of motion relevant to impulse and power development in gymnastics.
- Identify the angular kinetics subject to force application, torque and moment.
- Value the importance of the centre of gravity and moment of inertia in gymnastics practice.
- Recognize the principle of the conservation of angular momentum in gymnastics, in particular Newton's laws for angular analogue.
- Apply body inverse dynamics concepts to appreciate the forces applied at the floor exercises and at the vault table.

## Objectives (*Monèm Jemni*)

Biomechanics has always frightened students as they see it somehow as a complicated science that applies many mathematical and physical concepts. The overall objective of this part is to render "gymnastics biomechanics" an enjoyable and easy-to-understand fundamental science in coaching this sport. Basic to advanced concepts will be explained in order to help readers to understand how biomechanics could help coaches achieve their training targets.

## Introduction (*William A. Sands*)

### *What is biomechanics?*

Biomechanics is a part of physics that studies the mechanical or physical principles as they apply to the movement of living things. Biomechanics serves gymnastics by applying the principles and techniques of physics and mechanics to the movement of a gymnast and the apparatuses.

### *Why biomechanics?*

Biomechanics is an important part of all sports, not just gymnastics. Biomechanics relies on physics, which in turn relies on the character of physical law (Feynman, 1965). Physical laws are important for a variety of reasons and rely heavily on mathematics for unambiguous expression. One of the most important concepts behind physical laws is that once a law is determined, then further laws can be derived from the original law. For example, if we have a stopwatch and a tape measure and we know that a gymnast running on vaulting covered a distance of 10 meters in 2 seconds, then the average speed over that distance was 5 metres per second (10m/2s=5m/s). If we have a radar gun instead and we know that the gymnast traveled 5 meters per second for 2 seconds then we can determine that the gymnast traveled 10 meters (5m/s $\star$ 2s = 10m).

This may be overly simplistic, but the beauty of the mathematics in physical law is that we can use these concepts repeatedly, with modification, to learn new things.

### *Units of biomechanics*

When compared to the other scientific areas of sport and exercise, biomechanics has relatively few independent concepts and is thereby much easier to grasp. Table III.1 shows most of the units of mechanics that are used in gymnastics biomechanics.

Note, there are only four fundamental units, and all derived units are simply combinations of these fundamental units. This simplicity comprises some of the elegance of physical law and the application of physical law to a sport like gymnastics.

**TABLE III.1** Units of mechanics

| Fundamental units | Derived units | Units with special names |
|---|---|---|
| Time (s) | Velocity (m/s) | Force (newton, N = kg.m/s$^2$) |
| Displacement (m) | Acceleration (m/s$^2$) | Pressure and stress (pascal, P = N/m$^2$) |
| Mass (kg) | Momentum (kg (m/s)) | Energy and work (joule, J = Nm) |
| Temperature (Celsius, °C) (Kelvin° − 273.15) | Torque (Nm⊥) | Power (watt, W = J/s) |
| | Moment of force (Nm⊥) | Angle (radians or degrees) |
| | Density (kg/m$^3$) | |
| | Area (m$^2$) | |

The ⊥ symbol indicates that the distance is measured perpendicularly from the line of application of force rather than collinearly, as in energy and work.

In addition to a relatively small number of independent concepts than say physiology or psychology, the biomechanics of gymnastics can be further subdivided based on the type of motion involved and whether we are interested in just describing the motion or in knowing the origins of the motion.

### What is the role of biomechanics in gymnastics?

Biomechanics is the study of technique. Techniques are commonly studied and evolve over time. When addressing the technical errors involved in gymnastics, often the instructor sees the technique fault but is forced to rely on trial and error teaching in order to fix the fault. Moreover, the actual fault is often performed long before the technique problem becomes visible to the naked eye. For example, when a gymnast is airborne in a somersault and shows a fault in technique, that fault is almost always traceable to the take off or skills preceding the take off. Biomechanics is the science which can discover the source of the technical fault rather than simply identify the obvious.

### Conclusion

We further reduce the number of concepts in biomechanics by dividing our thinking into four areas: linear kinematics, angular kinematics, linear kinetics, and angular kinetics. Kinematics refers to a description of motion via the related concepts of position, displacement, velocity, and acceleration. Kinematics can furnish a nearly complete description of the motion with just these concepts and most coaches and teachers are experts at kinematics because this is the most visible part of performance.

Kinematics can be "seen" by the athlete, teacher, or coach, and thereby forms the bulk of the information used for the qualitative assessment of motion. Kinetics adds the concept of the genesis of the movement by including the forces that cause or underlie the motion that occurs. Forces are not visible; we can only infer forces by virtue of "seeing" a motion take place or by other forms of measurement.

# 8

# LINEAR KINEMATICS

*William A. Sands*

Taking a "divide and conquer" approach and moving from simpler concepts to those with more complexity, we begin with linear kinematics. Linear kinematics refers to a description of motion that is in a straight line or nearly straight line (called curvilinear).

A simple example of an object moving in a straight line is a body falling straight down due to gravity. Linear or near-linear movement in gymnastics is seen most obviously during the vault run. We can also see linear movement in various phases of a skill.

An image of a vaulter during the take off phase from the vault board is shown in Figure 8.1. The "stick figure" diagram of Figure 8.1 shows individual frames of movement as captured by high-speed video and then rendered by a computer and specialized software. Note that as the apex of the head moves right and upward, the points are in a nearly linear arrangement and that, although similarly spaced, the points get closer together as the movement proceeds.

Biomechanics of gymnastics uses this type of kinematic analysis. Most skills can be rendered in small enough segments of time that the small increments of motion are linear or nearly linear.

## 8.1 Distance and speed

As with other branches of science, biomechanics has special definitions of terms and the reader should be familiar with these terms and their applications.

When a body moves from one location to another, the *distance* of the movement is the path that the body followed. Distance, by definition, does not involve direction. Thus, distance is simply the path along which the body or body part moves.

If you express the distance covered relative to the time involved to cover that distance you have a new quantity called "speed", in this case the "average" speed. The term "average" is added because we don't know if the speed changed at any point along the path of the movement, thus we have a simple average. Speed is described mathematically in Equation 8.1. Average speed is also defined as the rate of change of position.

$$average\ speed\ (m/s) = \frac{distance\ (m)}{time\ (s)} \tag{8.1}$$

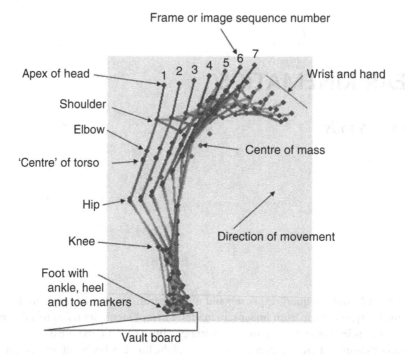

**FIGURE 8.1** Computer rendering of a gymnast taking off from a vault board towards the vault table.

Note: The motion is from left to right. Note that the paths of different joints and identified body parts are straight or slightly curved lines.

A simple example of speed can be seen when observing the vault run up. If a vault has a run up to the vault board of 15 metres (i.e., starts the run at 15 metres from the vault board) and the gymnast requires 2.5 seconds to run from the start to the vault board then the average speed would be:

$$average\ speed = \frac{distance}{time}$$

$$average\ speed = \frac{15m}{2.5s}$$

$$average\ speed = 6m/s$$

Although the average speed can tell us about the run up in its entirety, we know that the vaulter starts from a still position and then runs faster and faster until he/she reaches the vault board. The vaulter will often change his/her speed by speeding up in the beginning and then slowing down again just before board contact (Sands, 2000d; Sands & Cheetham, 1986; Sands & McNeal, 1995a; Sands & McNeal, 1999a,b,c).

Figure 8.2 shows the average speeds of a sprint test conducted on talent identified female gymnasts of 9–12 years of age (Sands & McNeal, 1999a,b,c). Note that the average speeds at the various distances changed. Although the overall average speed for this test was 5.5

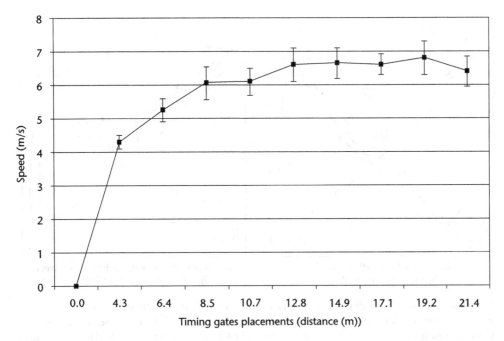

**FIGURE 8.2**   Speed vs. distance in gymnastics sprints.

m/s, we can see that there were probably very few, if any, recorded speeds at the average value. Moreover, the average value does not give a good idea of how the run ups were actually performed by this group.

The information shown in Figure 8.2 leads us to consider how often we might want to sample the gymnast's movement in order to get a better idea of the subtler characteristics of his/her skill performance. Some movements that occur slowly may be analysed with relatively few samples of the movement while very rapid movements such as a tumbling or a vaulting take off may require analyses with samples acquired every thousandth of a second or faster.

## 8.2 Displacement and velocity

Most lay people find the difference between the concepts of speed and velocity to be trivial. However, in biomechanics and physics the difference between the two is important. Speed is called a "scalar" quantity which means that the value has a magnitude only. Direction is not considered in a scalar value. Velocity is a vector quantity and includes a direction as well as magnitude. You can think of the speed of a movement as the magnitude of a velocity while the direction must also be determined or considered. In order to talk about velocity, you have to think of two things simultaneously, displacement (similar to distance) and velocity (similar to speed).

An example of the difference between the two concepts can be brought home by looking at a 400-metre run. In terms of speed, the distance over which the runner goes if he/she starts and finishes at the starting line (same place), the runner has covered a distance of 400 metres. If we know the time it took to cover the 400 metres then we can

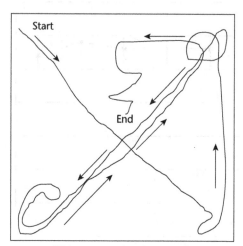
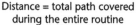
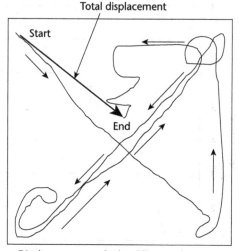

| Distance = total path covered during the entire routine | Displacement = only the difference between the initial position and the final position |

**FIGURE 8.3** Examples of distance and displacement. Each figure describes the path of a gymnast through a floor exercise routine. The left figure shows distance as measured along the entire path the gymnast follows, while the right figure shows the total displacement – the straight-line distance and direction from the start to the finish.

calculate an average speed. However, if the 400-metre runner starts and stops in the same place and we just look at the overall movement of the runner, then we would conclude that his/her displacement was zero metres because the athlete started and ended in the same place, thus he/she had zero or no displacement. The idea of the difference between distance and displacement, speed and velocity can be further illustrated by a floor exercise routine and the movement path of the gymnast around the floor exercise area, Figure 8.3.

Average velocity is the quotient of the change in displacement and the change in time. See Equations 8.2 and 8.3.

$$average\ velocity\ (m/s) = \frac{displacement\ (m)}{time\ (s)} \qquad (8.2)$$

$$average\ velocity\ (m/s) = \frac{final\ position\ (m) - initial\ position\ (m)}{final\ time\ (s) - initial\ time\ (s)} \qquad (8.3)$$

It may seem that the idea of deciding what "counts" as an initial position and a final position could be arbitrary. The biomechanist determines initial and final positions and initial and final times with considerable care in order to ensure that he/she is examining the portion of movement that is of interest. Moreover, we again confront the problem that an average speed or velocity is probably less insightful than smaller units of speed or velocity in examining any movement. When looking closely at speed and velocity we tend to divide the movement into smaller and smaller time segments as a means of determining an "instantaneous" speed or velocity. The term instantaneous refers to small segments of time and thus movement, and helps break a movement into smaller segments that reveal greater detail. An example of this process is shown in Figure 8.1.

## 8.3 Acceleration

A gymnast's ability to be explosive in his/her movements is often dependent on the ability to accelerate. Although rapid movement may be desirable, the gymnast is constrained by time limits of foot contact with the floor or apparatus, the correct moment for a release or other movement during a swing, and others.

The rate of change of velocity is called acceleration. Average acceleration is shown mathematically in Equation 8.4.

$$average\ acceleration\ (m/s^2) = \frac{final\ velocity\,(m/s) - initial\ velocity\,(m/s)}{final\ time\,(s) - initial\ time\,(s)} \tag{8.4}$$

Clearly, Equation 8.4 shows that if the velocity is constant (final velocity = initial velocity) then their subtraction equals zero, and thus there is no change in velocity and no acceleration.

Interestingly, an object at rest and an object in straight-line uniform (i.e., unchanging) motion have equal accelerations – zero.

As in the case of speed and velocity, acceleration can also be determined as an "instantaneous" value. If you look closer at Equation 8.4, it is apparent that there can be positive, negative and zero acceleration based on the values in the numerator.

When motions occur in a straight line, the positive and negative values provide little problem and people commonly refer to a positive acceleration as simple "acceleration" and a negative acceleration as a "deceleration". In biomechanics, however, the word deceleration is seldom used and the positive and negative designations are preferred. When the motion is not in a uniform straight line, such as when a body moves in one direction and then reverses and moves in the opposite direction, we can face the problem of the interaction of direction and change in velocity. For example, a body may be moving in a positive direction (e.g., to the right) but slowing down indicating a negative acceleration and vice versa. Keeping track of the positive and negative natures of direction and acceleration can prove difficult.

## 8.4 Linear kinematics units of measurement

Linear kinematics uses units that are relatively familiar to most people. Distance and displacement are measured in feet, inches, metres, centimetres, kilometres, miles and so forth. Time is measured in seconds, minutes, hours, days and others. Logically, if a displacement is known, let's say in metres, and we divide that distance by the time required to cover it, we get velocity in metres per second. Other familiar speed and velocity measurements are feet per second, miles per hour and kilometres per hour. The intuitive nature of these units often breaks down when we come to acceleration.

As you look at Equation 8.4, you should note that the number is the difference between final and initial velocities. The denominator is in seconds. Thus, average acceleration is measured in metres per second per second, feet per second per second, or metres per second squared – all of these are units of average acceleration.

When translated to simpler language, acceleration is a measure of the rate of change of velocity. For example, the speedometer in your car shows your current velocity. If you suddenly step on the gas and begin to move forward faster you'll see the speedometer needle or digital display show a rapid movement or a rapid change of digits. The speed of the movement of the speedometer is a measure of acceleration. If you suddenly hit the brakes while moving forward you'll note the speedometer shows a decreasing velocity. The more rapidly the speedometer changes, the larger the negative acceleration – you're slowing down.

All of these units of linear kinematics are shown in Figure 8.4, which shows a stroboscopic view of a gymnast performing a Roche vault (top image). The multiple images can provide

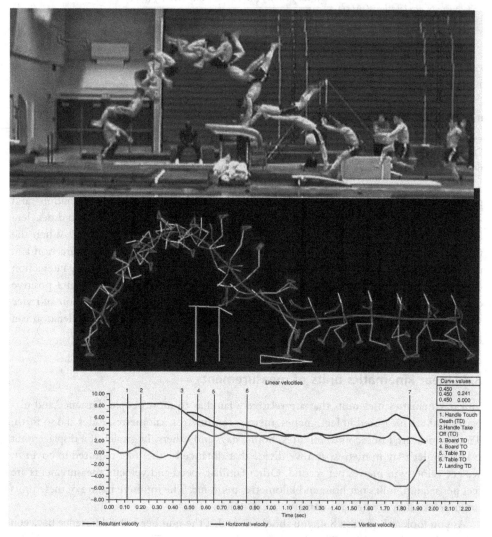

**FIGURE 8.4** Roche vault performed by an elite male gymnast. Note that the uppermost figure shows the performance as seen in a typical video format. The middle image shows a computer rendering (stick figures) along with the path of the centre of mass. Finally, the lower image shows the velocities of the centre of mass in the horizontal, vertical and resultant directions, phases of the skill and durations.

a great deal of information about the gymnast's performance. The middle section of Figure 8.4 shows the gymnast's body rendered by a computer to simplify the motion and provide yet another view of the skill, and shows the more linear component of movement that is the path of the centre of mass. Finally, the lower segment of Figure 8.4 shows the horizontal, vertical and resultant velocities of the centre of mass of the gymnast through the skill (Cormie, Sands, & Smith, 2004).

## 8.5 Frames of reference (*Patrice Holvoet*)

The determination of the frame of reference within which the motion of the body and its segments occurs is an important part of the motion analysis process. Indeed, the frame of reference is necessary as the background on which motion is "mapped". Therefore, the frame of reference must be precisely determined in order to maintain precision in further analyses.

The frame of reference is defined by a point of reference and axes relative to which the motion is described. The gymnast's change in location can be described relative to a stationary environment $R_O(O, xO, zO)$ as the apparatus, for example, or relative to another arbitrarily defined reference $R_G(G, xG, zG)$ or $R_S(S, xS, zS)$, so that movements of specific points can be referenced to a fixed external "framework" such that position, displacement, velocity and acceleration can be determined. For example, the movement of the athlete's centre of mass must occur within and about some frame of reference in order to have any physical meaning (Figure 8.6).

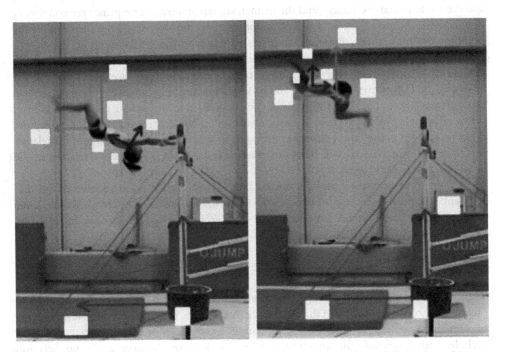

**FIGURE 8.5** Frames of reference usually used for analysing the motion of the body and its segments while performing gymnastic skills.

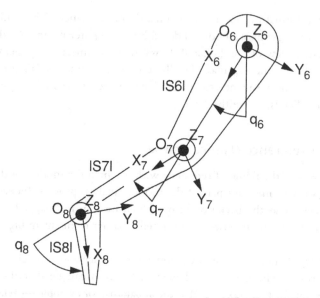

**FIGURE 8.6** Two-dimensional local frames of reference used for analysing segmental motions relative to the centre of joint. The lower limb motion is described with q8, q7 and q6 relative respectively to each local frame: R8 (O8, X8, Y8); R7 (O7, X7, Y7); and R6 (O6, X6, Y6).

The three spatial axes – horizontal relative to the ground (xO), the vertical axis (zO) and the mediolateral (yO) axis – and the frontal, sagittal and transverse planes are commonly designated as the stationary axes and planes of reference for describing the direction of the body's motion in three-dimensional space. Whenever the orientation of the body's segments is moving in space, the spatial axes do not change and are fixed relative to the ground or relative to an observer. This stationary frame of reference is appropriate for analysing the trajectories of the body and its segments during gymnastic skills executed on the different apparatus. When performing aerial skills, the movement can be analysed in terms of accuracy of the segmental motion relative to the centre of gravity G during the flight phase. When the biomechanical purpose is to describe the segmental motion relative to the centre of gravity, it will be appropriate to use a frame of reference relative to the centre of gravity defined by $R_G$(G, xG, zG).

In other cases in which the description of the segmental motion relative to another segment is required, it is appropriate to define a moving frame of reference by the three principle axes (mediolateral, longitudinal and anteroposterior axes) passing through the centre of a joint (Figure 8.6).

## 8.6 Vectors and scalars

Thus far, we have included velocity and direction and mentioned the terms "vector" and "scalar", but we haven't defined the terms fully. A scalar has been discussed as the magnitude of something with no concern for direction. A vector is more complicated. A vector includes both magnitude and direction, and, most important, the magnitude and direction are considered simultaneously.

Velocity, acceleration and many other values in biomechanics are vector quantities. Vectors are often depicted visually as an arrow from a point on the body at the beginning of the motion under consideration. The length of the arrow is scaled to the magnitude of the vector quantity and the direction in which movement occurs is depicted by the direction of the arrow. Figure 8.7 shows some examples of vectors.

Figures 8.7 and 8.8 show the parallelogram method for graphically displaying vector resolution. There are a number of interesting properties of vectors that will lead us to flight trajectories of bodies free in space, such as the flight phase of a tumbling skill, dismount, and so forth.

A trajectory is shown in Figure 8.4 for a vault sequence. Note that the line that follows the gymnast at about hip level depicts the path of the centre of mass during support and airborne phases. The path of this line is a trajectory. Vectors can be used to determine the nature of flight trajectories. Moreover, knowing the horizontal and vertical component velocity vectors can tell you most things you need to know about the subsequent flight trajectory.

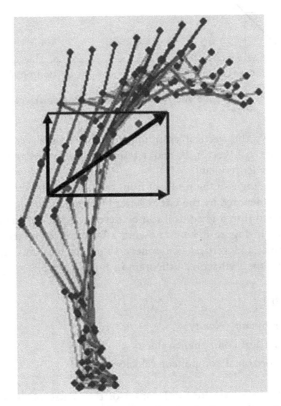

**FIGURE 8.7** Estimated horizontal, vertical and resultant velocity vectors of a vault take off. The horizontal velocity is greater than the vertical velocity due to the rapid run prior to the vault board contact. The vertical velocity component is shown as the vertical arrow. Both horizontal and vertical velocities have their origins in this case at the centre of mass of the gymnast. The angled vector is the resultant velocity of both the horizontal and vertical component velocities.

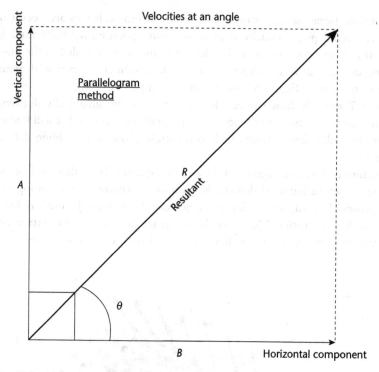

**FIGURE 8.8** Velocities at an angle and resolution of perpendicular velocity vectors.

Figure 8.8 shows the vertical component velocity as a vector $A$, the horizontal component velocity as a vector $B$, and a resultant (the actual resultant velocity of the object) as the vector $R$ (resultant).

Traditionally, the angle of the resultant from the horizontal is called the angle of launch or projection and is denoted by the Greek letter $\Theta$ (theta). The angle between lines $A$ and $B$ is a right angle and thus a great deal can be determined about the characteristics of the diagram of vectors in Figure 8.8 by knowing a little trigonometry. For example, if we know the horizontal and vertical component velocities we can determine the resultant simply by invoking the Pythagorean theorem in Equation 8.5.

$$R = \sqrt{(A^2 + B^2)} \tag{8.5}$$

Where: $R$ = resultant velocity

$\quad\quad A$ = vertical component velocity

$\quad\quad B$ = horizontal component velocity

More frequently, we estimate the location of the centre of mass (covered later), we know the resultant velocity from direct measurement and can determine the angle of launch by observation of the first few frames of video during a take off or release. From this knowledge and some elementary trigonometry we can determine the horizontal and vertical component velocities (Equations 8.6 and 8.7).

*vertical component velocity (line A) $= R \sin(\theta)$* $\tag{8.6}$

Where: $R$ = resultant velocity

$\sin(\theta)$ = the trigonometric sine of the angle $\theta$ (theta)

$A$ = the vertical component velocity shown in Figure 8.8

*horizontal component velocity (line B)* $= R\cos(\theta)$ (8.7)

Where: $R$ = resultant velocity

$\cos(\theta)$ = the trigonometric cosine of the angle $\theta$ (theta)

$B$ = the horizontal component velocity shown in Figure 8.8

To calculate the angle $\theta$, you need to know the horizontal and vertical component velocities and another trigonometric function called an arctangent. The arctangent of a number x is simply the angle whose tangent is x. Equation 8.8 shows the method for determining the angle of launch when the two component velocities are known.

$$\theta = \arctan\frac{A}{B}$$ (8.8)

Where: $\theta$ = the angle between the horizontal component velocity and the resultant

arctan = the trigonometric arctangent of the ratio of line $A$ and line $B$

$A$ = vertical component velocity

$B$ = horizontal component velocity

Of course, not all resultant vectors can be reduced to two vectors at right angles and a resultant velocity in the middle (Figure 8.9). In the situation where the components are not at right angles the resolution of these component velocities and determination of the resultant velocity requires a different approach and a trigonometric identity called the cosine rule. Equation 8.9 shows the means of calculating a resultant with an angle $\beta$ (beta) that is not a right angle (i.e., the angle between vectors $A$ and $B$).

$$resultant(R) = \sqrt{A^2 + B^2 + 2AB\cos\beta}$$ (8.9)

The direction in which the resultant acts can be determined by Equation 8.10.

$$\theta = \arctan\frac{A\sin\beta}{B + A\cos\beta}$$ (8.10)

Both Figures 8.8 and 8.9 show depictions of a resultant velocity vector and its components. Equations have been listed such that components can be calculated from a known resultant and a known angle of projection. Or, one can solve for the resultant by knowing the two components and the angle of projection. The figures also show completed parallelograms that include lines that parallel the $A$ and $B$ vectors. These parallel lines that complete a parallelogram can be used to graphically solve for any of the components and the resultant. Figure 8.10 shows an example of the "graphical method" that is used to solve for a resultant or the components. The horizontal component is 7 m/s, the vertical component is 3 m/s, the resultant is approximately 7 and approximately 1/2 m/s based simply on visual inspection. The calculated answer for the resultant is 7.6 m/s.

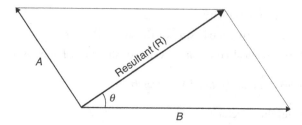

**FIGURE 8.9**   Velocities at a non-perpendicular angle and the resultant velocity.

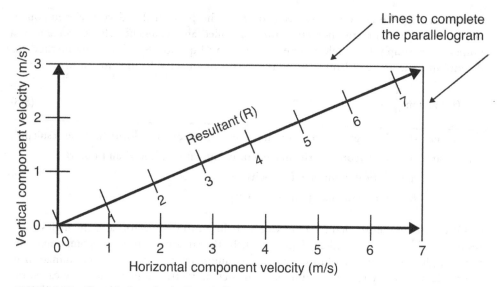

**FIGURE 8.10**   Graphical method of resolving vectors, and the resultant.

## 8.7 Taking flight – the kinematics of falling bodies and trajectories

Gymnasts often fall. One of the major factors in falling gymnasts is catching them and reducing injury (i.e., spotting). Interestingly, we can calculate the velocity at any point in a gymnast's falls, the time required to fall from some known height to the floor or any point between, and given a certain launch angle and velocity (i.e., take off) we can determine nearly all aspects of the flight of the gymnast. These kinds of calculations become important in such things as spotting. For example, if a gymnast unexpectedly falls from a known height, can a human spotter intervene and catch the gymnast before he/she hits the ground and/or apparatus and perhaps becomes injured?

Sands (1996, 2000a, 2000c) described the problem of catching a falling gymnast when the physical principles of a falling body in the Earth's gravitation are coupled with the limited ability of humans to react to a stimulus through the three reaction stages of stimulus identification, response selection and response programming. These three stages occur within the brain and perceptual systems – then you have to consider movement time to some effective position to catch the gymnast. It turns out that in many situations in which a gymnast falls, catching him/her is nearly impossible due to the combination of acceleration of the fall due to gravity and the limited ability of humans to process and act on the information they see.

How can we be so sure of the information in the previous paragraph? The principles of freely falling bodies and trajectories have been known for centuries. Let's begin with a body falling straight down from some height.

Because we live on the Earth with a nearly uniform gravitational field exerting a force (on everybody on its surface and in space around the planet) that is predictable and thus known we can characterize a great deal about a gymnast's fall.

Three equations are all that are necessary to handle the characteristics of a freely falling body.

Velocity at any moment of the fall = initial velocity + acceleration ⋆ time

$$v = u + at \qquad (8.11)$$

Distance of a fall = initial velocity ⋆ time + ½ acceleration ⋆ time$^2$

$$d = ut + at^2 \qquad (8.12)$$

Velocity$^2$ = initial velocity$^2$ + 2 ⋆ acceleration ⋆ distance of fall

$$v^2 = u^2 + 2ad \qquad (8.13)$$

Solving for time of the fall:

$$d = ut + \frac{1}{2}at^2 \qquad (8.14)$$

Let's assume that initial velocity is zero, therefore:

$$d = \frac{a}{2}t^2 \qquad (8.15)$$

$$t = \sqrt{(d/(a/2))} \qquad (8.16)$$

In most gymnastics falls and trajectories, the influence of air resistance is so small that it can be safely ignored. As an example of these equations and a falling gymnast, let's imagine that we have a gymnast standing on a balance beam (height = 1.25 m). For simplicity let's say that the gymnast just steps off the beam and falls to the floor. We'll focus on the feet, although in most biomechanics settings the interest would be in the centre of mass (we'll come to that later).

$$time\ of\ the\ fall = \sqrt{(d/(a/2))}$$

$$t = \sqrt{1.25/(9.806/2)}$$

$$t = \sqrt{1.25/4.903}$$

$$t = 0.50\ seconds$$

The calculations show that the time required for the feet to descend to the floor after stepping off the balance beam is approximately 0.50 seconds. Typical simple reaction time is around 0.25 seconds just to see a light flash and push a button (Henry & Rogers, 1960; Hodgkins, 1963; Stein, 1998). Unfortunately, trying to figure out something that will be effective in rescuing the falling gymnast takes a lot longer than 0.5 seconds (Woodson, Tillman, & Tillman, 1992).

For another example, let's assume that we have video tape of a dismount from the horizontal bar. We can plainly see when the athlete reaches the peak of his/her dismount flight trajectory. If we count frames from this point to the landing on the floor we can determine the duration of the descent. Each frame in standard video is taken in 1/30th of a second (0.03333 s) U.S. (NTSC standard) video and 1/50th of a second using European (PAL) video. Let's assume the descent of the gymnast required 24 frames at 1/30th of a second each. This means that the athlete required 0.8 seconds to fall from the peak of his/her dismount flight trajectory.

To determine how far he/she fell, or the height of the dismount, we approximate by using this information (at the peak of the trajectory the vertical component velocity is zero):

$$d = ut + \frac{1}{2}at^2$$

$$d = (0 \star t) + \frac{1}{2}(9.806) \star (0.8)^2$$

$$d = 4.903 \star 0.64$$

$$d = 3.14 \; metres$$

Now, let's determine how fast the gymnast is going (only in a downward direction) at landing. If the time of the fall is 0.8 seconds, then to determine an approximation of the impact velocity all we need do is multiply the time that the acceleration acts on the falling body. At this point we should identify where the constant 9.806 comes from. The acceleration due to gravity is 9.806 m/s/s, also written as 9.806 m/s$^2$. This means that the velocity of a falling body increases at a rate of 9.806 metres per second for each second in which it falls. Thus, the longer the time of the fall the greater the final velocity (disregarding air resistance). However, note that the value for the acceleration is constant or uniform. This is a very important idea.

$$v = u + at$$

$$v = 0 + (9.806 \star 0.80)$$

$$v = 7.8 \; m/s$$

Thus, the athlete's velocity at impact is 7.8 m/s.

There is an important caveat to the equations for uniformly accelerated motion – these equations only apply to situations in which the change in velocity is constant. Running, jumping (not the flight phase), swinging and so forth do change velocity, but the change or the acceleration is not constant. Thus, using these equations for anything but "uniformly" accelerated motion is a serious error. Fortunately, gravity provides us with a readily available source of uniformly accelerated motion via freely falling bodies.

Gymnasts rarely fall straight down; they usually rise from a take off or release and then fall to the ground or descend to regrasp a bar. Gymnasts can jump or rise straight up and then fall straight down. However, usually there is some horizontal travel during the flight as well. Interestingly, and simplifying understanding, it doesn't matter whether the gymnast moves horizontally or not, the principles of this type of flight are the same – we call this type of flight a "trajectory".

Figures 8.11 and 8.12 show a trajectory from a tumbling somersault. This particular somersault also included an attempt at four twists. The sequence runs from left to right and also includes the latter phase of the round-off and the complete back handspring (flic flac) along with the somersault. The trajectory of interest is the path of the centre of mass. The centre of mass is shown as a single dot in Figure 8.11 and as the series of dots in Figure 8.12.

Both Figures 8.11 and 8.12 show the two-dimensional velocities of horizontal and vertical components and the resultant velocity. Note the correspondence between the velocity changes that occur in the components and resultant along with the changes in motion and the flight trajectory.

The path of the centre of mass in Figure 8.12 shows a little "noise" or bumpiness that is largely due to digitizing error which is an artefact of the hand digitizing process that took place from the raw video footage of three cameras. However, the general shape of the curve of the flight path of the gymnast is parabolic and we will refer to these two figures in the following. You can also see a flight trajectory in Figure 8.4 when the gymnast leaves the vault table and launches into the post flight somersault phase of the vault. Once the athlete leaves the ground and undertakes flight, he/she is a projectile – just like a baseball, football or bullet.

Two things are of primary interest in characterizing a trajectory: time of flight (height) and horizontal range (horizontal distance of travel). We can learn the time required to reach the top of the flight trajectory by knowing the resultant velocity and the angle of projection.

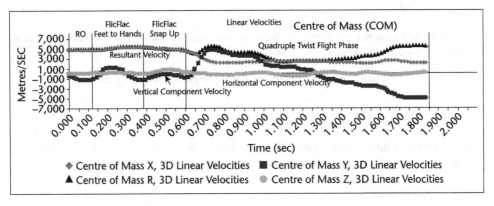

**FIGURE 8.11**  A stick-figure tumbling sequence of the snap-down of the round-off, back handspring (flic flac) and a high layout somersault with four twists. Note that the centre of mass in this figure is marked by a small triangle that is not connected to the stick-figure. The lower half of the figure shows the horizontal forward–backward and side-to-side, vertical and resultant velocities of the centre of mass.

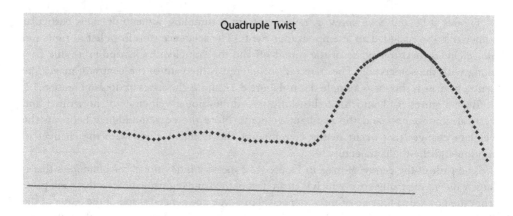

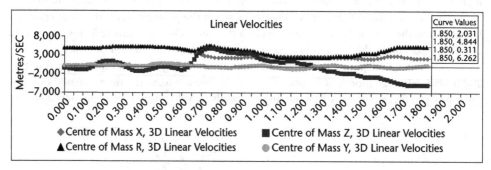

**FIGURE 8.12** Centre of mass path. The stick-figure is removed from this image, leaving the movement of the centre of mass as shown by the small triangles. The rightmost movement of the centre of mass shows the large bump in the trajectory of the gymnast during flight.

$$time_{up} = \text{resultant velocity at take off} \star \sin \theta$$

$$t_{up} = Rsin\theta \tag{8.17}$$

In order to know the total time of flight we also have to know the time from the peak of the trajectory to the landing:

$$time_{down} = \text{resultant velocity at take off} \, / \, g \text{ (acceleration due to gravity)}$$

$$t_{down} = \frac{Rsin\theta}{g} \tag{8.18}$$

In order to obtain the total time of flight we simply add the two terms. Or, by combining the equations we obtain the following:

$$T = t_{up} + t_{down} \tag{8.19}$$

$$T = \frac{Rsin\theta}{g} + \frac{v_{up}}{g}$$

$$T = \frac{Rsin\theta + v_{up}}{g}$$

An important caveat of the foregoing is that if the launch and landing are at the same level then:

$$t_{up} = t_{down}$$

Which can also be written as:

$$\frac{R\sin\theta}{g} = \frac{v_{up}}{g}$$

And, total time can be determined by the following:

$$T = \frac{2R\sin\theta}{g} \tag{8.20}$$

These relationships, when launch and landing are at the same height, indicate several things:

1.  The time from take off to the peak of the trajectory is exactly the same as the time from the peak of the trajectory to the landing. (This assumes that the centre of mass rises and falls the same distance.)
2.  The time of flight or height of flight are completely determined by the vertical velocity of take off. Thus, you will note that there is no term in Equations 8.17–8.20 for time of flight that includes a horizontal component velocity.
3.  In order to maximize time of flight or height, the gymnast must maximize his/her take off velocity which involves considerable technique and strength/power considerations leading up to the take off.
4.  Once in the air, the gymnast can do absolutely nothing to alter the parabolic flight path that his/her centre of mass is following.
5.  The height and time of flight and the horizontal range of the projectile are completely independent of each other in terms of the physics. However, we know that it is considerably harder to go high than far in a jump. The negative acceleration that is caused by gravity can be thought of as "robbing" the vertical component velocity of 9.806 m/s each second of flight. Thus, if you leave the ground with a vertical component velocity of 9.806 m/s, you will reach your trajectory peak in 1 second.

The preceding has involved an assumption that the take off and the landing occur at the same level. Often this assumption can provide fairly close approximations of flight time and height. However, there are circumstances where the gymnast launches from a position that is quite a bit higher than the landing (e.g., vaulting post flight, dismounts from the apparatus, etc.). Moreover, there are also circumstances when the take off point is below the landing position such as a mount to the balance beam, uneven bar, parallel bars and so forth. In these situations determining the total time of flight is more complicated and the equation a bit more difficult:

$$T = \frac{R\sin\theta + \sqrt{(R\sin\theta)2 = 2gh}}{g} \tag{8.21}$$

In this case, again $R$ refers to the resultant velocity, $\theta$ is the angle of launch and $g$ is the acceleration due to gravity. The only additional term is $h$ which is the height of the object from the landing point at the moment of launch. One should keep in mind that the height referred to here is often the height of the centre of mass, not simply the height of the object from which the launch occurred. Various body position distortions can occur and naked-eye appraisals of take off and landing positions are often deceptive.

Thus far the primary characteristic of interest has been time. What about displacement? Displacement occurs in two dimensions forward-backward and up-down. However, we usually just concern ourselves with a horizontal range from launch to landing and a vertical height of the trajectory. Height of flight is the easiest characteristic to determine, all vertical displacements are assumed to begin at the point of launch and the calculation is shown in Equation 8.22.

$$height\ of\ flight = \frac{(R sin\theta)^2}{2g} \tag{8.22}$$

You can also think of the height of flight as:

$$H = \frac{v_{up}^2}{2g} \tag{8.23}$$

Again, it should be obvious that the only things that determine the height of flight of any projectile are the negative acceleration due to gravity and the vertical velocity at launch. Although the athlete can perform a variety of airborne movements (e.g., swinging arms, tucking the legs, arching, piking, etc.), the only thing under the athlete's control in achieving a high flight trajectory is his/her vertical velocity at take off or launch.

Horizontal displacement is simultaneously simple and complex. Because the horizontal component velocity and the vertical component velocity are completely independent, one need only know the horizontal component velocity of the object or body and the time of flight. As such, the object or body gets to travel in a horizontal direction only so long as it's in the air; gravity has no influence on horizontal velocity. Therefore:

$$horizontal\ range = average\ velocity \star time\ of\ flight \tag{8.24}$$

Or:

$$s = vt \tag{8.25}$$

Where: $s$ = horizontal range
$v$ = horizontal component velocity
$t$ = total time of flight

Horizontal displacement becomes more problematic to calculate because we must determine the time of flight (determined by vertical velocity at take off) and the horizontal component velocity. Both of these values can be decomposed from the resultant, however until now we've not shown the horizontal component velocity as calculated from a known resultant velocity and the angle of projection ($\theta$).

Starting with Equation 8.24, we can substitute Equation 8.20 for time of flight. Then we must determine the average horizontal component velocity which is given by:

$$horizontal\ component\ velocity = R cos\theta \qquad (8.26)$$

Thus:

$$s = R cos\theta \star \frac{2R sin\theta}{g}$$

Equation 8.26 can be simplified by using a trigonometric identity:

$$sin2\theta = 2sin\theta cos\theta$$

which then gives us:

$$s = \frac{R^2 sin2\theta}{g} \qquad (8.27)$$

Equation 8.27 shows that the horizontal range($s$) of a body or object is determined by the resultant velocity at launch and the angle of the projection. In Equation 8.27, $g$ is a constant acceleration due to gravity and thereby not within the control of the gymnast.

Of course, Equation 8.27 presumes that launch and landing are at the same height. If this is not the case, the equation becomes more complicated, but the terms should be easily recognizable. The equation is simply a fancier way of using average horizontal component velocity and knowing the total time in the air.

$$s = R cos\theta \star \frac{R sin\theta + \sqrt{(R sin\theta)^2 + 2gh}}{g} \qquad (8.28)$$

Equation 8.28 can be reduced to:

$$s = \frac{R^2 sin\theta cos\theta + R cos\theta \sqrt{(R sin\theta)^2 + 2gh}}{g}$$

In both circumstances − the same launch and landing and different launch and landing positions − velocity of launch or take off is important. In the case of launch and landing at the same level, the optimum angle of launch or take off is 45 degrees measured from the centre of mass. In the case of launch and landing at different heights, the problem becomes one of optimization. When trying to achieve a maximum range or distance of flight, the following rules apply:

1. Equal changes in either height of launch or speed of launch do not result in equal changes in the optimum angle or the horizontal range.
2. The best angle for launch is always less than 45 degrees.
3. For any particular height of launch, the greater the resultant velocity, the more closely the best angle of launch approaches 45 degrees.
4. For any particular resultant velocity of launch, the greater the height of the launch the lower the optimum angle (Hay, 1973).

For launch and landings that move from a lower to a higher position such as a vault board take off to the vault table, the optimum angle of departure varies based on the intent of the landing on the higher surface. For example, do you want to land and stop (balance beam mount) or do you want to preserve as much horizontal component velocity as possible (vaulting)?

## 8.8 CONCLUSION

Linear kinematics provides a foundation for understanding much of the following material. You will see positions, displacements, velocities and accelerations throughout the remainder of this section. Moreover, in the next chapter on angular (rotational) kinematics, you will see how the linear ideas are only slightly modified and an additional thought process is needed to make the shift from linear motion to angular motion. In a sense, you will have to hold two thoughts in mind simultaneously, linear ideas and additional information about their rotation.

# 9

# ANGULAR KINEMATICS

*William A. Sands*

When a gymnast moves from one position to another he/she may do so by translation (linear movement), rotation about an axis and through an angle (angular movement), or both. Angular movement is "science-speak" for "rotation". Somersaults, handsprings, giant swings, hip circles, pommel horse circles and so forth are all examples of skills that involve angular motion. Fortunately, there are angular analogues to linear motion making the understanding of angular motion simply an extension of the concepts from linear motion.

Different sports describe angular movement using terminology that fits sporting contexts and usually has origins in long traditions. Diving and gymnastics refer to somersaults as forward (face leading) or backward (back of the head leading). However, in gymnastics a backward giant swing has the face leading and a forward giant swing has the back leading. The terminology in these cases appears contradictory. Diving uses both forward and backward somersaults along with inward and gainer to indicate which way the diver is facing relative to the diving board or platform during the take off.

If we add twisting to rotation terminology then the direction is referred to as right or left. A right twist has the right shoulder leading rearward or the head turns to the right. A left twist shows the opposite. The magnitude of twists and somersaults are usually simply described as single, double, triple and fractions thereof. Biomechanics uses a different approach to determining the direction and magnitude of rotation.

## 9.1 Angular motion

Although it is perfectly acceptable to talk about angular motion in terms of degrees, the more common approach used in biomechanics is to refer to angular motion in terms of radians. The reason for this is that there is a serious mathematical problem with degrees when the rotation involves more than one complete rotation. For example, if a diver completes two and one half somersaults, then he/she has performed 900 degrees of rotation. If we watch the diver rotate and standardize vertical as 0 degrees, then as the diver completes a single rotation his/her position changes from 359 degrees to 0 degrees very suddenly and creates a difficult problem for mathematics because of the continuous change in position with a discontinuous change in the mapping of that position onto a 360-degree circle.

While degrees are terrific for navigational purposes, they are much more problematic in a mathematical, geometric and trigonometric world. As such, there is an easy way to determine the rotation of something by taking advantage of properties of a circle. For example, we know that the circumference of a circle is given by Equation 9.1:

$$circumference = 2\pi r \tag{9.1}$$

Where: $\pi$ = pi ≈ 3.1416 which is the ratio of the diameter of a circle to its circumference
$r$ = the radius of the circle, and two times the radius equals the diameter

Therefore, we can describe a trip around the circumference of a circle for any size circle simply by knowing its radius. If the diver performs a single somersault (360 degrees) then the diver also rotates 2 $\pi$ times, or approximately 6.28 radii.

There is a special name for this distance around the circumference of a circle – a radian. In a sense, a radian takes the "angle idea" of degrees and converts it into a distance around the circumference always relative to the radius of the circle. If you want to convert from degrees to radians then a radian ≈ 57.3 degrees, 0.16 rotations or about 60 degrees. From the world of gymnastics we might say that a gymnast performed a one-and-one-half twisting double somersault. In degrees this would be a 540-degree-twisting 720-degree somersault, or a 3 $\pi$ twisting 4 $\pi$.

## 9.2 Angular speed and angular velocity

Angular speed and angular velocity are terms used similar to their linear counterparts. The angle through which a body moves in a period of time is called angular speed if you're not concerned with direction and angular velocity if you are.

$$angular\ speed = \frac{angular\ distance}{time} \tag{9.2}$$

If you're concerned with both the magnitude of the angular distance and the direction then you use angular velocity as the correct term, usually given by the Greek letter $\omega$ (lower case omega).

$$\omega = \frac{\theta}{t} \tag{9.3}$$

For example: if a gymnast rotates counterclockwise 60 degrees in a giant swing (and you've defined clockwise as positive) in 0.45 seconds, then the result is:

$$\omega = \frac{-60\ degrees}{0.45\ seconds}$$

$$\omega = -133\ degrees/second$$

## 9.3 Angular acceleration

Angular acceleration works in the same way as linear acceleration. Angular acceleration is the rate at which angular velocity changes with respect to time. Or, angular acceleration is the final angular velocity minus the initial angular velocity per unit of (or divided by) time. Angular displacements, velocities and accelerations can be both average and instantaneous values just like their linear counterparts.

$$\alpha = \frac{\omega_f - \omega_i}{t_f - t_i} \tag{9.4}$$

A gymnast who is travelling in a clockwise direction at 100 degrees/second at one point in a giant swing and then 0.6 seconds later is travelling at 140 degrees/second then the average acceleration of the gymnast would be:

$$A = \frac{140 deg/sec - 100 deg/sec}{0.6 sec}$$

$$A = (66.7 deg/s)/s \; or \; 66.7 deg/s^2$$

As shown in Figures 8.7 and 8.8 it is customary to graphically depict motion as vectors representing magnitude and direction.

Angular motion is little different except it is difficult to graphically depict rotational motion on two-dimensional paper. Therefore, a convention for showing angular motion vectors has been derived, called the "right-hand thumb rule". In order to describe an angular motion vector you use your right hand with the fingers flexed in a fist and the thumb out in a sort of "thumbs up" gesture. The direction of the fingers indicates the direction of rotation of the object while the thumb represents the direction in which the vector is drawn.

Since angular velocity, acceleration and other quantities are vector quantities, the right-hand thumb rule can be applied to all of these. Figure 9.1 shows the take off for a quadruple twisting backward somersault. Considering only the twist the gymnast is turning to the left, or the gymnast's face is turning toward the reader and away from the page. If we apply the right-hand thumb rule to set up our description of the twist velocity it would look like Figure 9.2.

**FIGURE 9.1** Quadruple-twist take off. Note that the gymnast is turning so as to face to the left prior to the twist and then to face the reader during the pictured initial phase of the twist.

**FIGURE 9.2** The right-hand thumb rule. The direction of the fingers in the fist shows the direction of the turn, while the thumb shows the direction in which you would draw the velocity vector of the twist.

Angular kinematics has one more aspect that requires explanation. Although we've been concerned with angular motions so far, described and calculated as an angular change in the position, velocity or acceleration of a body, there is another characteristic of angular motion which is the linear speed or velocity of the object at any given point in the angular movement. For example, in a golf swing the head of the club, while in angular motion, also has a linear motion component that might apply to the head of the club as it strikes the ball on the tee.

Average angular speed is measured as the angular distance through which the body rotates in a period of time. For example if an object is swinging about an axis through an arc $AB$ then

$$v_{Tangential} = \frac{distance}{time} = \frac{arc\ AB}{t} \tag{9.5}$$

Where: $v_{Tangential}$ is the linear speed of any point on the object.

Since the angular distance is measured in radians, the number of radians is determined by dividing the arc $AB$ by the radius, in this case $r$.

$$\omega = \frac{v_T}{r} \tag{9.6}$$

Rearranging:

$$v_T = \omega r$$

Finally, you can consider angular motion to be composed of two parts, both alluded to in the previous sections. One is a radial component which is the component that causes the object to rotate about an axis. If you swing a ball on a string in a circle above your head, the string is representative of the radial component. The string constrains the movement of the ball and by virtue of making the ball constantly change direction, accelerates the ball.

If you release the string at any given moment, then the ball will initially fly away at a tangent (right angles to the radius of the circling motion, i.e., the string) in the plane of the circling ball and string. The tangential component refers to the tangent of a circle which is a line drawn at right angles to the radius and touches the circumference of the circle at only one point. The tangential component of angular motion is important because it perfectly predicts in what direction an object will move once the radial component is eliminated. The radial acceleration is given by:

$$\alpha = \frac{v_T^2}{r} \tag{9.7}$$

The term $v_T^2$ is the speed of the ball tangential to its circular path as it is swung around your head by the string. When the ball is released, the speed of the ball will be this term.

Moreover, we can determine that the flight path of the ball will be at a tangent to the circle in which it was spinning. The exact same principle applies to the direction of flight following a release on the horizontal bar or uneven bars from a giant swing. The rate at which the speed of the ball changes as it moves along its curved path is called the tangential acceleration and is calculated by:

$$\alpha = \frac{v_{Tf} - v_{Ti}}{t} \tag{9.8}$$

This information is particularly important in some sports or events such as the hammer or discus throw in track and field.

In gymnastics these ideas can be seen in the swing and release of a flyaway. Figure 9.3 shows a dangerous flyaway from the uneven bars where the release occurred when the tangential motion of the centre of mass of the athlete caused her to move toward the bar and nearly strike it. Figure 9.4 shows a superior flyaway where the release and tangential flight path caused the gymnast to move away from the bar (Sands et al., 2004b).

## 9.4 Application: understanding relations between angular and linear motions

Take, for example, the analysis of the forward swing on the high bar performed by a young gymnast. What about angular and linear velocities? Figure 9.5 shows that at the beginning of the downward phase ($t_0$=0s) the hip is located at a radius of gyration of 1.35m from the bar ($rHipt_0$) with values of angular and linear velocities equal to zero. When the gymnast is swinging just under the bar ($t_1$=0.56s), the hip has rotated through an angle ($\theta_1$) of 140 degrees (or 2.44 rad) with a radius of gyration equal to 1.65m. When the gymnast reaches the front horizontal level of the bar ($t_2$=1s), the hip has travelled through an angle of 248.5 degrees (or 4.33 rad) with a radius of gyration of 1.45m.

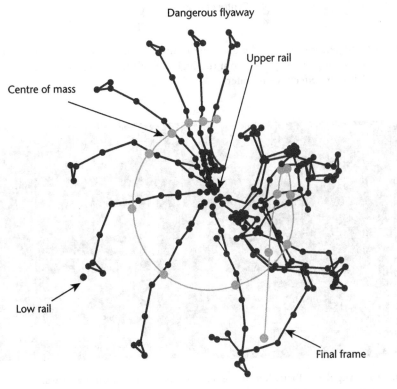

**FIGURE 9.3** Dangerous flyaway. The grey line depicting the path of the centre of mass of the gymnast shows that she released too late and her trajectory took her toward the high bar of the uneven bars.

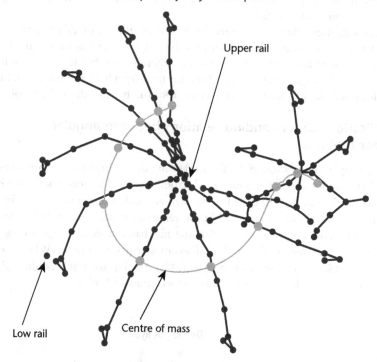

**FIGURE 9.4** Superior flyaway. The grey line in this case shows a release point that resulted in the gymnast moving up and away from the high bar of the uneven bars in a flight trajectory that was both more effective and safer.

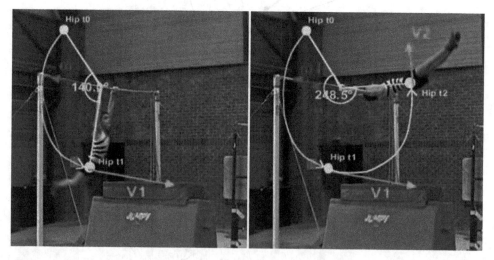

**FIGURE 9.5** Angular velocity and linear velocity of the hip during a forward swing around the high bar.

The axis of the angular velocity vectors $\omega_1$ and $\omega_2$ at $t_1$ and $t_2$ respectively is perpendicular to the plane of rotation and is along the bar. The direction of the angular velocity vectors along the bar is given by the right-hand thumb rule. The angular speeds $\omega_1$ and $\omega_2$ at $t_1$ and $t_2$ respectively are given by:

$$\omega_1 = \frac{\theta_1 - \theta_0}{t_1 - t_0} = \frac{140}{0.56} = 250°/s \text{ or } 4.36\,rad/s$$

$$\omega_2 = \frac{\theta_2 - \theta_1}{t_2 - t_1} = \frac{248.5 - 140}{1 - 0.56} = 246.5°/s \text{ or } 4.30\,rad/s$$

Knowing $\omega_1$ and $\omega_2$ and the radius of gyration of the hip from the bar at respectively $t_1$ and $t_2$, the linear velocities $v_1$ and $v_2$ can be calculated as follows:

$$v_1 = rHipt_1.\omega_1 = 1.65 \times 4.36 = 7.19\,m/s$$

$$v_2 = rHipt_2.\omega_2 = 1.45 \times 4.30 = 6.23\,m/s$$

The direction of the linear velocities, vectors $v_1$ and $v_2$, is along the tangent line passing through the curved path at $t_1$ and $t_2$.

The hip swinging action performed by the gymnast between $t_1$ and $t_2$ helps to avoid losing too much angular and linear speed during the upward phase.

Now let's determine how angular and linear accelerations are given during the swing.

Figure 9.6 shows that at the beginning of the upward phase $(t_1=t_{initial})$, the angular speed of the gymnast is equal to $\omega_1$. At $t_2$ $(t_2=t_{final})$ the gymnast reaches the front horizontal level of the bar with an angular speed equal to $\omega_2$. The average angular acceleration $(\alpha)$ acting during the upward phase is calculated as follows:

$$\alpha = \frac{\Delta\omega}{\Delta t} = \frac{\omega_2 - \omega_1}{t_2 - t_1} = \frac{4.30 - 4.36}{1 - 0.56} = -0.13\,rad/s^2 \text{ or } -7.95°/s^2$$

That negative angular acceleration is responsible for the deceleration of the angular motion of the gymnast during the upward phase.

It appears similarly in Figure 9.6 that at the beginning of the upward phase $(t_1=t_{initial})$, the linear speed of the gymnast is equal to $v_1$. At $t_2$ $(t_2=t_{final})$ the gymnast reaches the forward horizontal level of the bar with a linear speed equal to $v_2$. The average linear acceleration $(a)$ acting during the upward phase is constructed with the vector $V_2-V_1$ (Figure 9.6a) and is calculated as follows:

$$a = \frac{\Delta V}{\Delta t} = \frac{V_2 - V_1}{t_2 - t_1} \tag{9.9}$$

The vector drawing of the linear acceleration can be considered as the vector resulting from the addition of a tangential component vector $(aT)$ and a radial component vector $(aR)$ (Figure 9.6b).

The vector $aT$ determines the tangential acceleration and is acting along the tangential line passing through each point of the curved path. The vector $aR$ determines the radial or centripetal acceleration and is acting along the radial line connecting each point of the curve path to the centre of rotation.

Remember that the linear acceleration $(a)$ of a point rotating with a given angular acceleration $(\alpha)$ depends also on its radius of rotation $(r)$. The greater the radius of gyration,

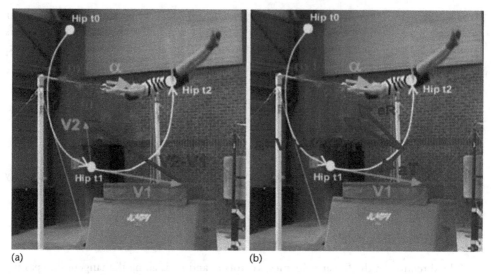

(a)                                                    (b)

**FIGURE 9.6** Angular acceleration and linear acceleration of the hip during the forward swing around the high bar.

the greater the linear acceleration of the rotating point. The relationship between linear and angular acceleration of a rotating point is given as follows:

$$a = r\alpha \tag{9.10}$$

With: $a$ *in* $m/s^2$
$r$ *in* $m$
$a$ *in* $rd/s^2$

If we suppose that during the downward phase of the swing (Figure 9.6) the radius of rotation of the hip is constant and is equal to $rHipt_1$, the average linear acceleration of the gymnast between $t_1$ and $t_0$ is given by:

$$a = r\alpha \ \ with \ r = 1.65m$$

$$\alpha = \frac{\omega_1 - \omega_0}{t_1 - t_0} = \frac{4.36}{0.56} = 7.78 \ rd/s^2$$

$$a = 1.65 \times 7.78 = 12.83 m/s^2$$

During the downward phase of the swing the linear acceleration of the gymnast is equal to 1.3 times the acceleration due to gravity.

## 9.5 CONCLUSION

Angular kinematics follows many of the same conventions as linear kinematics.

However, angular kinematics often involves several aspects of physical laws occurring simultaneously. A swing, for example, has both a tangential and a radial component. As we move into linear and angular kinetics, the importance of these differences will become even more pronounced.

# 10

# LINEAR KINETICS

*William A. Sands*

Kinematics is a description of motion: position, displacement, velocity and acceleration. There is no reference in kinematics to the underlying causes of motion.

Kinetics is the area of biomechanics where the source of motion is studied. The source of motion is a force.

Unlike the visibility of position, displacement, velocity and acceleration, forces are invisible. We can only infer that a force is present when we see or measure a body's acceleration. There are a few terms that should be defined before proceeding further. And one of them is important to the idea that an acceleration or a tendency to accelerate must be present in order for there to be a force.

## 10.1 Inertia

Inertia is a property of a body to remain in one place or to continue in uniform straight-line motion unless acted upon by an outside force. It may seem counterintuitive, but an object at rest (i.e., not moving) and an object in straight-line uniform motion are actually examples of the same thing – inertia.

Often, the word "reluctance" is used when describing inertia. It's important to realize that the biomechanist is not ascribing a property of consciousness to an object and that its reluctance comes from not "wanting" to move, but there are few words in English that suffice to describe this phenomenon without collapsing to circularity. Inertia is proportional to mass so a more massive object, whether at rest or in straight-line uniform motion, exhibits greater reluctance to move than a less massive object.

## 10.2 Mass

Mass is the quantity of matter in an object. Weight and mass are constantly confused. Weight is a measure of force, usually the force of gravity pulling on your mass. You can measure your weight using a scale. However, when you are in a freefall situation, such as while orbiting the Earth, you are weightless (or nearly so in microgravity), but you still have the same mass as when you were on the surface of the Earth. The distinction of weight and mass is one of the reasons we prefer to use "centre of mass" rather than "centre of gravity" when describing properties of a body or object, however here on Earth the ideas are interchangeable. The Imperial/English measurement system unit for mass is the "slug". The term sluggish comes from this term and implies the "reluctance" or resistance to motion that is described by the term "inertia". In the metric system, the unit of mass is the gram or kilogram.

## 10.3 Force

Forces are used to change a body's state of motion, either moving it from rest (accelerating it), changing its direction (such as radial acceleration discussed in angular kinematics) or a change in velocity while already moving (i.e., also an acceleration).

A force is defined as a push, pull or tendency to distort. The push and pull can be seen in the acceleration that results. The "tendency to distort" idea appears in situations where there is a force applied to something but the force is insufficient to actually accelerate the object. This might occur when the gymnast is trying to press to handstand (Figure 10.1) but is unable to, or when seated on a chair, the weight of the person's body is applying a force to the chair and if you put a scale under the legs of the chair you would find that there is a measurable force. However, the chair does not move in spite of the applied force. In the Imperial/English system a force of 1 lb will produce an acceleration of 1 foot per second squared. In the metric system the unit is the Newton. A Newton is the amount of force required to accelerate 1 kilogram by 1 metre per second squared.

Figure 10.1 shows force-measuring instruments that are required for collecting the forces exerted on the beam during the foot and hand support phases when performing a back walkover to handstand. Force transducers or force platforms convert the magnitude of force to an electric signal. Nowadays, these instruments are connected to computers which collect, process, plot and analyse the forces recorded.

## 10.4 Internal and external forces

Gymnasts also need to consider that they can produce forces from muscle tension (i.e., internal forces) and these forces can be seen in accelerating limbs and body. The body of the gymnast can also experience external forces, primarily gravity, but also the elastic forces of the apparatus such as springs in the spring floor and the bend and recoil of a rail or bar (Hay, 1973). For example, Figure 10.2 shows the flexion of the springboard during the takeoff executed by a gymnast to perform a vault. When the deformed springboard is recoiling, the recoil generates an external elastic force that moves the gymnast over the vault table. The recoil ability of the springboard depends on its coefficient of elasticity that is mainly determined by the nature of the materials from which the apparatus is made. The greater the degree of the springboard recoil, the greater its coefficient of elasticity.

## 10.5 Newton's laws of motion

Isaac Newton codified motion into laws. Newton's laws apply nicely to the world as we see it. However, Newton's laws have been theorized to break down when you consider the very large distances of outer space, very small distances of the subatomic and very fast motion of things moving at the speed of light. Fortunately, gymnastics movements fit nicely into the realm of Newton's laws and these laws have been well understood for hundreds of years. Some mysteries still remain. For example, Newton's first law has never

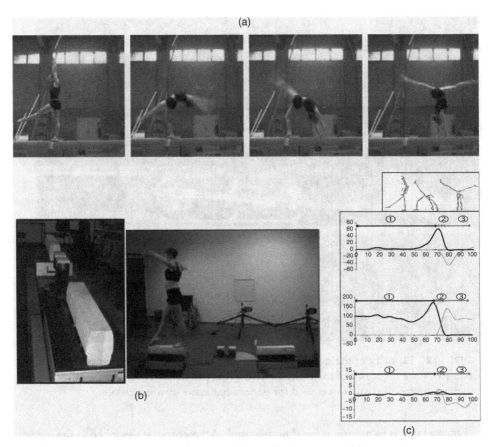

**FIGURE 10.1** Force-instrumented beam for collecting data on the forces exerted on the beam – for example, when performing a back and walkover handstand (a). Four pieces of beam are fixed on four Logabex force-plates so that they correspond to the foot and hand support target areas. The steel rail supporting the force measuring instruments can be raised to different heights (b). The different components of the reaction forces acting during the foot and hand support phases can be collected for the analysis of the dynamic balance control (c).

been proven directly because there isn't a situation in which a body has zero force applied to it. There are always forces being applied as a result of planetary rotation, orbital revolution and gravity (Hay, 1973).

## First law

> A body continues in its state of rest or motion in a straight line unless acted upon by an outside force.
>
> *Hay, 1973*

Newton's first law is also called the "law of inertia" or sometimes the "laziness law". Objects at rest tend to stay at rest. Objects in motion tend to stay in motion.

In either case, these are examples of an object's "reluctance" to change motion. An object's inertia is proportional to its mass, which is also proportional to its weight. A more massive body is harder to accelerate than a less massive body.

**FIGURE 10.2** Elasticity of the springboard. During takeoff the flexure of the springboard is indicated by the trajectory of the markers. The spring properties of the apparatus are an important source of elastic forces used for performing many gymnastic skills.

Once a body is in motion it has "momentum". Momentum is described as the quantity of motion in a body. Momentum is the product of the object's mass and its velocity. Increasing an object's mass and/or velocity increases the object's momentum and decreasing either term decreases the momentum.

## Second law

> *The rate of change of momentum of a body is proportional to the force applied to the body, is inversely proportional to the body's mass, and acts in the direction of the applied force.*

> *Hay, 1973*

A change in momentum per unit of time is shown in the following:

$$\text{force is proportional to } \frac{mv_f - mv_i}{t} \tag{10.1}$$

Where: $m$ = mass
$v$ = velocity
$t$ = time

When the mass of the body doesn't change, the statement above changes to:

$$\text{force is proportional to } m\frac{v_f - v_i}{t}$$

Of course, the second term (the division) is acceleration as you should recall from the section on linear kinematics. Multiplying one side of the equation by a constant (k) we get:

$$F = kma \tag{10.2}$$

The constant can be removed by noting:

$$1 \; Newton = k \times 1 kilogram \times 1$$

And then:

$$1 = k \times 1 \times 1$$

Thus, the equation for force becomes the more familiar:

$$F = ma \tag{10.3}$$

From a practical standpoint, a coach or instructor can often see the effects of a force on a gymnast when the coach or instructor is spotting. For example, a gymnast may be performing a tumbling sequence ending with a somersault. The coach assists the gymnast by "lifting" him/her slightly during the spot and thereby accelerates the gymnast in the direction of the applied force. Every spotter also knows intuitively that an older and probably heavier (more massive) athlete is more difficult to spot because it is more difficult to change the momentum of the heavier athlete.

## Third law

> *There are several ways of stating this law. For every force that is exerted by one body on another there is an equal, opposite, and simultaneous force exerted by the second body on the first.*
>
> *Hay, 1973*

According to Newton's third law, forces always work in pairs. A simple vertical jump can illustrate this. As a gymnast performs a jump he/she pushes against the Earth with a force exceeding the gymnast's weight (if he/she leaves the floor and travels upward) and the Earth pushes against the gymnast with an equal, opposite and simultaneous force. Of course, the mass of the gymnast is so utterly tiny compared to the mass of the Earth that the gymnast moves while the Earth's movement is imperceptible.

A word on the "simultaneous" term in the definition of Newton's third law is merited. Experience has shown that people often misunderstand the idea of this law by confusing aspects of jumps from elastic surfaces. If our vertical jumping gymnast was standing on a trampoline it appears to the eye that he/she pushes downward against the trampoline and then "waits" for a second or so for the trampoline to push back. The conceptual difference is that there is never a wait for the second force to arrive on the first body. The illusion provided by the trampoline is that some elastic energy is stored in the springs of the trampoline that is later applied or returned to the jumping gymnast. This is a separate pair of forces than those that occurred during the initial gymnast's downward push against the trampoline bed. One of the ways of stating Newton's third law is that for every action there is an equal, opposite and simultaneous reaction.

However, even in the statement above, there is no special meaning in mechanics for "action" and "reaction" and this makes the statement imprecise (Hay, 1973).

Sadly, the phraseology of this fundamental law of motion has been borrowed by other areas of study to demonstrate a conceptual point while ignoring the mathematical relationships. For example, the idea that for every action there is a reaction in a social setting may be partially true, but using Newton's law as justification for popular psychology is inappropriate.

## 10.6 Impulse

Impulse is the product of the force applied and the time in which force is applied. To increase impulse you can increase force, time or both. The unit of impulse is the pound-second in the Imperial/English system and the Newton-second in the metric system.

Impulse is an important concept in gymnastics because of the nature of how the principles of mechanics meet biology. Although it makes mechanical sense that you can increase impulse by increasing the time of force application, in gymnastics there is nearly always a limited window of time in which force can be applied in order to effectively perform the task. For example, the time of application of the downward and horizontal forces is limited by the mechanics of performing an effective takeoff. If the time is lengthened too long then the gymnast rotates about his/her feet to a position that can make the somersault trajectory "long and low" to a point where the takeoff becomes so dangerous that common sense would deem the increased time on the floor utterly unreasonable. There are also other constraints on impulse from biology such as the limited time in which muscle peak forces can be produced in terms of transmission of forces to bone, energy supply, neural recruitment, strength, pennation of muscle fibres, angles of joints relative to ideal lines of pull, unrecoverable and unusable transfer of muscle tension force to heat.

### Impulse-momentum relationship

Impulse is a change in momentum. If we begin with force:

$$F = ma \tag{10.4}$$

and then substitute the equation for acceleration, we get:

$$F = \frac{m(v_f - v_i)}{t}$$

Then distribute the mass:

$$F = \frac{mv_f - mv_i}{t}$$

rearranging:

$$Ft = mv_f - mv_i$$

A first example of this principle at work in gymnastics can be seen by gymnasts performing a standing back somersault. A typical performance error, usually caused by fear of failing to rotate backward effectively, results in a gymnast "pulling" his/her feet off the ground quickly and "early", reducing the time of force application. Based on Equation 10.4, the incomplete thrust of the legs also shows that the change in momentum will be suboptimal which usually causes a low flight trajectory for the somersault. A second example illustrates how different arm-swing techniques produce appropriate takeoff impulse for performing forward somersault on the floor (Figures 10.3 and 10.4).

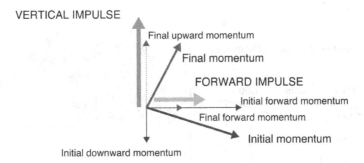

**FIGURE 10.3**   Linear impulse and momentum produced during the contact phase on the ground for performing a forward somersault.

Figure 10.3 shows that the gymnast's approach velocity produces at the beginning of the takeoff an initial momentum which is the result of great forward and downward velocity components. This existing momentum must be changed so that an appropriately directed momentum at the end of takeoff could be produced for performing an adequate flight trajectory during the aerial phase of the somersault. Thus, it is important to take into account the direction of the new momentum that was redirected. This last idea requires a decrease in the forward momentum and an increase in the upward directed momentum during a very short period of time – the impulse.

Figure 10.4(b-c-d) shows that different arm-swing techniques initiated during the support phase before takeoff can produce appropriate acceleration for generating a sufficient ground reaction force that is required to perform a forward somersault on the floor. Figure 10.4(a) shows the effect of the external forces exerted during the takeoff on the change in the gymnast's velocity. Performing a forward somersault requires generating high vertical peak force (18 times bodyweight) during a very short time of takeoff (0.10s). During this short period of time, because of the impulse due to the reaction of the ground in response to the takeoff force generated by the gymnast, there is a change in the direction of the body's momentum. Because the "overhead arm throw" technique produces weaker ground reaction peak force to execute a required takeoff, this technique may be appropriate to prevent gymnasts from ankle ligament sprains.

## 10.7 Work

Given that a force accelerates a body, what do we need to know about the actual motion of the body? Work is defined as the product of the magnitude of the force and the distance the body moves. The force and the displacement of the body are in line with each other, or collinear.

$$W = Fd \tag{10.5}$$

> Where: $W$ = work done by the force
> $F$ = magnitude of the force
> $d$ = relevant distance

When the force acts in the same direction as the motion of the body, work is considered positive. When the force acts in the opposite direction in which the body moves, the work

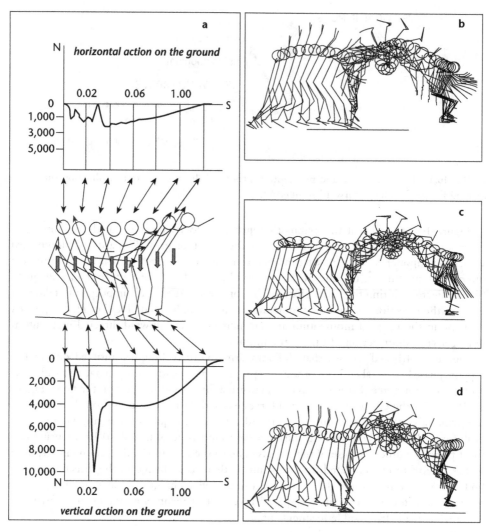

**FIGURE 10.4** External forces exerted during takeoff and change in the body's velocity when performing a forward somersault. The horizontal and vertical components of the ground reaction force (grey vector) and the gymnast's weight (double arrow vector) are plotted according to the stick-figure kinogramm (a). The effect of these two external forces is to produce the required variation of the gymnast's centre of gravity velocity (black vector) during the takeoff phase. Three forms of arm swing were commonly executed to produce the appropriate takeoff impulse:

(b)  backward/upward arm swing or reverse lift technique
(c)  classic forward/upward arm swing
(d)  overhead arm throw technique.

is considered negative. When a gymnast jumps to catch the high bar or rings, the work performed to jump upward is considered positive and the influence of gravity (weight of the gymnast) is considered negative work. For example, if the gymnast has a mass of 50 kg, and thus a weight of 490.3 N, and jumps 50 cm (raises his mass 0.5 m) to reach the bar by applying 850 N of force to the mat beneath the bar then:

$$work\ due\ to\ gravity = -(490.3 \star 0.5) = -245.15\ Nm$$

$$work\ done\ by\ gymnast = 850 \star 0.5 = 425\ Nm$$

$$total\ work = 425 - 245.15 = 179.85\ Nm$$

## 10.8 Power

Work is a relatively simple concept with no mention of how much time is required to perform the work. By simply computing the work (*Fd*) completed in a period of time (*t*) we have a new concept – power. Power is the rate of doing work:

$$power\ (watts) = \frac{work}{time} = \frac{Fd}{t} \qquad (10.6)$$

$$P = \frac{W}{t}$$

## 10.9 CONCLUSION

Linear kinetics involves a variety of concepts that refer to invisible forces and their visible effects on bodies at rest and in motion. A great deal of movement can be explained by kinematics and linear kinetics. Something that linear kinetics borrows from earlier sections is the idea that a force is a vector quantity which means that forces can be decomposed to components and combined into resultants.

Forces underlie all motion. A handy rule of thumb for coaches is that if you see an acceleration (the velocity of the body changes), then you can be sure that a force was involved. Moreover, you can be fairly certain that the direction of the acceleration is the result of a single force being applied or the sum of multiple forces that create a resultant in the direction you observe.

# 11

# ANGULAR KINETICS

*William A. Sands*

Angular kinetics is one of the features of gymnastics that best defines the sport.

Gymnastics is not so much about running, stamina, riding a conveyance, throwing something, hitting something, outsmarting a direct opponent or gaining ground on a standardized playing field – gymnastics is about spinning. Gymnasts flip and twist, rotate and revolve, and use flight time and the apparatuses to perform these skills. Angular kinetics is about the forces and torques, moments of force and inertia, conservation laws, Newton's angular analogues to linear kinetics, and other topics.

## 11.1 Eccentric force application

In order to produce angular motion you need a force, but not just any force. A force that rotates something has to be applied eccentrically. By eccentric we don't mean someone with bizarre behaviour but rather the application of force on the body somewhere away from its axis of rotation or its centre of mass.

Figure 11.1 shows how eccentric forces work. Assuming that the three blocks are free to move, let's say sitting on an ice rink or floating free in space (i.e., no friction), then the left-most block will translate and not rotate. The translation is due to the lack of an eccentric force. The force vector shown goes through the centre of mass resulting in translation only. The central block in Figure 11.1 shows that forces being applied are parallel to each other, in opposite (or non-collinear) directions, and not directed through the centre of mass. The middle block and its forces is a particular combination of eccentric forces called a "force couple". The force couple shown will just rotate, that is, the block will rotate but not translate. The right-most block shows the same kind of block and only one eccentric force. The right-most block will also rotate, but probably not precisely around the centre of mass. This third situation results in rotation because of the applied force and the inertia of the block on the upper portion of the block, however the right-most block and force situation will result in rotation and translation.

Figure 11.1 shows how eccentric forces work. However, note that all of the forces are applied in a way that the line of application of force is perpendicular to the axis of rotation – the centre of mass. What happens if the force(s) is/are applied obliquely?

Figure 11.2 shows a similar situation as shown in Figure 11.1.

## 11.2 Torque, moment, force couple

In order to understand Figure 11.2, we must introduce the concept of torque. A torque is not a force. A torque is the measure of the effectiveness of a force, and it consists of two parts. One of the parts is simply the magnitude of the force. A larger force has a greater tendency to rotate something than a smaller force. However, in angular kinetic settings, the placement or line of application of the force also matters. Figure 11.3 shows how forces of equal magnitude (middle diagram) have different levels of effectiveness in turning the object because one is applied closer to the axis of rotation than the other.

The perpendicular distance between the line of application of force (or resistance) is called a "moment". The "perpendicular distance" phrase is important. The line of application of a force is not simply where the applied force might be touching the object to be rotated; the line of force is the perpendicular distance between the line of application of force and the axis of rotation.

The right-most diagram in Figure 11.2 shows an eccentric force applied obliquely to the object being rotated. More importantly, the perpendicular distance between the line of application of force and the axis of rotation is also shown.

When determining the moment of a torque in the situation shown on the right of Figure 11.2, it is the perpendicular distance that must be determined. The equation for torque can be written as:

*torque = force x distance*

However, this confusing equation looks a lot like the equation we saw in Chapter 10 that calculated linear "work" (Equation 10.5). Interestingly, you can consider the equations quite similar in concept. Work is the measure of the effectiveness of a force in moving an object by determining the product of the force applied and the distance the object moved. In an angular setting the same idea applies. Torque is the measure of effectiveness of a force in turning an object. The confusing part comes from how the distance is measured. In a linear kinetic setting the distance moved is collinear with the force applied. In an angular

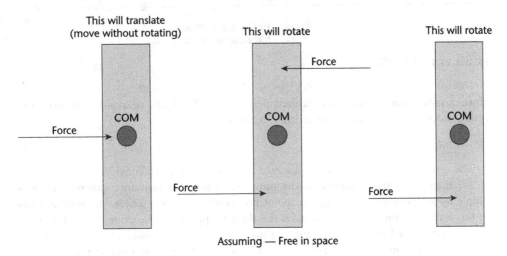

**FIGURE 11.1** Translation and eccentric forces.

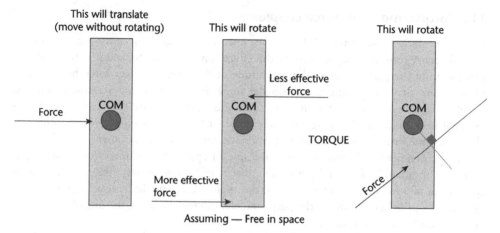

FIGURE 11.2  Moments of force.

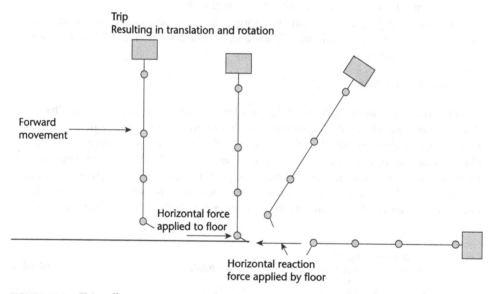

FIGURE 11.3  Trip effect.

kinetics setting the distance is measured perpendicularly. Thus, to avoid confusion, the equation for torque is sometimes written like this:

$$T = F \times d^{\perp}$$
(11.1)

The symbol $\perp$ indicates that the distance is measured perpendicularly.

An example of the influence of the moment of force or resistance can be seen in a simple door. Note that the hinge (or pivot point) of the door is where the door attaches to the wall. The force applied to open the door is applied to the door knob. The door knob is positioned far from the hinge, thus ensuring a large moment of force. If you apply a certain magnitude of force to open the door, the torque to open the door will be the product of the force applied and the perpendicular distance of the line of application of force (usually perpendicular to the door) and the hinge axis. Now let's imagine that the door and magnitude of force remain the same but the door knob is moved to the centre of the door. You will rapidly find that it is more difficult to swing, rotate or open the door. Now let's imagine that we put the door knob a couple of centimetres from the hinge. In this case the inertia of the door and the friction of the hinges may make swinging the door very difficult or impossible to do with the magnitude of force we used before.

In gymnastics we see the same issues arise when we want to create a somersault or salto in the air. For a given force a longer body is harder to rotate than a smaller body. This can be seen in the teaching progression that begins with learning a tuck somersault (knees and hips flexed) through to a pike somersault (hips flexed) to a layout somersault (straight body). We'll come back to this issue later when we consider moment of force and moment of inertia. These are somewhat arbitrary terms for the same thing – a force and a distance from an axis of rotation.

Both of these ideas are seen in a common gymnastics movement often called a "trip effect". A trip effect is shown in Figure 11.3. Note that as the gymnast runs forward and puts his/her feet together for a takeoff, the forward directed force (loosely defined here) will incur a backward directed (equal, opposite and simultaneous) force which will effectively stop the gymnast's feet while the inertia of the rest of his/her body will continue forward. These combined actions and forces will create a force couple that will serve to somersault the gymnast forward. In this case, rotating forward in a layout position to land on the front of the body on the floor, in a pit or on a trampoline.

As the speed of the run, the abruptness of stopping the feet and the distance of the fall increase, the amount of somersault achieved will increase.

## 11.3 Leverage

Levers are one of the classic simple machines that are used to magnify force or displacement. Levers consist of a force, resistance (both sometimes arbitrarily named), the distances of each of these from an axis of rotation, and an axis of rotation or a pivot point. There are three types or classes of levers based on where the pivot point is located relative to the force applied and the resistance (Figure 11.4).

The first type of lever is called a type I or a first-class lever and is commonly depicted as the teeter totter or seesaw found in parks where children play. A type II or second-class lever is usually illustrated by a wheelbarrow. A wheelbarrow has the pivot point at the wheel, the resistance is the load between the person lifting the handles and the wheel or pivot point, and the force is supplied by the person lifting the handles.

A type III or third-class lever is a classic example of most of the joints of the human body such as the human elbow when the elbow is performing active flexion against some load in the hand.

Figure 11.4 shows the three types of levers with the assumption that all of the levers are at equilibrium or that the products of moments of force and magnitudes of force equal the moments of resistance and the magnitude of the resistance. Note that the type I lever can be made to favour force by moving the pivot point closer to the resistance such as when someone is trying to pry a nail with a hammer or lift a stone with a bar placed on top of another smaller stone close to the stone to be moved. The type II lever will always favour force due to the moment of force, by definition, always being longer than the

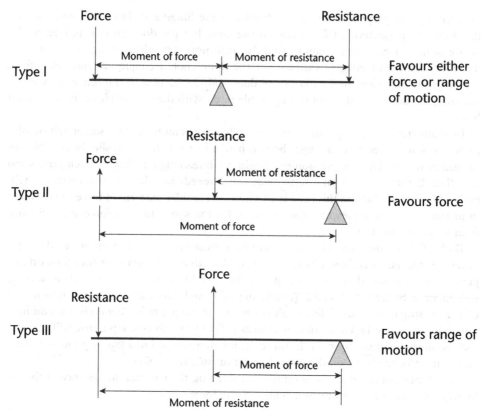

**FIGURE 11.4**   Lever types and moments.

moment of resistance. Type II levers can thereby lift a heavy resistance with a force that is less than the resistance.

Type III levers always favour range of motion due to their placements of applications of force and resistance. By favouring range of motion, we mean that for a given displacement of the point of application of force, the displacement of the point of application of resistance is larger. The contrasts in range of motion are shown in Figure 11.5.

Leverage does not apply as much to gymnastics apparatuses as to the functional motion of joints of the body. Moreover, classifying joints as particular lever types can be deceptive because load and axis placement may change depending on the specific joint and load configuration. For example, the ankle joint is shown schematically in Figure 11.6. Note the variations in lever type depend on the movement performed. The upper-left illustration shows a schematic of the ankle and the calf muscles that pull on the heel bone. The weight of the body is shown as the load placed on the shinbones and acting through the ankle and foot to the ground. In the upper-right illustration of Figure 11.6, the schematic is showing how the load remains acting through the shin and ankle to the foot. The pivot point during the motion of rising on the toes of the foot is accomplished by muscular shortening of the calf muscles, which in turn pull on the heel bone and raise the load (body weight). The motion described is like that of a wheelbarrow with the load in the middle the calf

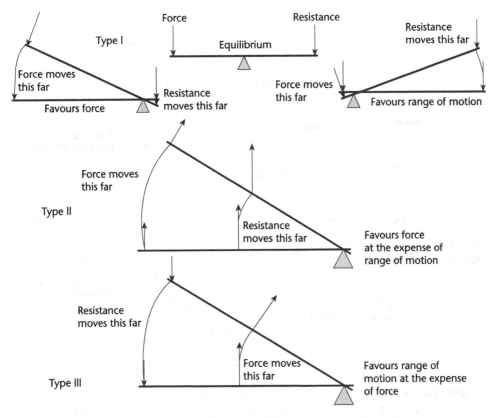

**FIGURE 11.5** Levers, showing ranges of motion of the components for each type.

muscles pulling upward on the handles of the wheelbarrow and the toes of the foot similar to the wheel of the wheelbarrow.

The lower illustration in Figure 11.6 shows the ankle as it would appear when an athlete is seated at a leg press machine and pushing with his/her toes against the foot platform. By pushing with the feet in the way shown by the lower illustration the load is coming from the leg press machine, the pivot point is at the ankle and the force is the contraction of the calf muscles. Thus, the ankle becomes the pivot point or axis making the ankle a type I lever.

## 11.4 Centre of gravity

The phrases "centre of gravity" and "centre of mass" have been used throughout preceding discussions without clarification. When a body is acted upon by gravity, we can use the thought experiment of considering that every particle of the body is separately attracted toward the Earth. If we summed the products of all the masses and gravitational attractions of each particle (weight = $\Sigma$ (mass x gravity)), we would get the total weight of the body. The gravitational attractions of all these particles would form a myriad of parallel lines going from each particle directly downward.

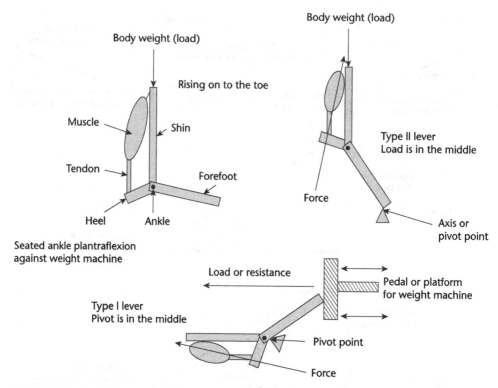

**FIGURE 11.6**   Ankle leverage characteristics. Note that rising on the toes exemplifies a type II lever, while pushing with the toes against a leg press machine to exercise the calf muscles exemplifies a type I lever.

However, all of the particles are not stacked on top of each other. Each line of gravity would be at some distance from the resultant of the summed products of each particle's mass and its gravitational attraction. Determining the resultant of all these particles and their masses requires also knowing the perpendicular distances of each particle from some unknown line of gravity of a resultant of all these forces. These perpendicular distances are the moments of the particles from the unknown line of gravity. By summing these moments we should achieve a perfect balance of all moments around a particular line of gravity. This new line of gravity originates at the centre of gravity of the body.

For illustration, let's consider a simpler body than the human body. If the body is an empty soda bottle and we lay the bottle on its side trying to balance the bottle on an outstretched finger we would find that there is a position of the bottle such that the bottle will balance on the finger. In the balanced position the sum of moments of the particles on either side of the finger will be equal. A plane of gravity projecting upward from the length of the finger through the bottle passes through the centre of gravity of the bottle.

If we then balance the bottle with the bottom on the finger, although more difficult, the plane of gravity projecting from line of the finger upward will go through the centre of gravity. Then to complete the location of the centre of gravity we need to turn the bottle 90 degrees along its long axis and repeat the procedure. Where all three planes intersect the resulting point of intersection is the centre of gravity.

The same thing can be done with a human body, but modern approaches to determine the centre of gravity of the human body rely on computing the moments of each body segment (e.g., thighs, shanks, upper arms, torso, etc.), using known or models of the masses of each segment and each segment's centre of gravity to determine the location of the body's centre of gravity. The centres of mass/gravity are shown in numerous illustrations in preceding figures.

Knowing the precise location of the centre of mass in the gymnast's body is probably unnecessary. However, practitioners should know that the centre of mass in a standing body lies slightly below the navel and about half way from front to back. Moreover, the centre of mass location changes due to the changing distribution of body segments and the shape of the gymnast's body position.

For example, if the gymnast raises his/her arms above the head then the centre of mass moves upward to reflect the redistribution of mass above the head. If the gymnast starts in a standing position and bends forward to touch the toes, then the centre of mass moves forward and downward relative to the body position. If the gymnast stays in balance over his/her feet, then the centre of mass will always be over the feet (base of support) while lowering. From a standing position, if the gymnast raises one arm up to a horizontal side position, then the centre of mass moves upward and toward the raised arm. Knowing the location of the centre of mass of the body helps determine how the body will move in angular motion settings.

## 11.5 Moment of inertia

The angular equivalent to linear inertia is the moment of inertia. Inertia is a measure of a body's resistance to changes in motion. In an angular setting, the moment of inertia is the body's resistance to rotation. The moment of inertia is the product of the mass of a body and its perpendicular distance from the axis of rotation, squared:

*moment of inertia = mass × radius²*

Since a rotating body has particles or segments in three dimensions around an axis of rotation, the determination of the moment of inertia requires the summing of all the masses and their individual radii about the axis of rotation.

$$I = \sum mr^2 \tag{11.2}$$

Where $I$ = moment of inertia

$m$ = mass of each particle or segment

$r$ = the radius of rotation of the particle or segment about the axis of rotation

The moment of inertia of a body can be determined in a variety of ways. If the body is shaped in some known geometric shape, then its moment of inertia can be easily determined. The same cannot be said for a human body that not only consists of multiple unusually shaped segments, but the segments are also free to move about. Thus, mass does not typically change, but the position of the segments does change.

## 11.6 Angular momentum

Angular momentum is analogous to linear momentum with the exception that the mass term is replaced by moment of inertia and velocity is replaced by angular velocity:

*angular momentum = moment of inertia × angular velocity* (11.3)

*angular momentum = $I\omega$*

## 11.7 Newton's angular analogues

Like Newton's three linear laws, there are three angular laws.

### *Principle of the conservation of angular momentum*

The first angular law can be summarized in that a rotating body will continue to rotate about an axis with a constant angular momentum unless acted upon by an external force. The first angular law is also known as the principle of the conservation of angular momentum. This conservation law is of particular importance to gymnasts.

If you consider that angular momentum is fixed at takeoff during a tumbling somersault and the gymnast is free to change shape during the airborne phase, the gymnast can change his/her angular velocity (somersault or twisting velocities) by changing the moment of inertia.

If the moment of inertia is decreased (e.g., move from a layout to a tuck, or from arms outstretched to arms close to the body), then the angular velocity of the somersault, twist or both will increase. Conversely, when the gymnast wants to slow his/her rate of somersaulting or twisting then the gymnast need only increase his/her moment of inertia about the axis of rotation – the centre of mass. The increase in moment of inertia will result in a corresponding decrease in angular velocity. However, since the angular momentum must remain constant or be "conserved" there is no net change in angular momentum.

### *Angular analogue for Newton's second law*

The second angular analogue states that the rate of change of angular momentum of a body is proportional to the applied torque and acts in the direction of the torque. This results in an equation for torque:

*torque = moment of inertia × angular acceleration* (11.4)

$T = I\alpha$

The second angular law is shown nicely by a gymnast swinging on the horizontal bar. Figure 11.7 shows an example of how torque and moments interact to cause rotation. Figure 11.7 shows that the torque of the gymnast in position *A* is larger than the torque found in position *B*. Although the gymnast's weight is unchanged in each position, the moment arms show that when the gymnast is horizontal his/her torque is larger and thus

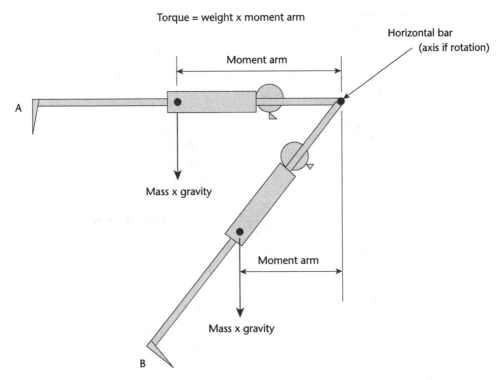

Torque = weight x moment arm

**FIGURE 11.7** Torque comparisons in swinging positions.

the gymnast is more effectively rotated in this position. In order to comply with the second angular law, position $A$ will result in a greater angular velocity during the downswing than position $B$ due to the differences in moment arms.

## Angular analogue for Newton's third law

The angular analogue for Newton's third law states that for every torque that is exerted by one body on another, there is an equal, opposite and simultaneous torque exerted by the second body on the first. Gymnasts often experience this "action–reaction" torque problem when they try to change positions in the air. For example, when a gymnast is free in space in a layout position and then pikes, the upper body and lower body will move toward each other.

Unfortunately, the gymnast is often rotating at the time so the illusion that one half of the body is moving more than the other half is exaggerated. This is shown in Figure 11.8.

Gymnasts often use the angular version of action–reaction in order to correct body position while in flight. Qualitative analysis of motion should include knowledge of angular action–reaction because the gymnast who must make large body position changes while airborne usually has serious technique problems. Interestingly, the actual error usually occurs before the visible large body position change, but knowledge of Newton's angular analogues can serve the practitioner by alerting him/her to problems and where to begin the search for performance (i.e., technique) solutions.

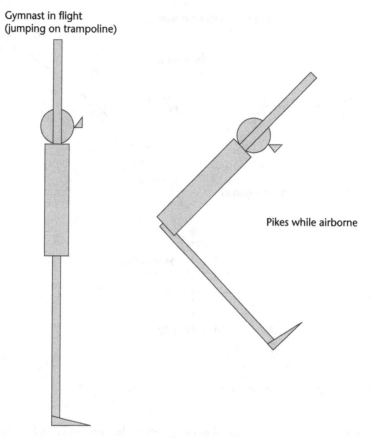

Gymnast in flight
(jumping on trampoline)

Pikes while airborne

**FIGURE 11.8**   Action–reaction in an angular setting.

## 11.8 CONCLUSION

Gymnastics biomechanics is the application of mechanical principles to gymnastics movements. In spite of the style and artistry of gymnastics performances, all gymnasts and their apparatuses must obey the laws of physics. Familiarity with these laws and their application can be of considerable assistance in teaching and coaching gymnastics by identifying those motions that represent errors.

# 12

# USE OF BODY INVERSE DYNAMICS TO EVALUATE REACTION FORCE DURING VAULT AND FLOOR SOMERSAULTS IN ARTISTIC GYMNASTICS

*Bessem Mkaouer*

## 12.1 Introduction and objective

Understanding the implementing rules of acrobatic skills presupposes the knowledge of the forces that create and/or modify the movement. Understanding forces, speed and take-off angle enables optimizing performance. Through information and communications technology, in particular video analysis, the calculation of speed, displacement and joint angles has become very accessible to coaches (smart phones, tablets with free biomechanics and motion analysis applications). On the other hand, the calculation of reaction forces, whether directly (i.e., via force plate) or indirectly (i.e., via inverse body dynamic analysis), remains quite difficult for coaches, considering the complication of the procedure and the required expertise.

The objective of this chapter is to provide readers with real applications of the above theories and concepts from the field of practice. Examples of studies will be taken from the floor and the vault.

To our knowledge, there are only a few studies that estimated gymnasts' reaction forces at take-off during acrobatics series and/or vault table by inverse body dynamics. Mkaouer et al. (2013) used a concomitant dynamic and kinematic analysis to evaluate reaction force during acrobatic series at the floor exercise. Sano et al. (2007) analysed the reaction force on springboard, using five Kistler® force plates, during handspring vault. Seeley and Bressel (2005) used a concomitant dynamic and kinematic analysis to compare upper-extremity reaction forces between the Yurchenko vault and floor exercises. Sands et al. (2005) and Coventry, Sands and Smith (2006) used magnetic and infrared sensors with kinematic analysis to calculate the compression of the springboard during the take-off phase during the handspring vault. Smith (1983) analysed the backward salto via kinematic analysis using inverse body dynamics.

In this context, we suggest a simple method for calculating the take-off reaction force during floor exercises and/or balance beam and vault table based on Smith's calculation method (1983).

## 12.2 Application at the floor exercises (and/or balance beam)

Smith's equations (1983) are based on the simple law of biomechanics (Equation 12.1a and 12.1b). They could be used either during the standing acrobatic skills (Figure 12.1a and 12.1b) or during the acrobatic series (Figure 12.2a and 12.2b) at the floor exercise and/or the balance beam.

$$(a) \ F_x = m \cdot \left(\frac{v_2 - v_1}{t_1 + t_2}\right) \qquad\qquad (b) \ F_y = m \cdot \frac{v_3}{t_2} \qquad\qquad (12.1)$$

Where: $F_x$ = horizontal force

$\quad\quad F_y$ = vertical force

$\quad\quad m$ = mass of the athlete

$\quad\quad t_1$ = braking time

$\quad\quad t_2$ = propulsion time

$\quad\quad V_1$ = initial horizontal velocity "beginning of the preparation phase"

$\quad\quad V_2$ = final horizontal velocity "end of the propulsion phase"

$\quad\quad V_3$ = final vertical velocity "end of the propulsion phase"

Six elite-level male gymnasts volunteered to take part in this application (age 23.70 ± 1.94 years; body mass 58.67 ± 8.24 kg; height 1.66 ± 0.06 m). They performed two back somersaults in a dual approach study (i.e., kinematic and dynamic); the first start from the standing position (SBS) and the second from an acrobatic series (i.e., round-off flic-flac back somersault "RFBS"). Direct kinetic data of the take-off for back somersaults were

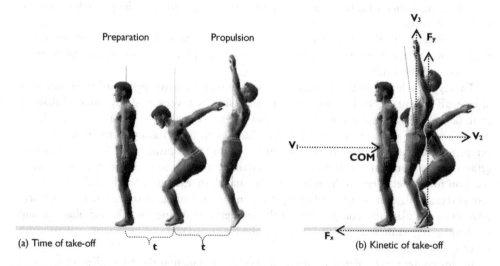

**FIGURE 12.1** Calculating method related to Smith (1983) during standing acrobatic movements.

Note: (COM) centre of mass; ($F_x$) horizontal force; ($F_y$) vertical force; ($t_1$) preparation time; ($t_2$) propulsion time; ($V_1$) initial horizontal velocity "$V_0$ at standing position"; ($V_2$) final horizontal velocity "end of the propulsion phase"; ($V_3$) final vertical velocity "end of the propulsion phase".

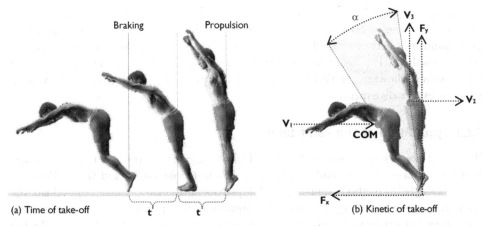

**FIGURE 12.2** Calculating method related to Smith (1983) during acrobatic series.

Note: (COM) centre of mass; ($F_x$) horizontal force; ($F_y$) vertical force; ($t_1$) braking time; ($t_2$) propulsion time; ($V_1$) initial horizontal velocity "beginning of the braking phase"; ($V_2$) final horizontal velocity "end of the propulsion phase"; ($V_3$) final vertical velocity "end of the propulsion phase".

measured using a Kistler® QuattroJump force plate (QJ) and kinematic data using SkillSpector® motion analysis software (sampling rate 50Hz) through inverse body dynamics analysis.

The baseline device, a portable force plate QJ, recorded the vertical ground reaction force at the take-off in two different acrobatic series RFBS and SBS (6874.40 ± 1204.70 N and 1794.66 ± 141.97 N, respectively). The SkillSpector® motion analysis programme recorded similar vertical reaction force values to the baseline, using rigid body inverse dynamics procedure, as described by the Smith (1983) equation (SE), in both acrobatic series (7361.71 ± 1427.93 N and 1816.58 ± 145.75 N, respectively, for RFBS and SBS). The difference was not statistically significant ($p > 0.05$) and the effect size was acceptable (Table 12.1).

The comparative study showed no significant difference between the two methods, whether for standing and/or dynamic performance (i.e., SBS and RFBS).

The results of this study suggest that the kinematics analysis via SkillSpector® software, using rigid body inverse dynamics (Smith's (1983) equation), measures similar vertical reaction force values when compared to the baseline method (i.e., using QuattroJump

**TABLE 12.1** Comparison of body inverse dynamics reaction force measurement with the criterion portable force plate device

| R1 vs. R2 | Wilcoxon Test (P) | r | $R^2$ | CV(%) | ICC | Effect size (dz) |
|---|---|---|---|---|---|---|
| RFBS | 0.062 | 0.99 | 0.99 | 2.08 | 0.99 | 2.18 |
| SBS | 0.062 | 0.99 | 0.99 | 3.46 | 0.99 | 2.76 |

(RFBS) round-off flic-flac back somersault; (SBS) standing back somersault.

RFBS $F_{max}$ = 681.45 + 0.84*$F_{SkillSpector®}$ (ESE = 0.02).

SBS $F_{max}$ = -38.15 + 1.01*$F_{SkillSpector®}$ (ESE = 0.02).

portable force plate). Lescura and Bagesteiro (2011) conducted a spatio-temporal analysis of human walking in 2D with SkillSpector® and in 3D with Vicon system. They found great similarities between the displacement curves of the hip and knee when assessed by the two methods. It is worth noting that the advantage of SkillSpector® is its free cost (open source) allowing similar motion analysis compared to other software (i.e., Vicon), which is extremely expensive.

## 12.3 Application at the vault table

We would like to suggest two methods to calculate the gymnasts' reaction forces on springboard at the take-off of the vault routine according to Hooke (1678) and Smith (1983).

Hooke's (1678) method combined with kinematic analysis was applied to evaluate the reaction force by recording the spring compression of the springboard. Progressive loads (i.e., 10 to 300 kg) up to a complete total compression of the springboard were applied to determine the constant $k$ (Figure 12.3). $F_{max}$ and $k$ are calculated according to Equation 12.2a and 12.2b. See the Smith (1983) methods already detailed earlier in section 12.2.

$$\text{(a) } F = k \cdot \Delta l \qquad \text{(b) } k = F / \Delta l \qquad\qquad (12.2)$$

Where: F = reaction force

$k$ = springboard constant

Dl = displacement of springboard

The springboard (i.e., AAI, Stratum® Springboard, 1.20 m × 0.60 m × 0.22 m) used in the study presents a compression constant $k$ equivalent to 47.5 N·m$^{-1}$.

Ten elite female gymnasts (age 20.5 ± 2.1 years; body mass 56.5 ± 2.4 kg; height 1.62 ± 0.04 m) have participated in this study. A 2D kinematic analysis (50Hz) of the Yurchenko and Handspring vaults were performed using the SkillSpector® video analysis software.

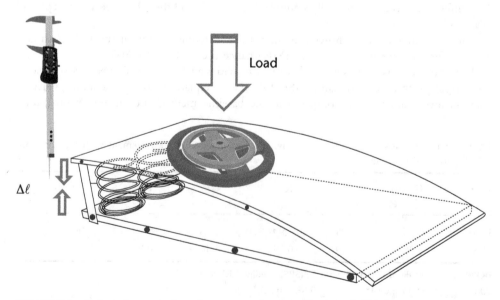

**FIGURE 12.3** Calculating springboard constant related to Hooke's (1678) law.

**TABLE 12.2** Comparison methods Hooke (1678) vs. Smith (1983)

| Variables | t-Test | r | $R^2$ | CV(%) | ICC | Effect size (dz) |
|---|---|---|---|---|---|---|
| Handspring | 0.01 | 0.99 | 0.99 | 6.35 | 0.99 | 2,61 |
| Yurchenko | 0.01 | 0.99 | 0.99 | 6.32 | 0.99 | 4,48 |

The results show that the two methods of calculation (i.e., Hooke and Smith) showed significant difference at $p < 0.01$ with a $D_\%= 6.3\%$ for the two vault jumps in favour of the Hooke method (1678) [Handspring: $3054.23 \pm 615.76$ N and $2852.79 \pm 553.45$ N; Yurchenko: $4969.72 \pm 628.53$ N and $4655.17 \pm 638.71$ N, respectively] (see Table 12.2).

The reaction force estimated at the handspring vault by the Smith method (1983) is comparable to that reported by Sano et al. (2007) calculated directly on a force plate (i.e., $2852.79 \pm 553.45$ N and $2693 \pm 485$ N, respectively). Similarly, the deflections of the springboard recorded during the kinematic analysis are comparable to that presented by Coventry, Sands and Smith (2006) and Sands et al. (2005) for female gymnasts (i.e., $0.074 \pm 0.014$ m, $0.08 \pm 0.016$ m and $0.068 \pm 0.018$ m, respectively).

## 12.4 CONCLUSION

The indirect estimation of the gymnast's reaction force, through the inverse dynamics, whether during the acrobatic series and/or on the springboard, could be very helpful to coaches and scientists to optimize their gymnasts' performance and to prevent injuries. The Smith (1983) method seems to be very effective during the floor exercises whereas the Hooke method (1678) seems to be more adaptable to the vault table to optimize the jumps.

# PART III REVIEW QUESTIONS

Q1. A body is measured rotating at 320 degrees per second at the first measurement and 360 degrees per second at the second measurement. The time between measurements is 0.1 second. What is the average angular acceleration of this object? Give your answer in degrees per second per second.

Q2. If an object is projected at 30 degrees from the horizontal at 30 m/s, what are the horizontal and vertical component velocities? (*2 decimal places in metres per second.*)

Q3. If an object is launched and lands at the same level at 10m/s horizontal component velocity, and has a flight time of 4 seconds, how far does the object go (disregarding air resistance)? (*Answer in metres.*)

Q4. When a figure skater brings her arms close to her body's vertical axis during a spin, she spins faster. Explain why this is.

Q5. The muscular force of a muscle is 160 pounds, and the muscle is pulling on a bone at an angle of 15 degrees. What are the vertical and horizontal components of this force? (*3 decimal places, answer in pounds.*)

Q6. Draw two stick-figure diagrams of the lower extremity (thigh, shank and foot) and show in Figure 1 a large moment of inertia of the leg during the recovery phase of a running stride. Then draw Figure 2 to show a smaller moment of inertia of the recovery leg than Figure 1. Label the hip, knee and ankle in both diagrams. What could be the implication of the smaller moment of inertia in running?

Q7. What is the distance of a fall that lasted 2.5 seconds (disregard air resistance)? What is the final velocity at impact for this fall? (*Answers in metres and metres per second.*)

Q8. If an object is launched at 35 degrees from the horizontal at 12m/s to land at the same level, what is the peak height of the trajectory and what is the horizontal range?

Q9. If a gymnast's vertical jump is 0.25 m, what was his vertical velocity at take-off?

Q10.  When a gymnast performs a giant swing, the distance from his hands to his feet is approximately 2.1 metres. If he swings at 270 degrees per second at the bottom of the swing, how fast are his feet going – linearly?

Q11.  A gymnast (mass = 57kg) performs a tucked forward somersault on the floor. The following list shows the initial and final conditions of the centre of gravity during the take off:

| | |
|---|---|
| horizontal velocity at touchdown | 4.15 m/s |
| vertical velocity at touchdown | −1.85 m/s |
| horizontal velocity at take-off | 2.22 m/s |
| vertical velocity at take-off | 3.01 m/s |
| height at take off | 1.12 m |

Use appropriate data/equations to calculate:

- the magnitude of the resultant velocity of the centre of gravity $(v_R)$
- the angle of $v_R$ $(\theta)$
- the horizontal impulse $(p_H)$
- the vertical impulse $(p_V)$
- the peak height of the flight $(H)$

Q12.  Explain why the body's moment of inertia is:

- smaller when performing a tucked front somersault than when performing a straight one?
- greater when performing a straight front somersault than when performing a vertical jump with full twist?

Q13.  (True/False) By skilfully changing body position in the air (i.e., completely free of support), the gymnast cannot change his/her flight time.

Q14.  (True/False) If I push on a gymnast (spot) while the gymnast is in flight and the direction of my push is directly in line with his/her centre of mass, the gymnast will rotate faster.

Q15.  (True/False) The centre of mass of a gymnast is a fixed point and does not move during a gymnast's changes in body position.

Q16.  (True/False) A torque is the same as a force.

Q17.  (True/False) If the gymnast is somersaulting forward, the vector is drawn from the gymnast's centre of mass toward the gymnast's left.

Q18.  (True/False) By increasing the horizontal velocity of the gymnast, when the gymnast stops his/her feet suddenly together on the floor, the "trip effect" will be increased and the magnitude of the angular momentum experienced by the gymnast will decrease.

Q19.  (True/False) Distance and displacement are the same thing.

Q20.  (True/False) If the gymnast's centre of mass is 1.1 metres from the top of the balance beam and she departs the beam from a run at that height and later departs the beam by simply stepping off and falling straight down, the time of descent in both cases is the same.

Q21.  (True/False) The primary determinant of success in vaulting is the horizontal component velocity achieved during the run-up.

Q22.  (True/False) When comparing height and distance of a dismount from the horizontal bar or the uneven bars, it is much more difficult to achieve a large range or horizontal distance than a high peak flight of the dismount trajectory.

Q23.  Solve the following scenario using the kinematic data, the mechanical laws and the Smith equation (1983):

A video analysis (50Hz) of a gymnast (mass 70kg) who performed a salto forward on the floor (see Figure III.1) shows that he took 4 metres of backswing [from position (a) to position (b); image interval from i1 to i51]. The take-off angle is 75°, the braking time is from position (b) to position (c) [from i51 to i53] and the propulsion time is from position (c) to (d) [from i53 to i55]. The rise time during the salto is from position (d) to (e) [from i55 to i75] and the decent time is from position (e) to (f) [from i75 to i95]. The landing is from position (f) to position (g) [from i95 to i145].

a.  Calculate the backswing time.
b.  Calculate the initial velocity.
c.  Calculate the vertical velocity.
d.  Calculate the horizontal velocity.
e.  Calculate the braking time.
f.  Calculate the propulsion time.
g.  Calculate the flight time.
h.  Calculate the vertical displacement.
i.  Calculate the horizontal displacement.
j.  Calculate the horizontal component of force.
k.  Calculate the vertical component of force.

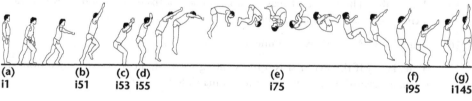

(a)         (b)         (c) (d)                          (e)                          (f)         (g)
i1          i51         i53 i55                          i75                          i95         i145

**FIGURE III.1**

# PART IV

# Psychology for gymnastics

# PART IV

# Psychology for gymnastics

## Learning outcomes (*Monèm Jemni*)

After reading this part, you will be able to:

- Analyse the difference between men's and women's events' task demands.
- Discuss the different models of expertise development throughout a gymnasts' career.
- Distinguish the roles of the coaches and parents in gymnasts' development.
- Distinguish the difference between the foundation skills, the psychosomatic skills, the cognitive skills and emotional states.
- Examine how human senses intervene in the analysis of the movement-related tasks.
- Illustrate how visual perception is an important concept in the learning process for a gymnast and for a coach.

## Introduction and objectives (*Monèm Jemni*)

The objectives of this part on sport psychology, mental training and perception are to outline the research on the learnable, and thus, teachable mental and sensorial skills related to gymnastics, and more importantly to consider how each of the concepts that has been developed and published can be applied in actual training and coaching competitive gymnastics. Every theme has an "implications" section in which the personal, related experiences from observations, interviews, research and interventions related to each concept in practice are developed, and then put into practice. Thus the beauty of this part is to bring together both theory and practice in sport psychology, mental training and perception, as has never previously been accomplished for the sport of gymnastics and is aimed at educated undergraduate or graduate students, gymnasts or coaches, all of whom would find appropriate concepts to support their learning process.

Consideration will also be given to the specific task demands of gymnastics, the various stages of learning throughout a gymnast's career from different authors' perspectives, ages and genders and the roles of parents, coaches and mental trainers during these various learning and performing processes. Scientifically based concepts to movement perception of complex movements will be updated with the latest research and practical recommendations will be provided.

# 13

# TASK DEMANDS AND CAREER TRANSITIONS IN GYMNASTICS

## From novices to experts and the stages of learning across the career

*John H. Salmela*

## 13.1 Performance task demands in gymnastics

If you were asked to establish a physical and mental training program for shot putters, the task would be relatively straight forward. They must propel a 16 lb iron ball as far as possible after having shifted themselves rapidly across a 7-foot circle. Biomechanically, you would suggest that they start from a low position with the shot on their neck and in a position opposite to the intended direction of the throw. They would thrust themselves in a low backwards position, and then explosively, with the transfer of energy from their legs, to the twisting torso and finally to the arms and fingers, finally propel the shot during the release. Physiologically speaking, you would train strength and speed of the legs, developing torque speed and arm power prior to the release. Psychologically, you would train the athlete to get pumped up emotionally and to keep in mind self-talk words such as "drive" and "explode." It is pretty straight forward stuff, right?

However in gymnastics there are six men's and four women's events, each with specific strength, power, balance and skill dimensions requiring somersaulting, twisting and remaining upside down for many of the events. Also, the complex dismounts from the horizontal, parallel and uneven bars have inherent elements of danger and fear. In addition, some events such as the pommel horse require many rapidly performed movements demanding fast hand changes and balance, while the rings are performed at a more leisurely pace, but require great power and strength.

Thus, John Salmela (1976) and his students did a time-motion analysis of the 1972 Munich Olympics for the final of the men's routines, as well as a classification of the various performed skills. The results for the men were obviously more complex than the analysis of the shot putter's task (Figures 13.1 and 13.2).

The most striking dimension of this data was the diversity of the number of elements and the relative importance of each of these for each event. For example, for floor exercises the women performed 46 elements, although some were small and balletic, while the men performed only 31.2. On the pommel, or side horse, the gymnasts executed a total of 29 movements while there were only 13.4 on the rings, and in the same light, on the pommels there were 47 manual releases and obviously only one on the dismount from

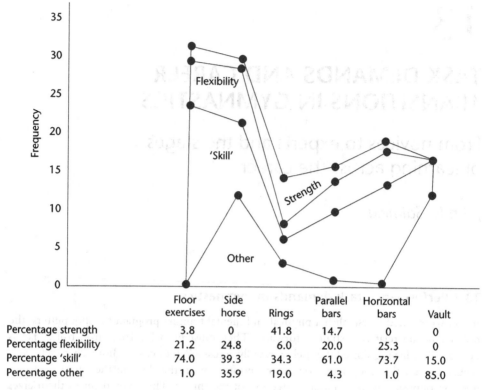

| | Floor exercises | Side horse | Rings | Parallel bars | Horizontal bars | Vault |
|---|---|---|---|---|---|---|
| Percentage strength | 3.8 | 0 | 41.8 | 14.7 | 0 | 0 |
| Percentage flexibility | 21.2 | 24.8 | 6.0 | 20.0 | 25.3 | 0 |
| Percentage 'skill' | 74.0 | 39.3 | 34.3 | 61.0 | 73.7 | 15.0 |
| Percentage other | 1.0 | 35.9 | 19.0 | 4.3 | 1.0 | 85.0 |

**FIGURE 13.1** Frequencies and types of gymnastic components for the men's gymnastic finalists at the Munich Olympics.

(Salmela, 1976)

the rings. Given the evolution of the sport over more than 37 years since Munich, it is obvious that the frequencies and numbers of all of these measures would be greatly increased. In recently reviewing the films from Munich, it appeared to me that the men were performing in slow motion, especially on the horizontal bar, as compared to today's very dynamic performances.

If the same analyses were carried out in rhythmic gymnastics (RG), it seems that the patterns would be similar to those of the women's floor exercises, with a greater emphasis on changes in visual orientations and flexibility. RG is highly skillful, and arguably more intricate than artistic gymnastics, since the gymnasts must also perform an open skill task of catching an object. They perform their rolls and twists which must be timed with the hoop, ball, clubs and ribbon, with no tumbling, but with demonstrations of extreme flexibility.

From a mental training perspective, it is evident that the tasks for a sport psychologist or mental trainer would vary between events. Emotionally, the gymnasts performing on the pommel horse or balance beam would require an alert but more relaxed mental state, while on the floor and vault, for both sexes, the gymnasts must be firing on all cylinders, in that vaulting and tumbling require maximum speed to accomplish the somersaulting and twisting movements.

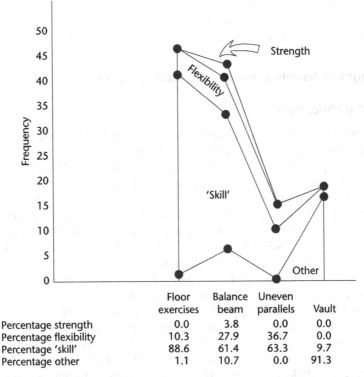

| | Floor exercises | Balance beam | Uneven parallels | Vault |
|---|---|---|---|---|
| Percentage strength | 0.0 | 3.8 | 0.0 | 0.0 |
| Percentage flexibility | 10.3 | 27.9 | 36.7 | 0.0 |
| Percentage 'skill' | 88.6 | 61.4 | 63.3 | 9.7 |
| Percentage other | 1.1 | 10.7 | 0.0 | 91.3 |

**FIGURE 13.2**    Frequencies and types of gymnastic components for the women's gymnastic finalists at the Munich Olympics.

(Salmela, 1976)

## 13.2 Implications of gymnastic task demands for learning and performing in gymnastics

It is essential to understand the nature of the tasks which must be trained before beginning to coach, either in artistic or RG. In men's gymnastics, the floor exercises and the vault have some common properties such as the powerful running and tumbling elements; however, the pommel horse and the rings have nothing in common, while the horizontal and parallel bars have some similar swing elements.

In women's artistic gymnastics, the floor exercises, vault and balance beam have some identical elements, although the tolerance for errors is less on the beam which is only 10 cm wide. For women, learning a proper round-off back handspring covers many elements on the above three events. The uneven parallel bars are quite similar to the men's horizontal bar, and for this reason, men often will coach this event as well as the floor and vault, where they are stronger for the spotting of the performers.

In RG, there are extensive spinning and rolling movements which must be executed in relation to a thrown set of clubs, ribbons, balls and hoops. I would imagine that one of the primary task demands is to be able to launch an instrument into the air in a predictable manner.

Gymnastics technique is continuously changing and, as I have seen while giving mental training courses for the FIG, gymnastics from a technical viewpoint does not at all resemble what I was taught more than 50 years ago. In fact, the great British gymnastics champion

Nik Stuart published a book in the 1960s and at the end of the book he outlined a series of men's routines that he dreamed of and envisaged in the future. Now almost every element that he dreamed of can be realized by a 12-year-old boy!

## 13.3 Stages of learning across a gymnast's career

### The fixed abilities view

There are at least two ways of assessing the career structure of developing gymnasts. Ogilvie and Tutko (1966) were probably the first sport psychologists to study "problem athletes," using paper and pencil tests. But to understand a population of athletes is to first assess all gymnasts cross-sectionally within each age group, using a variety of valid sport science measures, which I did as the research chairman of the Canadian Gymnastics Federation from 1976–1985. The other method is to continuously and personally intervene with them as young gymnasts until their adult career using "softer" methods, such as mental training, which I also did from 1985–1995 (Salmela, 1989). Both approaches have their strengths and weaknesses, and in many ways are complementary.

In the 1970s, all Canadian sports were obliged to follow the highly successful sporting model of talent identification of the East Germans who won an extraordinary number of medals with a small country of only 17 million inhabitants. Thus the Canadian Gymnastics Federation, under my guidance as research chairman, created the Testing for National Talent (TNT) program for men's gymnastics. I consulted with sport scientists who had carried out research in gymnastics, as well as national and international coaches in Canada to come up with a battery of multidisciplinary tests that might be related to identifying gymnastic talent. The resulting TNT test battery can be seen in Table 13.1. A portable testing package was constructed and was administered in every Canadian province, where each male competitive gymnast from 10–25 years old was evaluated, the data were collated and the measures were correlated with their provincial championship scores (Régnier & Salmela, 1987).

The results were astounding with these 236 evaluated gymnasts, since for each age division the relative contributions of the physical, organic, perceptual and psychological families, which included all variables, changed drastically across the six age groups with the perceptual and morphological variables usually being dominant (Figure 13.3.). It is clear that weight, flexibility, power and all of the perceptual variables are trainable, while limb lengths, height and most selected personality psychological traits are not. What was of

**TABLE 13.1** List of potential determinants of gymnastic performance in the TNT test

| Morphological | Organic | Perceptual | Psychological |
|---|---|---|---|
| Skin folds (4) | Flexibility (19) | Coordination | Anxiety (2) |
| Breadths (2) | Strength | Kinaesthetic sense (2) | Pain tolerance (2) |
| Girths (6) | Shoulder power (6) | Rotation sense | Neuroticism–stability |
| Lengths | Leg power (3) | Foot balance (2) | Extroversion–introversion |
| Height | Speed | Hand balance | |
| Weight | | Time estimation | |
| Morphology (3) | | | |

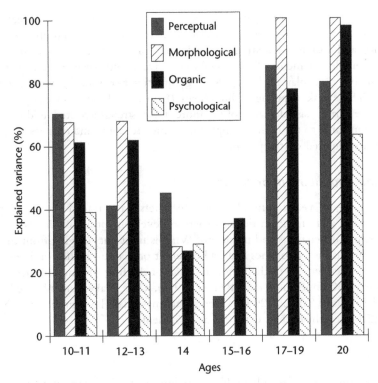

**FIGURE 13.3** Explained performance variance using all of the possible variables in each family of determinants.

(Régnier & Salmela, 1987)

interest was that the fixed psychological variables almost always explained the least of the performance variance. Years later, it became clear to me that with the exception of the pain tolerance variable, all of the other psychological evaluations were assessed by paper and pencil instruments which measured fixed personality abilities and did not take into account the learning across the gymnasts' careers.

However, when the number of tests was reduced to the two best predictors, the perceptual variables, requiring balance and spatial orientation, were predominant; but only in the 14–16 age groups did the psychological tests emerge from explaining the least variance. Also the morphological and organic variables of height, strength, power and speed were consistently good predictors, but to varying degrees in each age range. Based upon these data, it would appear that psychology, as measured by personality tests, at least, was not important in learning, performing and winning, which is contrary to what many gymnasts and coaches report regarding psychology, in general (Cogan & Vidmar, 2000). This also confirmed Singer's (1988) view on the usefulness of paper and pencil psychological evaluations.

The implications of fixed abilities for learning and performing in gymnastics are great. For example, body height cannot be trained, and most gymnasts have a short physical stature! According to Ericsson, Krampe and Tesch-Römer (1993), height is one of the few physical characteristics that cannot be trained, even with extensive practice with the support of expert coaches and nurturing parents. It is for this reason that the perceptual variables remained very substantial contributors, especially over the last two ages groups (Figure 13.2), since balance and spatial orientating are malleable and continue to contribute to the increasingly complex elements of somersaulting, twisting and balance. Perhaps there is a better alternative of explaining progress in gymnastics performance within a learning perspective. These task dependent differences (Figure 13.1) are especially true when considering the various task demands of the more stable, strength-related hold positions on the rings, as compared to the highly complex movements, rapid hand changes and dynamic balance that are required on the pommel horse.

## The learning and intervention views

Over the last 25–30 years, a number of researchers have traced the developmental milestones, causes and characteristics of expertise development across the careers of not only athletes, but of scientists and musicians. What is most encouraging from the results, which used various methodologies and had different objectives, is the consistency of the findings and the refinement of their models. Each of the four outlined studies of Bloom (1985), Ericsson, Krampe and Tesch-Römer (1993), Côté, Baker and Abernethy (2003) and Durand-Bush and Salmela (2002) added sometimes small, but significant, elements to Bloom's initial research on their expertise development.

### Bloom's view

The nature–nurture issue of whether athletes are born or made is currently tending to side with the environmental perspectives based upon the ecological research of Bloom (1985) and the cognitive research of Ericsson (2007). Bloom wrote an influential book entitled *Developing Talent in Young People* (1985), in which they interviewed 120 world experts in science, the arts and sport, as well as their parents and teachers/coaches. It was shown how the delicate interplay between the athletes evolved with their parents and coaches across three phases of their careers. As they progressed through the *early*, *middle* and *late* stages, their attitudes, activity patterns and goals evolved in a somewhat predictable manner.

In addition, Bloom and his colleagues pointed out that the types of coaching received often changed across the various stages of the performers' careers. Caring, pupil-centered interventions were used early on in tennis and swimming to increase their intrinsic interest in training, and then nurtured them to love practicing their sport activity until they were "hooked" and decided to progress to the next level. The middle phase was more performance-driven, and demanded of the athlete more work and less play, and was supervised by a more disciplined, task-centered coach. During the late stage, sport performance perfection of the sport was required, but this was done so in a more collegial manner between the athlete and the coach than in the earlier stages. Bloom also discussed the roles of the parents during the first two stages where they were required to make family sacrifices regarding meal times and transportation of the athletes to practice, especially during the early stages.

### Ericsson, Krampe and Tesch-Römers' views

The acquisition of exceptional performance of athletes has paralleled the development of other expert performers. Within this developmental framework, Ericsson has shown that expert musicians, athletes and other successful performers carried out significantly more deliberate practice, or effortful practice with the intent of specifically improving current performance levels. The fact that exceptional performers did not excel in laboratory tasks which supposedly measured innate capacities led to the conclusions that exceptional performances were driven by environmental factors rather than biological or genetic ones, as was the case in the TNT gymnastic project. The main constraints which limited deliberate practice were those of effort, motivation and the availability of human and physical resources. In addition, a metric for the minimum amount of deliberate practice was calculated as at least 10,000 hours, or 20 hours per week, normally spread over a minimum of a ten-year period. Young and Salmela (2002) evaluated Canadian middle-distance runners using their daily running logs and found, contrary to Ericsson et al.'s finding, that the effortful training activities were also enjoyable. Ericsson (2007) and his colleagues then transformed Bloom's useful qualitative framework by using quantitative methods, which detailed the practice patterns, the number of repetitions and practice duration of expert performers and developed a metric which is applicable in various domains. The necessary required effort and enjoyment levels were also quantified, as well as the presence of physical and personal resources, such as practice facilities and teachers or coaches, to help the performers endure and profit from the necessary but grueling quantities of practice.

## Côté, Baker and Abernethys' views

Côté, Baker and Abernethy (2003) again extended the work of Bloom (1985) and Ericsson, Krampe and Tesch-Römer (1993) regarding the early developmental steps for attaining expertise in sport. This is particularly applicable in gymnastics, and especially for girls, since it is essential that training begins at a younger age than for the boys, since it is easier to provide spotting when they are lighter. They were able to identify and label three developmental stages based upon interviews with various athletes, parents and coaches. The earliest stages for the sports of rowing and tennis were designated as the *sampling years* (ages 6–12), but could begin as young as five years old. The *specializing years* (ages 13–15) might include 20 to 30 hours of training and could increase up to 40 hours a week during the *investment years* (aged 16 years and older) (Cogan, 2006). Côté, Baker and Abernethy (2003), working only in sport, further developed the previous research findings of Bloom and Ericsson and developed age-specific guidelines for the stages that were previously outlined.

## The sampling years

Côté, Baker and Abernethy (2003) showed that both successful Canadian and Australian athletes evolved through an initial period termed the sampling years, where children experimented with a number of sports, some of which could be related to gymnastics, such as diving or trampoline. What differed with Côté et al.'s research was that many of these athletes did not specialize early in one sport, as was found by Bloom (1985) in tennis and swimming, nor by Ericsson (2007) in music, but they experimented with various sports and physical activities. Thus their total number of hours practiced in one sport did not attain the magical criterion of 10,000 hours. In addition, these individuals did not all grow up in "athletic greenhouses," as was the case in many former and present socialist countries; they lived normal social lives, and most went on to finish their university careers.

In more governmentally controlled societies, the early interventions were almost entirely monitored by the coach, whose mind-set was not directed towards the exciting experiences of practicing gymnastics, i.e., being upside-down, spinning, rolling, hanging and swinging, and being higher in the air than is normal, and then landing safely on a soft and secure gymnastic mat. Rather, many socialist coaches begin, early on, by making students practice deliberately and engage in work, rather than play.

## The specializing years

During this period one interesting phenomenon that I have observed when working with the Canadian national age group teams, as they developed into world competitors or Olympians, is what I have called the period of *mental changing of gears*, from being boys to men gymnasts. When they are from 13 to 15 years of age, anything that the coaches or I said as a sport psychologist was digested by the gymnasts as the TRUTH! And if they remained at a medium level of performance, this belief in the coach remained fairly constant. Nevertheless, when they then began a six- to eight-year period of deliberate practice, they learned and perfected the required skills to perform at an international level. However, there was a so far unidentified moment in a skilled gymnast's development during adolescence when this mental changing of gears occurred, and it was real! The parents, coaches and sport psychologist no longer held all of the keys to the universe in gymnastics, or to their lives, and they may rebel and disregard advice. Many parents and coaches hate this period, but personally I loved it! They were now autonomous, or at least semi-autonomous, gymnasts.

## The investment years

Once the gymnasts have committed themselves to a concerted attempt at succeeding at the highest levels, the serious deliberate practice phase begins where the intent is to improve on a daily basis and to refine partially learned skills into highly polished ones. Since the gymnasts are now in their middle to late adolescence, they realize that their investment in the sport requires that they now make choices with the time spent with friends and even in education. The parents are no longer the driving force (nor the taxi drivers!), 148 the coach now plays a central role in the daily activities of structuring training and practice. As Ericsson (2007) points out, this is not necessarily enjoyable because of the need for the constant repetition of some of the daring gymnastic elements.

### Durand-Bush and Salmela's views

Durand-Bush and Salmela (2002) added another dimension to the above three perspectives with qualitative approaches similar to those of Bloom (1985) and Côté, Baker and Abernethy (2003). The main difference was that all of the athletes were part of a unique population, since they were either World or Olympic champions who had repeated their gold medal achievements at these major championships in different years.

### The maintaining years

Following on from the previously cited research on the career stages of experts in sport, the present study extended the three career phases to a fourth stage (Figure 13.4). Four men and six women who had won gold medals in at least two separate Olympics or World Championships were interviewed, and the term the *maintaining years* was coined.

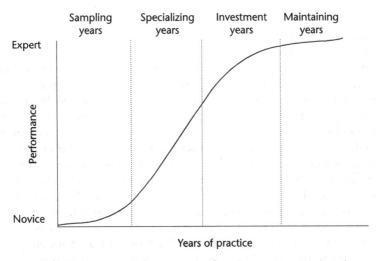

**FIGURE 13.4** Stages of learning and changes in performance across a gymnast's career.

No gymnasts were included in the sample, since most gymnasts retire after their first Olympics and get on with their education or obtaining a job. During these years, the athletes trained less but smarter, and they allocated more time for recuperation, since their bodies were battered from years of training and often numerous injuries.

## Implications of child development for learning and performing in gymnastics

In this section, consistent but evolving findings have been outlined across the careers of expert performers, which have profound implications for athletic development in gymnastics. The pioneering work of Bloom (1985) over 25 years ago traced the pathways and provided profound implications for the progression of high level performers.

But another dimension that is often forgotten is in regards to the practice of gymnastics during free and deliberate play (Côté, Baker & Abernethy, 2003) during the sampling years. During these early years of a gymnast's career, it is essential that one or both of the parents devote some of their daily routine to becoming a taxi driver, sometimes for two-a-day workouts. Durand-Bush and Salmela (2002) reported that parents of multiple Olympic or World champions rarely had the energy to devote the same amount of time to other siblings and had to find other performance niches in their lives, such as in the arts or music.

Once the gymnasts pass through the specializing or investment years, they still love or respect you, and hopefully both, but now the rules have changed! You as a coach, teacher or parent do not have the same total control. They now can call some of the shots, or at least participate in the planning of training and competitions with their coach as a colleague. The extreme stages of deliberate practice, which when viewed from the outside appear repetitive and boring, is not always the case.

I invited a world renowned Serbian concert violinist, Dragan Rodosavljevic, to a graduate seminar in Brazil. He remarked that when others watched him repeatedly practice the same piece over many hours, it appeared to them to be boring. He explained that he was curious about minute variations in his techniques and these small changes were not apparent to non-violinists. After 25 years, he was always experimenting with slightly different approaches. It is interesting that the term curiosity is not found in the sport psychology literature regarding adherence to deliberate practice.

Durand-Bush and Salmela (2002) outlined how multiple Olympic and World champions maintained their training with meticulous planning and commitment, but did not require the same quantity for achieving their first gold medal, but they improved the quality of training for the second one. They now had more free time and were able to plan for their future employment possibilities when their sport careers ended, since the ten athletes then were either currently attending high school or university. They reported that imagery, relaxation and self-talk were important skills that they used during this maintaining period and also outlined the importance of their physical and mental recovery to prolong their careers. There are a number of factors that may have caused the underestimation of how individuals become champions in gymnastics, and the athletes do not always attribute it to the number of hours of deliberate practice, but sometimes to the genetically driven notion of *natural talent*.

One of my childhood gymnastic heroes was Abie Grossfield, an American gymnast who competed in the 1956, 1960 and 1964 Olympic Games and was also a long-time coach for the USA men's team. In 1995, I had the opportunity of talking to him at the 1995 Sabae Worlds and I asked why he was such a good gymnast at such a young age, since he made his first Olympic team at 18. He said that he learned all of his gymnastics from 15–18 years of age!

Thinking of Ericsson's 10,000 hours of deliberate practice, I asked what he did prior to these three years. He said, "Nothing just five years of rope climbing and trampoline twisting and somersaulting." Now, this is what the Russians are currently doing: massive conditioning at a young age and learning spatial orientation skills on the trampoline. Figure 13.2 reveals the high levels of the contributions of the organic (muscular strength and flexibility) and perceptual skills (balance and spatial orientation) during the final stages of gymnastic development. It seems that Abie was wrong in his self-assessment and that the developmental theory of learning was more appropriate!

Durand-Bush and Salmela (2002) reported the remarkable feats of athletes who became multiple Olympic or World champions in separate Games. However, a number of Russian gymnasts, like Boris Shaklin and Yuri Titov, participated in multiple Olympics and World championships in the 1950s and 1960s for the Soviet Union. What was more impressive was Tanaka's (1987) report which outlined over a 28-year period what he termed the "Japanese Golden Era of Gymnastics," when 32 Japanese gymnasts competed in multiple Olympics or Worlds! Ono and Kenmotsu competed in seven consecutive games, between 1952–1964 and 1968–1979, respectively; an amazing record of longevity in such a demanding sport.

Two other examples from athletics and one from gymnastics spring to mind regarding passing from the sampling to the investment years, and in the case of gymnastics, remaining in the investment years in gymnastics. The first regards Wilma Rudolph, the three-times Olympic champion in the sprints in athletics (100 m, 200 m, and 4 x 100 m relay) at the 1960 Rome Games and the second is Lee Evans, world record holder in the 400 m and in the 4 x 400 m relay at the 1968 Mexico Games. I was fortunate enough to meet them both at a conference in Perugia, Italy, in 1991. While thinking of Bloom's (1985) stages of development I asked them both: "When did you first think that you were *good* in athletics?" and "When did you think that you were *great*?"

Wilma was the first to respond and said that she never thought that she would be good at anything in sport, since as a child she experienced no sampling years since she had polio, scarlet fever and pneumonia, and for some years lost the use of one leg. But she did say that as a high school student she trained at the Tennessee State University with the famed "Tiger Bells," and was beating their times in practice. These successes in the intervening years transformed her from being good to great in her progress towards the Rome Olympics, where she was great!

Lee Evans was another story, since he was always in good shape, and when he began training as a freshman at San Jose State University, in repeated 400 m runs he noted that his more experienced teammates were laying on the track gasping for air while he was ready for the next series. He said to himself: "Maybe I do have a future in this sport, and maybe I can become great." His world record of 43.86 seconds lasted for almost 20 years.

But perhaps the most amazing gymnastics-related event which occurred during the investment and maintaining years was Dmitri Bilozerchev from the USSR, who won the European Championships at the unheard-of age of 16 in 1985. Shortly afterwards, he was involved in a car accident in which he broke one leg in 40 places. But the Russian medical staff was able to reconstruct his leg with pins and straps and two years later, after an incredible rehabilitation program during which he worked on his upper body, in 1987 he won the World gymnastic all-around title! This went beyond the maintaining into the *reconstructive* years! In that year at the Rotterdam World's, his tough mental attitude developed during his recovery and performance levels led to him being labeled as the *foreman*, since he always looked in charge of everything.

## 13.4 CONCLUSION

Gymnastics for men includes six events, with four for women. Each event has various and sometimes extremely different task demands. Some apparatus requires balance, others strength and power and others highly complex and difficult elements. Thus, sport science must be multidisciplinary and specific for each apparatus

Knowing what are the task-demands for each apparatus: each apparatus has different task-demands. The balance beam requires women gymnasts to perform on a beam that is 10 cm wide, and the movements are often equivalent to those performed on floor exercises on a 12 by 12 m surface. The vault, for both men and women, requires a rapid sprint towards the horse, contact with the horse with their hands at an appropriate angle, somersaulting and/or twisting in the air, "spotting" the landing point then "sticking" the landing with no further movement. The other events require gymnasts to perform movements which are not natural for human beings including being upside-down, twisting in the air, being supported by only one arm (uneven and horizontal bars), performing movements requiring extreme flexibility (floor and balance beam), dynamic balance (pommel horse) or strength (rings). These elements can be assessed through some sport science methods, but more frequently through the eyes of experienced coaches.

Understand the evolution of the various points of view and methodologies of researchers in gymnastics since 1985. In 1985, Bloom interviewed the best scientists, musicians and athletes in the United States. He found that they followed very similar career paths. During their initial stages of involvement in their activity, they received very caring interventions both by their parents and teachers, and became "hooked" on the activity. This led them into the middle stage, which required stricter training by the coaches and more parental support. In the final stage, after many years of practice, they worked with the coach in a more collegial way, during which decisions regarding training and performance were decided upon together.

In 1993, Ericsson and his collaborators used a quantitative approach to understand how musicians evolved from novices to experts using a quantitative approach. Using diaries and questionnaires, they found that expert performers completed, at least, 10,000 hours of what was termed "deliberate practice," or practice that was goal-related, but was not necessarily enjoyable, and thus they were able to put a metric on Bloom's work.

Côté, Baker and Abernethy (2003) replicated Bloom's study and labeled the early stage the "sampling years," when the athletes tried a variety of sport activities, and then selected the one in which they would specialize in a given activity for eight to ten years, and this was termed the "specializing years." This was followed by another period, called the "investment years," during which the athletes devoted all of their time to practicing and competing in their selected domain.

Durand-Bush and Salmela (2002) added another dimension to Bloom's model by interviewing multiple World or Olympic champions, with special attention given to the period between their first and second gold medals, which was termed the "maintaining years." Their practice was less intense and more focused, and rest, mental practice and recuperation were emphasized.

In summary, the models and guidelines first suggested by Bloom provided a qualitative framework for expertise development which intertwined the career phases of athletes, coaches and parents (Bloom, 1985). Côté and Hay (2002) then later labeled the sub-components of the various activities that young, successful athletes were engaged in prior to moving into the deliberate practice regimes during the specializing years. Finally, Durand-Bush and Salmela (2002) considered the development of expertise from the other end of the continuum, when athletes had already achieved Olympic or World Championship gold medals and continued to achieve others during the next cycle, during the maintaining years.

# 14

# COACHING AND PARENTING

*John H. Salmela*

## 14.1 Coaching

The notion of an autonomous, self-directed athlete's propulsion to world class standards has been put into perspective by the research of Salmela (1996). Expert coaches reported on how they helped shape the learning environments of young athletes by judiciously creating and monitoring achievement goals, facilitating the acquisition and maintenance of exceptional performance in training and competition by minimizing the constraints which limited their practice, competition and skill development.

Some dimensions that have been so far omitted are the central roles that coaches play, not only in gymnastics, but in other sports. Success in gymnastic achievements cannot occur in isolation, nor does it in any other academic pursuits. I spent many hours with great former Soviet Union coaches abroad, such as Leonid Archiaev and Edouard Iarov, and many foreign and native gymnastics coaches living in Canada. Of course, when comparing the best in the world to myself, my shortcomings in understanding the rapidly evolving gymnastics coaching techniques quickly became embarrassingly obvious, and also the fact that I was now part of the very old school of coaching.

Rabelo (2001) discovered that young Brazilian soccer players, usually from lower socio-economic classes, received only minimal coaching until they reached the professional ranks as juniors. But they played and practiced for an enormous amount of time – at each break in their classes, before and after school and in nearly all of their leisure time. In contrast, all Brazilian gymnasts received specialized coaching from the beginning of their careers, but during their leisure time did other activities apart from gymnastics (Moraes et al., 2004). This is in agreement with Durand-Bush, Salmela and Green-Demers (2001) who reported that both multiple Canadian Olympic and World champions spent their youth discovering a variety of other domains and in only a few cases were totally devoted to their selected activities.

Côté and Hay (2002) discussed the importance of two different types of activities in which coaches could facilitate in the progression from the deliberate play, sampling years through the investment, and deliberate practice years. Deliberate play involves the child's active participation, is voluntary and pleasurable, provides immediate gratification and is driven by intrinsic motivation. On the contrary, deliberate practice is not as enjoyable, requires effort and involves the delayed gratification of rewards (Ericsson, Krampe, & Tesch-Römer, 1993).

A coach's decision about the kinds of learning activities to provide and how to decide whether children are capable of profiting from them becomes an important aspect of successful performance. These issues are often discussed, while keeping in mind that a coach's duty, at any level of development, is to stimulate the early fruition of skill and sustain the athletes' motivation for learning and improving. Csikszentmihalyi, Rathunde and Whalen (1993) conducted a four-year study of talented teenagers from the arts, sport and science who would be in their specializing years, and reported some damning commentaries from the most talented group regarding their teachers/coaches: "Previous research suggests that teenagers are singularly uninspired by the lives of most adults" (p. 184). This is a strong message for young, aspiring coaches.

However, their most successful mentors were both inspired by the subject matter that they were teaching and reinforced the individual progress and interest of their students. It is not difficult to make this transition to successful gymnastic coaches. Within a positive learning environment, successful gymnasts relished both the hardships and the challenges of training and competition. However, there is a price to pay, since to become great through extensive practice, they have to better manage their time than their normal friends. They were also more often alone than their cohorts, which was not always enjoyable. But, when involved in the activities which they enjoyed, they were more concentrated and happy. However, when these tasks were completed, they reverted more to mindless tasks such as watching TV or going on the computer than did their colleagues (Csikszentmihalyi, Rathunde & Whalen, 1993). After interviewing 22 expert international Canadian team coaches (Salmela, 1996), I realized that my investment in training and coaching graduate students in achieving international academic success was greater, more intense and better planned than my meager gymnastics coaching skills.

To date, the psychological differences between men, women or girl gymnasts have not been discussed. One of the first studies on this subject was by Jerome et al. (1987) who evaluated a sample of 50 young girls involved in an elite Canadian gymnastics program. The intention of the study was to determine the factors which differentiated between gymnasts who remained in the program and those who dropped out. Twice-yearly evaluations were carried out using psychological instruments, as well as others in motor behavior, physiology and anthropometry.

From my perspective, the most disturbing results were found in the psychological domains. The girls who remained were significantly more obedient and conforming, more introverted, had lower social comprehension or intelligence, and higher trait anxiety (p. 98). In some ways, these characteristics are similar to those presented by Csikszentmihalyi, Rathunde and Whalen (1993) which indicated that talented teenagers often reported fewer daily positive episodes and psychological states, since they had to sacrifice their social lives for a period of their adolescence to achieve excellence. Actually, this profile made these gymnasts extremely coachable, but not every parent would like to have this mental profile for their normal child.

## 14.2 Implications of coaching for learning and performing in gymnastics

Young, talented women are no longer allowed to compete internationally in gymnastics at 13 years of age, but now only at 16. One of my good friends, Dave Arnold, with whom

I worked for 20 years, both as a gymnast and as the National Coach, had some interesting insights about coaching young boys and girls. He said that young Canadian girls began training seriously when they were 8–10 years of age, and often skipped the sampling years. He said that the coaches and their assistants imposed intensive conditioning and had them repeatedly practice the same elements with little chance for play and experimentation. The result was that after years of this monotonous training, at the end of their careers, they often totally abandoned the sport, hated their coach and rarely became either a judge or a gymnastics coach.

Because of the necessity by the men of awaiting post-pubescence to develop sufficient strength and power, the coach still conditioned and trained them, but let them play around on the trampoline, get involved with improvised gymnastic games and joke with their teammates and the coach. The coaches' task was to make them love gymnastics and continue to practice into their twenties. Later on, many of these gymnasts went on to become judges and/or coaches at various levels. I am still in contact with my university gymnasts whom I first enjoyably coached in the seventies, and I believe that the fact that we enjoyed each other's presence, liked traveling together and had fun was central. In numerous international competitions, when the Canadian men and the girls were together, the latter often commented upon the fact that the men, with whom I only worked, seemed to be so relaxed and happy compared to how seriously, and almost grimly, they viewed their gymnastic training, competition and life

## 14.3 Parenting

Bloom (1985) was the first to outline the roles of parents in the development of expert athletes simply by being supportive of their achievement activities and permitting them to choose between practicing formally, or to just continue playing. During the middle period, there was more dedication by both the parents and the athletes to continue in their deliberate practice, while during the later stages their support was less physical and emotional, but more financial.

Côté (1999) studied the family environments of elite junior Canadian athletes across their careers. The roles of the parents changed from a leadership role during the sampling years to a follower/supporter role during the investment years. Over all of the years of their child's participation in sport, the parents tried not to pressure them within their given sport. During the investment years, the parents backed off, provided financial support when necessary, but tried not to create additional demands or pressures.

Soberlak and Côté (2003) interviewed professional ice hockey players regarding their progress from the sampling through the investment years. Parents did facilitate their *deliberate play*, for example, by building a backyard ice hockey rink, but avoided providing technical instruction. During the investment years, they observed the progress in their son's performance and provided them with positive, but non-technical, feedback. One common thread that characterized all of the studies of Bloom (1985), Soberlak and Côté (2003) and Côté (1999) were the transformations of the roles of parents from the sampling to the investment years. During the sampling years the parents had greater direct involvement in their child's sporting activities by teaching and playing with them. This direct involvement, however, decreased as the young athletes moved from the specializing to the investment years, and occurred even less during the maintaining years when the athletes were young, semi-autonomous adults.

Cogan (2006) convincingly argued that most parents of gymnasts want what is best for their children. But in women's gymnastics, especially, they may exert unnecessarily high expectations on their young athletes, which may have effects that are the reverse of those intended:

> Parents can exert too much pressure if they become overly invested in their child's sport. Sometimes parents can get so absorbed in their own goals and desires for their child, that they are not aware that the child is losing interest in the sport. Parents can try to live vicariously through their child if they never achieved their sport goals and the child shows potential to excel. In this instance, the children may feel pressure to perform well and keep training when they would rather do something else.
>
> *pp. 646–7*

The practice and play of aspiring soccer players in Brazil was largely unsupervised by parents and often without coaches (Salmela, Marques, & Machado, 2004). Rabelo (2001) reported that 78.3% of the families reported that their son's activity in soccer did not change any aspect of their daily life routines, while only 17% reported making adjustments to allow them to participate in sport. This was in contrast to the interviewed parents of, at least, a middle class standing in Canada (Côté & Hay, 2002). However, Moraes et al. (2004) found that 24% of middle class and upper middle class parents of tennis players took time to interact with the coaches regarding their children's progress. Finally, Vianna (2002) showed that the tennis parents understandably maintained daily contact with their children at all levels of performance, while 50% of the soccer parents only saw their sons every one to three months at the time of the interviews. Thus the financial status of parents and athletes determined the quality of the interactions between the athletes and parents, and while play, although potentially financially rewarding in the long run in Brazil, their support played a lesser role than putting food on the table and a roof over their heads (Salmela & Moraes, 2003).

## 14.4 Implications of parenting for learning and performing in gymnastics

Social status and parenting in sport are intimately linked. For example, it was shown earlier that soccer players from poor families from the interior of Brazil received almost no social or financial support from their parents compared to the middle and upper middle class athletes in private clubs, in gymnastics, swimming and tennis. One poor football player was forbidden to play in a championship final by his parents, but jumped out of the window of his house to play the game anyway and then when he returned home after the victory he was beaten by his parents (Rabelo, 2001)!

I came from a working class family in Montreal and both of my parents worked, and I think that my parents were happy that I was out of the house. They attended only one gymnastics meet, little league baseball game or Canadian football game. I have known many Montreal gymnasts who made the national team who were poorer than me, and they walked to the gym without their parents. While traveling in China, Japan, Russia and Ukraine, I never saw a parent watching a gymnastics training session, so obviously socio-economic elements play both positive and negative roles in gymnastics.

In North America, some girls' gymnastic clubs either keep the parents away with limited possibilities of seeing them train, or place them behind a glass enclosure so that they cannot hear the coach/athlete interactions. If not, the parents would complain that their daughter did not receive the same attention or feedback as the other gymnasts, and they then would complain to the club president. So there is such a thing as *over-parenting*, where the parents become somewhat familiar with the gymnastics terminology and begin advising the coaches on what they should be doing and exerting unnecessary pressure on their child, or even worse, take a judging course and begin judging with little or no practical experience in the sport.

Therefore, parents are essential for the provision of a nurturing environment for the introduction into sport of their children, and then providing them with the necessary resources to practice their sport, which sometimes means driving them to the facilities and to competitions, as well as financing the purchase of equipment and uniforms. Very often, in middle and upper class families, this often means rescheduling their daily routines, such as meal times, to accommodate training regimens. Caution, however, should be taken not to over-parent their children, which could take away from their intrinsic enjoyment of the sport.

## 14.5 CONCLUSION

In a highly technical sport, such as gymnastics, coaches play a central role for the development of athletes. However, in the developed, non-socialist world they function in very different ways across the careers of gymnasts. As Bloom (1985) pointed out, during the early years they must be process and athlete-centered, otherwise the athletes who have many life choices will drop out of the sport. During the specializing years, they must be more demanding and require that the gymnasts receive and accept strict conditioning, longer hours of practice and confront often dangerous and challenging tasks. Later in the gymnasts' career, during the investment years, they can act more as consultants since the majority of the training has been accomplished over, at least, ten years of training.

Parents, however, play central roles during the early years of a gymnast's career since they have to introduce them to the sport and often have to transport them to the training sites. When they are in the specializing years, they also might have to invest in their travel and equipment, but this often diminishes during the investment years where they are supported by their team or national federations.

# 15

# MENTAL SKILL DEVELOPMENT AND VARIATION IN GYMNASTS

*John H. Salmela*

## 15.1 Mental skill learning for enhancing performance

Many approaches have been developed over the years to assess the nature and effects of mental skills for performance enhancement in sport, and to identify those factors that were most critical for sporting excellence. Some of these approaches which could be useful in gymnastics included the following: consulting with mental trainers or sport psychologists, coaches and/or colleagues using existing psychological instruments (Bernier & Fournier, 2007; Mahoney et al. 2009; Mahoney, Gabriel & Perkins, 1987; Nideffer, 1987), interviewing of athletes (Orlick & Partington, 1988; Ravizza & Rotella, 1982) and observing gymnasts' behaviors in sport settings (Salmela et al., 1980).

The use of psychological inventories has received much attention by researchers. Fogarty (1995) made a number of useful comments on the administration of psychological instruments within sport settings. He stated that:

> Many of the tests appear to be new, often developed for the purpose of a single study. Furthermore, because most of these new tests are not fully validated, they are not released for commercial publication and consequently do not find their way into the major distribution channels. More importantly, they are not subjected to the formal review processes which most commercial tests have to undergo.
>
> *p.167*

In gymnastics, little sport psychology research has been carried out besides the personality research of Jerome et al. (1987), which did not consider levels of expertise, but of dropouts and non-dropouts in an elite club in Canada. The only other study compared the competitive behaviors of top ranked gymnasts with lower classed individuals, concerning their actual competitive behaviors (Salmela et al., 1980). It seems obvious that the levels of expertise of performers within any domain need to be considered since national and international Canadian athletes in a number of sports, and international Iranian athletes in all sports who qualified for the Asian Games, and those who did not, all had different mental skill levels which narrowed in number as they became more skilled

(Salmela et al., 2009). Thus, the understanding of these various paradigms is important to understanding gymnastics.

## 15.2 How each mental skill affects the others

One thing is clear when assessing, teaching or applying mental skills in sports. First of all, the mental skills should incorporate the three most general conceptual fields in psychology, that of *behaviors* or actions, *cognitions* or thinking and *psychosomatic* elements or feelings. Second, it must be made clear that these general families and their more specific sub-components are interactive in nature and that changes in one of the components will always affect the other two. For example, if a gymnast receives a knee injury when tumbling (the behavior), this will immediately affect his/her current emotional state of anger or sadness (psychosomatic conditions), which in turn will affect their thinking, such as they are not worthy and may not make the team (cognitions).

## 15.3 The development of the OMSAT-3

As the result of research in the field of sport psychology, it has become evident that mental skills play an important role in achieving excellence in sport (Mahoney et al., 1987; Nideffer, 1987; Orlick & Partington, 1988). Orlick and Partington (1988) found that the most successful interviewed Canadian Olympians, who participated in the 1984 Games in Sarajevo and Los Angeles, were seriously committed individuals who believed in themselves and had the determination to accomplish their specific goals. According to Orlick (2008), being committed to excel through the good times and the bad, and believing in one's ability to succeed and reach personal goals, are fundamental elements that gymnasts needed to perfect to achieve exceptional performance. Ericsson, Krampe and Tesch-Römer (1993) also demonstrated the essence of developing high levels of commitment to overcome effort and motivational constraints associated with daily deliberate practice. Commitment is simply indispensable in the pursuit of expert performance as achieving such a goal.

Furthermore, there have been many attempts to distinguish between those mental skills that are most critical for performance enhancement. Based upon substantial consulting experiences, active interactions with well-known applied mental trainers such as Terry Orlick, Ken Ravizza, Bob Rotella and Len Zaichkowsky, the Ottawa Mental Skills Assessment Tool (OMSAT-3) was developed over a decade and validated using confirmatory factor analysis, a powerful statistical tool which is rarely used to evaluate the validation of the psychological scales in relation to actual sport performance (Durand-Bush, Salmela, & Green-Demers, 2001).

It should be noted that from now on in the text, the term of *sport psychologist* will be changed to *mental trainer*, since the profession of psychology in many countries is protected by law. But, the term of mental trainer with teams is comparable to conditioning trainer, strength trainer or tactical trainer, and what will be outlined in the rest of this book will relate to mental training. The research in many theoretical aspects of sport psychology will be used, but it is believed that mental training relates mainly to the application of mental skills training for sport performance in a more holistic and interactive manner.

The OMSAT-3 is made up of 12 scales which have been regrouped into the three broader categories of Foundation (*goal-setting, commitment and self-confidence*), Psychosomatic (*stress control, fear control, relaxation and activation*) and Cognitive skills (*focusing, refocusing, imagery, mental practice and competition planning*).

Fournier et al. (2005), using the OMSAT-3, found that refocusing skills improved significantly after a ten-month mental training program, while stress control showed no significant changes. Thus, refocusing appears to be more state-like, i.e., capable of being taught and learned, and it is thus a true mental skill. Stress control, however, might be more trait-like, in that it showed the lowest scale scores and resisted mental training, which may have been genetically inherent in these elite athletes.

## 15.4 Variations in the mental skill patterns over levels of expertise

In the initial development of the OMSAT-3 (Durand-Bush, Salmela & Green-Demers, 2001), over 200 national and international experienced Canadian athletes were administered the instrument (Figure 15.1). The results from the sport, psychology and sport psychology literature and common sense were confirmed. Just as would be expected with international athletes as compared to the national level ones, their mental skill levels paralleled their advanced physical skills, and the scores were statistically higher for the international than for the national group on 10 of the 12 mental skills. These differences were statistically significant with the exception of the stress control and refocusing scales, which were higher.

The variables of stress control were particularly intriguing given the findings in the other OMSAT-3 studies which included only high level, international athletes competing in Iran. The significant dimensions which marked the Iranian studies (Salmela et al., 2009) compared to the Durand-Bush, Salmela and Green-Demers (2001) study were that all of the Iranian athletes who were assessed were already performing at the international level. However, those who were selected for the Asian Games differed from the non-selected athletes on the OMSAT-3, but only on two scales, i.e., those for stress control and

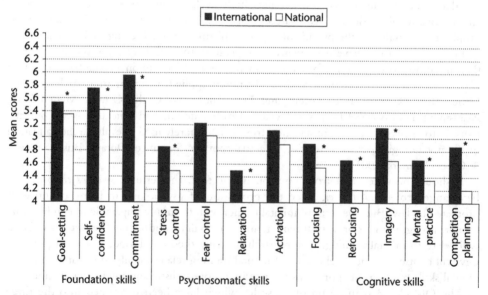

FIGURE 15.1 Profiles of Canadian national and international level athletes on the OMSAT-3 scales. The '★' represents statistically significant difference (p < 0.05) between the groups.

(Salmela et al., 2009)

refocusing. These scales significantly differentiated between the two groups and relaxation skill differences approached significant levels. The reduction in the number of scales which differentiated between the groups of international athletes seems logical. As Durand-Bush, Salmela and Green-Demers (2001) pointed out, higher scores on the three foundation skills of goal-setting, self-confidence and commitment were central to exceptional performance in all sports. All members of this sample reported high levels on these three foundation scales with values greater than those reported by Durand-Bush and Salmela (2002) with Canadian national and international athletes (Figure 15.2).

In the second phase, the Iranian medalists were found to be different from their non-medalists on the single scale of stress control. This again revealed that there was a consistent hierarchy of mental skill scores across levels of expertise which paralleled their sport perform-ance levels. The present sample of athletes demonstrated the evolution from the ten OMSAT-3 scale differences in the Durand-Bush, Salmela and Green-Demers (2001) study to only two scales of stress control and refocusing for the selected and non-selected inter-national Iranians, and then only one, stress reactions, differentiated between the medalists and non-medalists. These interactions between expertise and mental skill levels are unprecedented in the mental skills literature (Figure 15.3).

What was of particular interest was the consistency of the findings between the two comparisons, with stress control acting in a fixed trait-like manner, while refocusing showed significant learning over a ten-month training period, and while stress reactions showed no improvements with mental training and acted in a stable manner (Fournier et al., 2005). What was most striking was the fact that stress reactions and refocusing were consistent among the most highly correlated scales in both the Durand-Bush et al. (2001) study with international and national Canadian athletes (.53), as well as in the present one

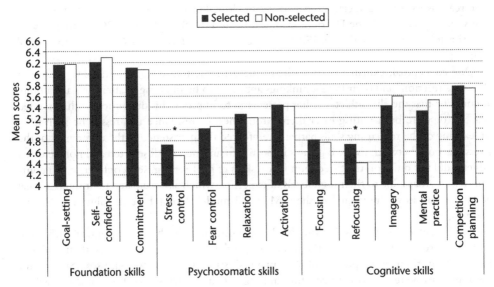

FIGURE 15.2   Mean OMSAT-3 scores for Iranian athletes selected and non-selected for the 2006 Asian Games. The '★' indicates statistically significant differences ($p < 0.05$) between the groups.

(Salmela et al., 2009)

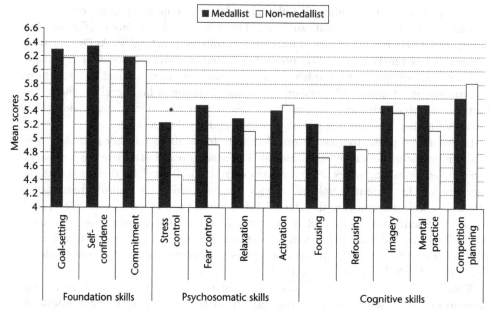

**FIGURE 15.3** Mean OMSAT-3 scores for Iranian medallists and non-medallists in the 2006 Asian Games. The '*' indicates statistically significant differences ($p < 0.05$) between the groups.

(Salmela et al., 2009)

for Iranian international athletes for both the selected (.57) and non-selected (.49) athletes. The reasons for these associations are as yet unclear, but it is interesting that the levels of stress reactions for both the Fournier et al. (2005) and the Salmela et al. (2009) studies were also the lowest of all of the OMSAT-3 scale values, and were most resistant to mental training.

One conceptual, explanatory hypothesis might be the following: the trait-like nature of the stress reactions and their lowest rankings appeared to be unchanged after mental training. It could be that the elite athletes either persevered to improve this dimension with extended practice or were born with greater mental toughness or resistance to stress. The high level of learning of the refocusing skills partially reinforced the persistence of the differences of the stress reactions between the two groups in both Iranian studies. All selected and non-selected athletes did not differ on the three foundation skills of goal-setting, self-confidence and commitment, perhaps due to their high levels of expertise, which also validated their international athletic status, as was predicted in the OMSAT-3 model (Durand-Bush, Salmela & Green-Demers, 2001).

Of particular interest was that the five Iranian coaches who had daily contact with these athletes were also asked to rank the importance of the same mental skills of their athletes, based upon their perceptions of the players' strengths and weaknesses. Significant discrepancies were found between the rankings of the athletes and those of the coaches. The coaches perceived athletes to be skilled at maintaining focus and self-confidence, while these athletes perceived just the opposite. Coaches viewed their players to be poor at controlling their emotions and tension, while the athletes reported themselves as being proficient in these skills.

## 15.5 CONCLUSION

Mental skills have been shown to be better predictors of sport performance compared to fixed, innate psychological parameters, such as personality tests or IQ. The OMSAT-3 was developed to assess a variety of mental skills, which were shown in the sport psychology literature to be essential for expert performance. The essential ingredient was that these skills could be taught and learned by developing athletes and coaches, sometimes with the help of a mental trainer.

The basic elements were the foundation skills, without which expertise in gymnastics, or any other domain in sport, science or music, would be impossible. These three skills included goal-setting or having high expectations for potential achievements. The second category included psychosomatic or emotional skills, such as stress and fear control, relaxation and activation, some of which would be used to different degrees in training and competition. The third category was cognitive or thinking skills, which included imagery, mental practice, focusing, refocusing and competition planning.

All of these mental skills were submitted to athletes of different levels, and some skills differentiated successful athletes at different performance levels from national, international and medalists from non-medalists at international competitions (Durand-Bush, Salmela & Green-Demers, 2001; Durand-Bush & Salmela, 2002; Salmela et al., 2009).

# 16

# OMSAT-3 MENTAL SKILLS ASSESSMENT

*John H. Salmela*

## 16.1 Goal-setting

Goal-setting is perceived to be an essential performance enhancement skill by many researchers (Burton, 1993; Harris & Williams, 1993). More precisely, it is believed to help athletes focus their attention, remain intense and persistent, increase their self-confidence, and control anxiety (Burton, 1993). Gould (1998) also suggested that athletes set specific, measurable goals which are difficult, but realistic to achieve, to maximize their effects. Others have revealed that athletes' performances can be additionally increased if they set short- and long-term goals (Harris & Williams, 1993) as well as performance goals, such as increasing their number of successful routines or perfecting problematic elements, as opposed to outcome goals, like defeating a given opponent or obtaining a certain score on an event (Burton, 1993).

"I set difficult but achievable goals" in the goal-setting scale of the OMSAT-3 (Ottawa Mental Skills Assessment Tool) was an excellent discriminating item between the Canadian national and international athletes, which suggested that elite athletes set goals, but not any type of goal; rather, they set challenging goals that they believed they could achieve (Durand-Bush, Salmela, & Green-Demers, 2001).

## Implications of goal-setting for learning and performing in gymnastics

It would be unrealistic to assume that aspiring gymnasts would set high level goals on a daily basis to be proficient in such a multi-faceted and difficult sport such as gymnastics. When young girls first saw on television Romania's Nadia Comeneci and the Soviet Union's Nelli Kim score perfect tens during the 1976 Montreal Olympics, or Mary Lou Retton who helped the USA win the team title in 1984 in Los Angeles, gymnastic club registrations sky rocketed across the world in women's gymnastics. However, even though goals in sport can maintain one's focus, persistence and self-confidence, this is not always sufficient. But, gymnasts for whom this is their only livelihood in financially depressed societies may persevere in their attainment of these lofty goals since they are dependent upon state support for successes and competition cash prizes for a good future life. This is also the case in Brazil, where successful athletes such as Deanna dos Santos and the Hypolito family can now attract lucrative, financial contracts.

I often ask my students in some FIG courses who would like to be seventh best in the world in anything. Most people raise their hands. I then reply that the men's all-around champion, Li Ning from China, the 1984 Los Angeles Olympic all-around champion, did not share this same goal. I was with the National Coach of Canada with Curtis Hibbert in the warm-up gym preparing for the horizontal bar finals at the 1987 Rotterdam World Championships. At that same time, the medals were being awarded for the pommel horse finals, and shortly afterwards all eight finalists entered the gym. Li Ning had fallen from the pommel horse and was awarded seventh place. When he walked into the warm-up hall he had his certificate in his hand, he threw it in the air in disgust and it landed close to where we were warming up. Since he was obviously going to leave it there, I went over and kept it as a goal-setting illustration for future classes and courses in gymnastics. Obviously, seventh place was not a goal for Li Ning!

Concerning the lifelong-term goal-setting of each university student, I often ask them in my classes, courses or workshops the following question: "How many of you would like to be an Olympic or World champion in your favorite sport?" Almost all of their hands immediately shoot up. But when I ask them: "How many of you are willing to work at your sport for six days a week, for 20–30 hours weekly for the next ten years?", almost all of the raised hands completely disappear. Again, goal-setting has to be accompanied by the two other foundation skills of self-confidence, in being able to realize these goals, and their commitment to doing what is necessary in the attainment of these goals, despite the time investment and possible setbacks and normal injuries.

This is what Ericsson (2007) termed deliberate practice, or effortful practice with the intent of improving current performance to the highest levels in the world. But to do so, extremely high goals are required. The normally cited minimum amount of deliberate practice in order to achieve expertise in sport, chess or music is 10,000 hours, although the number of hours with musicians can increase up to 55,000 hours if they continue during their golden years. During a seminar in Canada with elite gymnastics coaches, I asked them to estimate through-out their careers how many hours they had trained as a gymnast. The answers varied from 3,000 to 7,000 hours for the Canadian gymnasts/coaches with one exception: a former women's USSR team champion at the Montreal Olympics, Elvira Saadi, who is presently successfully coaching in Canada, reported an astounding 20,000 hours of training!

Personally, I cannot claim more than a bronze medal as a junior at the Canadian national gymnastics championships and a fifth placing in the all-around classification in 1961. I also captained the Canadian juvenile football championship team in 1964. What was interesting, in retrospect, was the fact that I did not consciously set any personal goals. I guess when you come from a working class family, your realistic life ambitions are to have a real job and to install telephones for Bell Canada, which I also did for two years. My father, until he died in 1991, told me that I should have stayed with Bell, not pursue a career as a gymnastics coach, because I might break an ankle!

One of the more humorous incidents concerning lifelong goals occurred when I was teaching a graduate sport psychology course in Brazil. I asked them, "When you die, what do you think your friends and colleagues would have inscribed upon your gravestone regarding your goals and accomplishments?" The class was stunned by the question and when I asked them all to give me a response, the responses were hilarious. They never said: "He worked harder than anyone on the team," or that "He had his dreams and realized them," but were much more superficial. Obviously, these were people who had never thought of far-reaching, lifelong goals, or exceptional goal-related accomplishments!

## 16.2 Self-confidence

Another element reported to be extremely important in the achievement of exceptional performance is belief, or self-confidence (Harris & Williams, 1993; Nideffer, 1987). As previously mentioned, self-confidence is another core skill that Orlick (2008) reported as being essential: "The highest levels of personal excellence are guided by belief in one's potential, belief in one's goal, belief in the significance or meaningfulness of one's goal, and belief in one's capacity to reach that goal" (p. 112).

Self-confidence has been reported to be cyclical and variable. Individuals can believe in themselves more on some days than on others. Orlick revealed that individuals manifesting high levels of self-confidence often had a solid support network, that is, other people and loved ones who believe in them, and maintained a positive attitude towards their performances. These individuals received much positive and constructive feedback, drew out constructive lessons from training and competition and regularly experienced improvements and successful performances.

It can be seen that commitment and belief in oneself are critical mental perspectives athletes need to develop and maintain their gymnastics performance to achieve their high level goals and skills. Moreover, it has been seen that goal-setting is an important skill athletes have used to enhance their self-confidence and commitment to excel. Other mental skills inherent to achieving success have also been identified in the literature, such as developing the *flow* state (Jackson & Cziksentmihalyi, 1999). First of all, flow is a state of consciousness where one becomes totally involved in the given activity, with the exclusion of all other negative thoughts and emotions. Flow has also been called the ideal performing state (Unestáhl, 1975), or being in the zone, and does not necessarily end upon winning.

The flow state occurs when gymnasts perceive a high level of challenge in a competition and in skill, and have performed and trained well to have the perceived competence and this is the situation where the state of flow *may* set in (Jackson & Cziksentmihalyi, 1999). This phenomenon does not always occur so often in gymnastics; however, it seems to increase with additional experiences. Gymnasts might feel that their performance is effortless, non-stressful and that time distortions may occur so that the performance from a psychological perspective is either in a slow dream-like state or happens very quickly, or in *flow*. The occurrence of flow is affected by many factors, such as motivation, appropriate psychosomatic mental states, appropriate pre-competitive and competitive plans, and physical and mental readiness (Weinberg & Gould, 1999). The good news is that both skill levels and challenges are teachable and learnable mental skills, which can change with practice. But the source of flow is not always evident to the performer. But the ultimate aim for a gymnast and a coach is to appropriately balance both physical and mental training, which can eventually enable them to enter this flow state.

### Implications of self-confidence for learning and performing in gymnastics

A fundamental element in the mental skills foundation triangle is self-confidence, or belief. Often non-gymnasts, when asked "What is an important mental concept central to success in gymnastics," often report the term of self-confidence. We can all remember in elementary school our colleagues who were picked out as being self-confident in math or reading and, usually the older boys or girls, in sports. Self-confidence is developed by having achieved success in gymnastics, often before their peers could achieve a given skill. Self-confidence is rewarded by coaches, which, in turn, increases gymnasts' belief in themselves.

Prior to elementary school, I was fascinated on TV by the weekly hour of professional wrestling, which included "midget" wrestlers. I began practicing and learned the single skill of a neck spring on our lawn, which was often done by the "midgets" to quickly kip up from their backs on the mat to their feet. It wasn't pretty and I must have looked like a four-year-old, one-trick, "midget" wrestler! But when I entered my Grade One physical education class, I repeatedly did them in front of the gym teacher on a wooden floor, much to his delight! I believe that this move was my first experience in developing my self-confidence in sport. But as is the case in any physical or academic activity, there are daily ebbs and flows of performance and it is for this reason we have peers or coaches to provide immediate feedback for less well performed movements and to reinforce what Orlick (2008) refers to as "highlights" or small, daily positive events to improve self-confidence for the next day.

Setbacks and losses of confidence can also occur at the highest levels of performance with World and Olympic champions. When I was accompanying the Iranian national delegation to the Asian Games in Qatar in 2006, the tae-kwon-do coach directed me to Youssef, an athlete, for a discussion regarding his confidence This young man had won the silver medal in the Athens Olympics and was twice World champion, yet he lost his confidence in the four bouts that were ahead of him in Doha. I met with him seven straight days and, even on the day of the bouts, I was becoming exasperated and told him that he only had 24 minutes left to fight and had already accumulated 14 years of high level training, and achieved many Olympic and World victories.

After the first fight, which he easily won, I approached him in the warm-up gym with a sign on which I wrote: "18 minutes," after which I got up and left the hall. After each successful bout, I returned with new signs stating "12 minutes" and then "Six minutes." By this time, the whole Iranian team was laughing uproariously, including him. He won the gold medal, but he needed a bit of a self-confidence shove!

An example of this state of flow in gymnastics is illustrated by a former World and Olympic champion in gymnastics who I personally observed and later commented about my perceptions during the 1976 Montreal Olympics. In 2004, I had the privilege of participating in an FIG Coaching Academy with Nelli Kim, double Olympic gold medalist on the vault and the floor exercises in Montreal. During the Games, I was sitting in the best seats with my graduate students doing behavioral observations of the Olympic gymnasts (Salmela et al., 1980). I recalled to her that she seemed in the psychological *zone* for exceptional performance from the flow literature.

She told me the exact characteristics of her flow state: effortlessness, total contact with the routine and distortions in time, as if competing in slow motion. I mentioned to her on the bus to the sport center in Kuala Lumpur that I would be talking about this flow state that afternoon. She seemed surprised and replied to me: "I always thought that this magical performance was because of God." I kept this alternative hypothesis in mind as a possibility, but as a scientist the flow hypothesis was more testable in scientific settings, rather than in religious musings.

In the preparation for the 1988 Seoul Olympics, during a three-week training camp, the coaches and I came up with a confidence building exercise where all seven members of the team had to successfully go through their routines twice without a major error, by anyone, or else everyone had to repeat the whole process again. We also told them that we could delay supper if there were many misses. Since each gymnast's performance affected the number of repetitions for their teammates, this exercise, both in self-confidence and commitment, permitted the whole team to go through 84 consecutive routines without a major error and on our rest day we played golf, and the gymnasts were still glowing. This stressful training exercise served us well and we achieved our best team finish ever in Seoul. Oudejans (2008) supported these methods and suggested that "… turning up the heat from the very first day of practice may be one of the most effective ways to immunize yourself against *blowing* it," or choking. He continued: "They're trained in how to play the game, but they don't train under pressure, so they will fail" (p. 40).

Another non-gymnastic, but remarkable, example of self-confidence was shown by a Canadian canoeist, Larry Kane. In 1983 he had printed up a business card on which was written, "Larry Kane, 1984 Los Angeles Olympic Canoe Champion," which he distributed to everyone, and he won! Now that is an example of public self-confidence!

## 16.3 Commitment

Goal-setting may be "one of the best performance enhancement techniques available in the behavioral sciences" (Burton, 1993, p. 469), but failure to develop an ongoing commitment to attain the set goals may diminish their enhancement capabilities. Orlick (2008) perceived an athlete's level of commitment to be a critical ingredient for success. Niemi (2009) showed strong evidence that placebo effects in both medicine and sport can be powerful, even though they are a sham, either medically or through psychological miss-information, like "You are the best ever," but can positively affect commitment.

In order to achieve excellence, individuals must possess or develop high levels of commitment, to the point where one's activity in a sport endeavor becomes their main life focus during a specific, and sometimes over a relatively short, period of time in sport (Orlick, 2008). However, being totally committed may also have detrimental effects on performance, leading to exhaustion or athletic burnout (Weinberg & Gould, 1999). To prevent overtraining, Orlick suggested balancing commitment with appropriate recovery periods, and joyful activities, not related to work.

Researchers have shown that there are several ways to increase the levels of commitment in their chosen activity. Having an inherent passion or love for a given sport leads to higher levels of commitment. Perceiving goals to be worthy and achievable, and believing in oneself, also have been shown to be characteristics of highly committed individuals (Orlick, 2008).

Orlick discussed the important roles it plays in enhancing athletes' self-confidence and commitment levels. Furthermore, according to Harris and Williams (1993), the levels of commitment of athletes can be heightened when certain sacrifices are made, when the time and effort invested is acknowledged and supported by their family and teammates, and also when their commitment is publicly displayed.

In the attempt to explain the attainment of expert performance, Ericsson, Krampe, and Tesch-Römer (1993) proposed a model which postulated that "The primary mechanism creating expert-level performance in a domain is deliberate practice." Deliberate practice was defined as an effortful activity that is motivated by the goal of improving performance. Of interest is that, unlike play, deliberate practice was perceived as not being inherently motivating, and unlike work, it does not lead to immediate social or monetary rewards. Furthermore, the length of time required to reach an expert level in a particular domain was estimated to be at least 10,000 hours of meaningful, goal-directed practice, which virtually defines commitment.

The three best discriminating commitment questions between the international and the national group on the OMSAT-3 were: "I am willing to sacrifice most other things to excel in my sport"; "I am committed to becoming an outstanding competitor" and "I feel more committed to improve in my sport than to anything else in my life" (Durand-Bush, Salmela, & Green-Demers, 2001). This is not surprising since, as has been previously mentioned, researchers have found that elite athletes are extremely dedicated individuals who are willing to do almost anything to become the best, even if this means sacrificing everything else that is important to them for a certain period of time; and unfortunately, as we have recently observed, has led to cheating with the use of illegal substances (Orlick, 2008; Orlick & Partington, 1988).

## Implications of commitment for learning and performing in gymnastics

Almost everyone always sets their New Year's Day resolutions or goals for the year, such as losing weight, increasing their exercise levels or stopping smoking. However, this commitment to practice and competition is not just the task of the gymnast, but considerations must be attributed to the parents and the siblings of aspiring gymnasts and to the coach. In large urban centers where most gymnastic centers are centralized, because of the high costs of equipment and coaching, young gymnasts are most often transported to and from the gyms by their parents, and if their parents are working, by caretakers or grandparents. If the child has siblings, it is rare that they will also participate in the same niche or activity since their time resources are already stretched thin, which makes the social dynamics and the transportation more complex (Côté, 1999). Thus the level of commitment is not just the decision of the aspiring gymnast, but also of the siblings and the parents.

At the Gymnix Club in Montreal, one of the city's most successful women's gymnastics clubs with multiple Olympians and world performers, a unique procedure was undertaken by the founding president and head coach, Nicole McDuff, during the 1970s for gymnasts' selection. She would invite aspiring young girls and their parents to a three–four-hour training session with her elite team, and after the training session she would independently ask both the parents and the young candidates whether they wished to endure the necessary parental transportation to and from the gym, six days a week, for three to four hours of training, as well as the weekend competitions over the next ten years! In more than 50% of the cases, this question regarding commitment was negative and this solved much wasted time which would have been spent with someone who would have eventually dropped out of the sport, whether it be the gymnast or the parent. To support this point of view, the best discriminating OMSAT-3 question of all was, "My sport is the most important thing in my life." As an adolescent or a young child, this may be true, but it seems obvious that this perspective would change with the nature of the professional and daily activities of the family.

But commitment extends at times beyond personal commitment, but to that of a gymnast's team or country. I was present at two events at the 1976 Montreal and the 1988 Seoul Olympics where this commitment level went way beyond what could have been imagined. In Montreal, the Japanese team had already created a dynasty in the men's team, with team gymnastics titles in the 1960s and 1970s. By the time of the Montreal Games, Japan had won four consecutive team gold medals. During the team finals, Shun Fujimoto broke his kneecap on floor exercises. Rather than withdrawing from the meet, he masked his injury on his final events, the pommel horse and rings. On the rings, Fujimoto scored a 9.7 after landing his full-twisting double back dismount with his broken knee cap on the other foot! His score helped the Japanese earn their fifth consecutive team gold, and he is still revered in Japan for his selfless commitment to the team. Another Japanese gymnast who demonstrated this same commitment was Sawao Kato at the 1974 World Championships in Varna. He was Olympic all-around champion in 1968 and 1972, but during his horizontal bar routine he dislocated his shoulder and his arm was out of its socket. Despite this, and with the encouragement of the fanatical Japanese fans, he attempted unsuccessfully to remount with one arm dangling from his shoulder. He tried, but logically he could not continue, and failed, not in his commitment or courage, but for obvious medical reasons.

Another example of what could be arguably called team commitment, mental toughness or stupidity, occurred with Bobby Baun in Canada during a 3–3 tie in the sixth game of the 1964 Stanley Cup ice hockey final. The Toronto Maple Leaf's defenseman was hit with a hard shot which broke his ankle. He returned to the trainer's clinic and they told him that it was surely broken. He said that it could not hurt more, so he told them to just inject him with pain killers and wrap it tightly so that he could return for the overtime. While not being known as a great goal scorer, he scored the winning goal and then pointed out that "This was the greatest *break* in my life!" Now you be the judge in which of the above categories he should be placed: either committed, tough minded or stupid.

The final instance was a bit closer to home for me, since I had been involved with the Canadian team and in mental skill consultations for more than ten years. I met Philippe Chartrand ten years previously when we were conducting the T–T, or talent identification program for men gymnasts across Canada, using more fixed and/or genetic variables. But honestly, anyone could have "eye-balled" out this strong, flexible and engaging gymnast as a "talent" in gymnastics without any other scientific information. Philippe progressed across his career and became the world champion on the horizontal bar in the 1983 Edmonton FISU Games at the age of 21. During the six-week preparation for the 1988 Seoul Olympics, we went through a number of competition contingency plans, one of which was what the team should do if one of their teammates was injured during the competition. The consensus was to step over the body and start focusing on the next event.

During the team finals at Seoul, Philippe dislocated his knee on the vault and everyone thought that he was through. But the team did as trained, ignored him, and went on to warm-up for the parallel bars. But Philippe was no ordinary gymnast, he was the team captain, made of steel and when the other gymnasts were warming up, he shook his leg a bit on the sidelines, put on his white pants and then preceded to the parallel bars, which stunned the whole team. Since he was the last and best man up for the Canadian team, he went through his full warm-up and finished with double and triple somersault dismounts on the parallel and horizontal bars, landing on one leg! Because of him, the team's personal worry diminished and Canada finished in ninth position, Canada's highest placing ever! Now this was commitment to both the team and the country.

## 16.4 Implications of the mental skills foundation triangle for learning and performing in gymnastics

A number of important lessons can be drawn by considering the foundation skills and their crucial interactions. First of all, unless the foundation skills are assessed as being relatively high, there is little chance of achieving success in gymnastics or any other realm of achievement. Concerning goal-setting, if you do not have a map before you, either drawn up by you, your coach or both of you together, chances are that you will end up somewhere else. Start out with what would be your ultimate dream, such as making the national squad, representing your country at the Worlds or Olympics, or winning a medal.

The second element of self-confidence or belief is the glue that brings together the three fundamental components. Later on, recent evidence on national, international and athletes who were medalists, including gymnasts, have nicely tied together these three concepts when considered across their levels of expertise in sport. I have met some lower skilled gymnasts whose dream was to make it to the Olympics, and when they got there, they either did not compete or faked an injury. Obviously, their goal-setting was inappropriately low, of just making team and going to the opening ceremonies, but their pride in their accomplishments was not very high. If they were committed to that dream and their goal was to make it to the top half of the draw or to one final, this dream goal would have been acceptable. It is clear that for exceptional performance in any domain to be achieved, the gymnasts must be committed to extensive practice over many years (Ericsson, 2007). Currently, the commitment levels to practicing are up to 30–36 hours a week of conditioning and training, with only short summer holidays.

## 16.5 Psychosomatic skills

### Stress control and its implications for learning and performing in gymnastics

Stress is an intrinsic component of training and competition. Research has shown that negative reactions to stress, or competitive pressure, can be detrimental to gymnastic performance and that, conversely, positive reactions to stress, arousal or nervousness can lead to enhanced results (Rotella & Lerner, 1993). Murray (1989) conducted a study in which athletes were asked about how they interpreted their pre-competitive arousal levels. Over 70% of them reported that they enjoyed the nervousness associated with competition, that it helped their performance and was a good indicator of their readiness to perform. Rotella and Lerner have thus stressed the importance of developing effective ways to respond to stressful situations which could potentially limit the achievement of a gymnast's goals.

Much research has been conducted on the topics of stress, anxiety and arousal (Gould & Krane, 1993). The three terms have often been used interchangeably in the sport psychology literature. Martens (1977) defined stress as:

> A process that involves the perception of a substantial imbalance between environmental demand and response capability, under conditions where failure to meet the demand is perceived as having important consequences and is responded to with increased levels of the A-state [state anxiety].
>
> *p. 9*

Spielberger (1966) made an important distinction between state and trait anxiety. State anxiety was defined as a situation-specific emotional state that reflects the perceived feelings of apprehension and tension, which are associated with increased or decreased arousal. Conversely, trait anxiety was defined as a stable behavioral predisposition to many situations which are perceived as threatening. Various theories have been postulated to explain the relationships between arousal and performance, or anxiety and performance (Gould & Krane, 1993). Recent findings have led researchers to believe that the multidimensional theory of anxiety contributes greatly to the understanding of the anxiety–performance relationships. This theory predicted that cognitive and somatic anxiety affect sport performance in different manners (Burton, 1988). More specifically, it suggested that there are strong negative linear relationships between cognitive state anxiety and performance, and a less powerful inverted-U relationship between somatic anxiety and performance. Gould and Krane raised a need to conduct more research in this area before valid inferences can be made from the assumptions of this multidimensional theory of anxiety.

Rotella and Lerner (1993) made some insightful comments when considering responses to competitive pressure:

> The more similar the abilities of the competitors and the greater the importance of the event, the more likely that high levels of pressure will be experienced by athletes facing that challenge. They must be able to consistently perform at or near peak levels when exposed to the highest levels of competition.
>
> *p. 528*

Murphy and Jowdy (1993) found that imagery and mental practice techniques are important components of stress management. Lazarus and Folkman (1984) found that subjects used imagery interventions to become more familiar with effective strategies for coping with stress. In other studies, imagery techniques were successful in reducing different types of anxiety, such as medical anxiety, test anxiety (Wine, 1971), and also in changing sports behavior.

Hardy (1990), who has worked with the British national gymnastics team, developed a model which explained the complex interactions between *physiological arousal* (muscular tension, sweating, increased heart and respiration rates) which has been shown to be beneficial for increases from a sleepy to an optimum emotional state of performance. But with further increases in arousal, it causes performance decreases. This has been termed the inverted-U hypothesis, a physiological state that has been well known for more than 100 years (Yerkes & Dodson, 1908). The inverted-U phenomena demonstrates physiological arousal states which can be captured by daily experienced life activities by yourself and others. For example, when you wake up in the morning, your physiological state of arousal is at its lowest, and this would not be a good time to do your income taxes since your performance would be necessarily low. However, once you get up and about, your physiological state would improve and you start thinking more clearly. If, however, you decide to go for a vigorous run, and then upon returning immediately do your taxes, this might also not be the optimal period because of the fatigue to carry out this activity. This physiological activity level would follow a bell-shaped curve, and if you had no further stresses, it would slowly return to the early morning state and you would fall asleep.

However, *cognitive anxiety* is another issue, since it deals with worry, doubt, fear and other external pressures. In gymnastics, it can be related to worrying about physical and mental readiness, the importance and consequences of doing well, making the team, considerations of strength of the competitors, or even the nature or nationality of the judges. However, when *physiological arousal* is combined with *cognitive anxiety*, or states of doubt or worry, a threshold is reached and the arousal curve drops dramatically, unlike the smooth physiological arousal bell curve – it is like a pool ball falling from the table. The shape of the curve looks like an oval piece of cheese which is halved, with a gradual increase in the curve and then a 90-degree fall in the center. This results in immediate performance decreases. Thus Hardy (1990) introduced the idea of the *catastrophe model*, which could provide another explanation for choking in stressful situations in sport. This state can be reduced after physical relation, and the mental restructuring of the gymnast's thoughts in which confidence and control are regained.

An example of the application of the catastrophe model could be seen if a gymnast was asked to do a kip to a perfect handstand on the bars, swing down and do a free hip circle back to the handstand position. In normal conditions, the gymnast would be in a calm state of *physiological anxiety*, and if asked to do a series of five or ten, each ending in a perfect handstand position, a well-trained gymnast could probably perform this with ease. If, however, the gymnast was instructed to do ten free hip circles to handstand perfectly, and failure to do so would mean their elimination from the Olympic team, *cognitive anxiety* would rise considerably and their performance might break down in a catastrophic manner because of the worry caused by this increased cognitive anxiety.

It is very difficult at this point to ascertain the nature of this scale of the OMSAT as being a true mental skill. In one instance it was found to discriminate well between athletes (Salmela, Monfared, & Mosayebi, 2009) and in another (Durand-Bush, Salmela, & Green-Demers, 2001) it did not. In both cases, it was found to have weak correlations with the other scales in the inventory, except for refocusing; however, the scale still yielded acceptable levels of internal consistency. This scale has not been found in any of the other multi-dimensional inventories, but this does not mean that it is not an important skill or capacity that athletes need to develop to achieve high performance levels. Examples of two items for the stress reaction scale were: "My body becomes unnecessarily tense in competitions" or "I have experienced problems in my performance because I was very nervous."

Another explanation for the inconclusive results could be that the scale is sport-specific, that is, it may apply more to athletes involved in sports like gymnastics, alpine skiing, luge, bobsleigh and whitewater kayaking, in which physical stress is a major factor in training and competition. Yet, items in this scale such as "I am afraid to lose" and "I am afraid to make mistakes" apply to every athlete, not only to those participating in stress inducing sports.

What must be remembered is that gymnastics is a sport that is unnatural for human beings. The human vestibular, or balance system in the inner ear, is designed for individuals to remain upright, balanced in a standing position with the head on top in the normal anatomical position. However, many gymnastic movements, with the handstand being the most obvious one, require that the gymnast performs upside down, which, as a beginner, is stress inducing. For this reason, many of the early photographs of Olympic champions from the 1920s until the 1950s showed the gymnasts in an arched position in the handstand position so that the head could remain in a somewhat normal, but less stressful, vertical position. Therefore, gymnastics cannot be considered, in relation to stress, as a normal human sport activity, such as running, walking, throwing or jumping. It is inherently, biologically, anatomically and psychologically, stressful!

## Fear control and its implications for learning and performing in gymnastics

On this subject, psychologist David Feigley reported his far-reaching observations on fear in gymnastics:

> For some gymnasts, fear is the major psychological barrier preventing learning and success. It prevents the learning of new skills and retards the improvement of already learned ones. It causes athletes to feel helpless and out of control of their lives. Fear erodes athletes' feelings of self-worth and competency and destroys the fun of the sport. Fear even ended the careers of otherwise successful gymnasts who quit because they were unable to cope with its constant presence.
>
> *Feigley, 1987, p. 13*

However, overcoming fear is one of the major reasons for certain types of individuals to be attracted to the sport of gymnastics. Feigley outlines that gymnasts must "… perceive that they can effectively assume individual responsibility and personal control over the fears they encounter in their sport. Successful gymnasts, feel in control of their lives" (p. 13). However, fear cannot be eliminated from high risk sports such as skiing, ski jumping, motor sports and gymnastics. It is fear that attracts the moth to the flame – close enough is fun, but too close can be a disaster. Many gymnasts try to eliminate fear, but it is a natural reaction to the risky skills that gymnasts perform and is actually a necessary component for the advancement of gymnastics.

Cogan (2006) related: "Fear helps a gymnast maintain enough focus to perform difficult skills safely. Therefore, a gymnast's goal should be to work with fear, rather than to eliminate it" (p. 644). Often fear is the result of carrying too much excess mental baggage, or negative thinking, from their daily lives by thinking too much. Gymnasts should learn to drop this luggage mentally, by opening their hands and lowering the luggage to the ground. Another strategy, suggested by an Australian, Peter Terry, is to maintain their favorite attention grabbing music on their iPods until just before the competition – a strategy that eight-time Olympic gold medalist, Michael Phelps, did during the 2008 Beijing Olympic Games in swimming.

Feigley (1987) astutely points out:

> A gymnast who is never afraid is either foolish or ignorant. Sophisticated gymnasts experience fear on a regular basis. Those who are successful at their sport have developed the means of dealing with their fears so that the risks are controlled. Unfortunately, some athletes try to deny their fears because they: a) view fear as a weakness; b) are afraid to look foolish; or c) believe that they are the only one in the group who is afraid.
>
> *p. 14*

Herein, lies the role of a skilled coach or mental trainer.

Kerr (1997) referred to this as *reversal theory*, by which it is not only the degree of fear or stress control that may affect performance in gymnastics; there are individual differences which individuals may sustain in their *interpretations* of fear or stress which may play even larger roles. We know that gymnasts may be highly activated, and initially this state could be interpreted as being negative and stress inducing. However, the coach or mental trainer can help reverse the perception of the gymnast's mind set from being fearful to one of pleasant excitement.

In 1989, I accompanied a group of 13–16-year-old male gymnasts in Lilleshall, England, with the coaches' tasks of submitting them to intense technical training, and I provided mental training for six coaches and 12 athletes who were our hopefuls for the 1988 Seoul Olympics. Besides the mental training sessions, the mission was to learn and develop the new prescribed skill difficulties to help qualify Canada for the Olympic Games.

Learning new elements necessarily meant dealing with new fears about being injured. When I did the theoretical part of dealing with the various types of fear, e.g., realistic versus unknown fear, the class became quiet. We specifically were dealing with the Tkatchev, a relatively new skill on the horizontal or bar. The movement was quite scary since it was initiated by an overhand giant swing and about three quarters from the initial vertical starting position, the gymnasts' body sharply arched and then flexed which resulted in a counter-rotation of the giant swing and propelled them backwards over the horizontal bar with no visual guidance and, hopefully, a successful re-grasp on the other side of the bar. The fearful elements were contacting the bar with their body. When I specifically asked about what their most fearful concern was regarding learning the Tkatchev, one bright gymnast said the following: "My teeth on the steel bar!"

Later that day, in this fully secure, modern training facility, our young hopefuls did the lead up exercises which included the pre-release "timers" of a beat and an arch, but no one let go of the bar! Then one lanky French-Canadian, Benôit, after doing 25 timers, went for it! He released a bit early and went about two meters above the bar, dropped and hit the back of his thighs on the bar, which flipped him harmlessly backwards into the foam landing pit, and he surfaced laughing and grinning.

The worst fear-inducing alternative had occurred, and he was still alive, happy and uninjured! Benôit had turned his fear into team excitement, and transformed these awfully negative emotions into a higher positive state of excitement within 10 seconds. Within the 20 minutes that remained before our lunch break, this transformation spread amongst all of Ben's teammates, and every one of our young squad either released their grasp over the bar and landed in the foam pit, or successfully re-grasped the bar for another attempt. Such was the beautiful mental link between fear and excitement.

However, there are also a number of non-rational fears involved in gymnastics which can be best resolved by the coach discussing either the biomechanics of the movement or its possible performance consequences:

> Another type of mental block is losing a skill that a gymnast has performed alone or in competition without difficulty … One day she feels a little disoriented and does not throw the double full on floor quite right, or her timing is off … From there. Her performance deteriorates until she cannot do the skill at all.
>
> *Cogan, 2006, p. 645*

So how does a skilled gymnast transform a well learned skill after years of practice to a fear-inducing element? To help illustrate one example of the above situation, I was invited to consult with a women's coach whom I knew, whose daughter was the provincial junior champion and a champion at the nationals in one month. She told me that her daughter was afraid of doing a back walkover on the beam, a skill that she had been doing in competition for eight years. When visiting the gym, I asked the gymnast to do a back walkover ten times on a mat. She did them perfectly. I then asked her to do the same on the low beam which was about 25 cm from the floor, which she also did wonderfully.

Then I heard a scream from across the gym from her mother, who was also her coach, telling her that she had a competition in one month, that she had done this movement thousands of time on the high beam, and that she needed this movement in the compulsory exercises. I stopped my intervention and approached the mother/coach and said: "Do you want to lose a gymnast or your daughter? She must be coached by someone else!"

Most gymnasts are proud when they accomplish a new element and then excitedly return home to share their success with their mother. But when the gymnast's coach is their mother, there is shame, fear and no place to hide. Ten years later, they both approached me at a national championships and thanked me for my very brief, but helpful, intervention. She was from then on coached by another colleague and received a full gymnastics scholarship to an American university, where she did very well, and she loved the experience. Such a mental skill learning effect for such a small, but important, intervention!

## Relaxation and its implications for learning and performing in gymnastics

Relaxation is a method often employed to decrease arousal or worry. Although several different types of relaxation techniques exist, they are all variations of those developed by Jacobson (1938). Zaichkowsky and Takenaka (1993) made the following statement regarding Jacobson's progressive muscular relaxation (PMR) technique:

> It is clear that the mastery of the Jacobsonian relaxation techniques results in reduced levels of anxiety, muscular tension, and physiological arousal. Research on the efficacy of the specific PMR technique for improving athlete performance, however, is still quite sparse.
>
> *p. 521*

Effects of relaxation have often been studied in combination with other arousal control techniques such as deep breathing, meditation, and cognitive techniques including imagery and self-talk (Zaichkowsky & Takenaka, 1993; Uneståhl, 1975). Relaxation techniques have not only been used to regulate arousal, they have also been used to control anger, reduce muscular tension, promote assertiveness, concentration and confidence. Relaxation techniques can be divided into two different categories: *muscle-to-mind* and *mind-to-muscle* techniques (Harris & Williams, 1993). Jacobson's PMR, which involves tensing all muscle groups progressively before relaxing them, falls into the former category. However, this method is lengthy and somewhat boring for gymnasts, and does not include essential cognitive rehearsal. On the other hand, transcendental meditation would be categorized as a mind-to-muscle technique.

According to Harris and Williams (1993), relaxation skills must be practiced on a regular basis. Although some individuals may take longer than others to develop these skills, most people are able to observe improvements after a couple of sessions of practice. Harris and Williams emphasized the importance of being able to relax completely and quickly. They reported that through deep relaxation, athletes can detach themselves from the environment, allow their central nervous system to regenerate physical, mental and emotional energy, and create a base for learning *quick* relaxation. They defined quick relaxation as the ability to relax within a short period of time. It can also be an effective strategy to regain a full focus during competition, and to return to a balanced, controlled state of mind after competition.

A tried and tested combined relaxation script has been successfully applied with thousands of athletes in a number of countries and it takes less than 20 minutes. It includes a brief muscular contraction, followed initially with a script that is spoken by either a coach or a mental trainer (Uneståhl, 1975). Initially, with this muscle-to-mind procedure, preferably the gymnasts should be lying down barefoot on a comfortable surface, such as the floor exercise mat, wearing comfortable, loose clothes with their feet apart and their eyes closed. This form of cognitive–behavioral relaxation has proven to be effective with single subjects or with large groups. It initially involves a single, but intense, muscular contraction of the non-dominant hand and the taking in of a long, deep breath which they must hold as instructed by the mental trainer in a loud, forceful voice while calling out "Harder, harder, hold it, hold it" (for 10–15 seconds) and then the command to relax is spoken in a lower and calmer tone.

The gymnasts are then introduced to the mind-to-muscle or cognitive or mental component and are instructed to: "Notice the differences between the contracted and the un-contracted arm." This directed suggestion element begins in a quiet voice. The script continues to further guidance to relax. The readers could find more about them by referring to Uneståhl (1975).

The mental skill of imagery is later introduced in the cognitive skills section, but for the continuity's sake, now is the time to include some other imagery skills. Another script is next introduced:

> Imagine that you are in front of a set of 10 stairs. As you descend the first stair you become even more and more relaxed. Now at the second one, you get deeper and deeper and you become more and more relaxed, in control, certain and comfortable.

Continue this until they have reached the tenth stair.

Now the phase of the gymnastic-specific imagery begins. This suggestion is then given:

> Look in front of you and you will see a large, sky blue bubble. This is your personal mental bubble which keeps stress, worries, and other distractors (e.g., other gymnasts, coaches or parents), away from you and you can use it either in training or competitions. It is nicely decorated and in the middle is a soft comfortable reclining chair where you can now sit, or lie down, and practice some elements in your routines. In your bubble, these movements are always carried out effortlessly and perfectly, without fear or anxiety.

Let them enjoy this experience, which they do, and then a number of gymnastic skills can be introduced which should help them imagine with confidence and certainty.

For the women, the sequences may include approaching the beam in a competition, mounting and performing with ease, on a beam which appears wider, their first difficult elements, or the final tumbling sequence with a stuck dismount, which they can easily repeat 10–20 times within a short period of time. For the men, it is usually the pommel horse where I start, and the instructions include executing flat, double leg circles and with fast hands, and finally ending with a high, controlled and stuck dismount.

I will then bring them back up the stairs with instructions to become increasingly awake and aware as they go up each successive stair. I will say "With each stair, you will become more and more awake and feel more comfortable, sure and confident." The number of each stair is given and they are told that they are becoming increasingly awake and aware, and from the eighth to tenth stair I ask them to move their fingers a bit, which is sometimes quite difficult because of the deep relaxation, then slowly move their ankles and bend their knees. They are then asked just to lie there and enjoy this altered state of calm control, and when they are ready, to sit up slowly. The whole process takes from 14–17 minutes and the gymnasts really enjoy it, which I always discovered in the short debriefing and feedback sessions held immediately afterwards.

After two to three sessions, I attempt to remove myself from the interventions because I do not want them to be dependent upon me, and I ask them just for the contraction, then have them imagine the red ball of light to relax their body, then to concentrate upon their breathing, go down the stairs and enter the bubble to rehearse. When they are ready, they should begin to reactivate themselves. I have done this whole process with over 200 athletes from 15 sports in Iran in a number of uncomfortable environments, such as sitting on hard chairs in a classroom, and there was only one athlete who could not, or would not, release, let go and relax.

It is also possible in critical situations to use the one *second relaxation technique*, and this is often seen in close-ups of players in ice hockey or soccer in penalty shoot-outs where, just before performing, the shooter or the goalkeeper takes a single deep breath and expires fully. While not as dramatic as the above techniques, it does physically relax the shoulders and helps adopt a positive focus, rather than one based upon worry. This is also clearly seen in gymnastics before mounting on an apparatus, or before a final tumbling run on the floor exercises.

### Activation and its implications for learning and performing in gymnastics

Sometimes athletes are under-aroused, or are mentally flat, before or during a competition. In these instances, energizing techniques would be most effective to increase their chances of achieving successful results in gymnastics. Many energizing techniques have been used by coaches and athletes (Anshel, 1990; Harris & Williams, 1993), despite the limited research on their efficacy (Weinberg & Gould, 1900).

Energizing techniques used in the past have included the use of rapid breathing techniques, running, stretching, yelling out loud, or listening to stimulating music or videos, or energizing imagery, verbal cues, pep talks and doing energetic pre-competitive workouts (Zaichkowsky & Takenaka, 1993). Athletes have also been known to *psych themselves up* by drawing energy from their environment, that is, from the crowd, their opponents, teammates, their flag or from hearing their national anthem. Zaichkowsky and Takenaka have suggested that athletes may also energize themselves by transferring negative emotions, such as anger, fear, disgust and contempt, into positive emotions such as excitement for their achievable challenges which injects positive energy for their performance goals.

It is believed that certain factors have to be considered before arousal-regulating techniques are implemented and effective. According to Zaichkowsky and Takenaka, it is important for both coaches and athletes to develop a sense of awareness that will allow them to detect if and when arousal levels need to be altered. Second, coaches and gymnasts need to be aware of techniques which are most effective for each gymnast. Coaches have to realize that there exist individual differences in gymnasts' responses to arousal-regulation techniques.

As has already been shown in Figure 15.1, the movement dynamics between the various events in gymnastics sometimes require a calm mental state, which should occur on the balance beam or the pommel horse, and then there are the efforts required for the vault and floor exercises. However, it was found with the OMSAT-3 assessments that most athletes had no problems activating in competitive situations, but needed to learn to relax, both physically and mentally.

For example, the sport of weightlifting provides some interesting examples, since the athletes need to exert maximum power within a couple of seconds. The coaches often scream aloud at their athletes, smack their faces and give them smelling salts to pep them up to lift the maximum weights. In gymnastics, these procedures may be effective on events which require power, but not so on those which require control of the finer movements for optimal performance.

I have seen a video of Mike Tyson and his famous coach, Gus D'Amado, prior to an amateur fight in "Iron Mikes's" boxing career, and he was crying in front of his coach. He was a boxer who would be feared by all of his rivals in professional boxing, but he could not get ready and activated for an amateur fight! I have also already described the lack of mental energy, or activation, of a two-time Iranian World champion in tae-kwon-do and a silver Olympic medalist prior to an international competition. So, activation levels cannot be taken for granted, even at the highest levels of sport!

## Implications of all psychosomatic skills for learning and performing in gymnastics

Emotional or psychosomatic skills are often the most dominating aspects of success in gymnastics. From my own perspective, fear was the greatest factor which limited my performance. During the 1960s and 1970s, there were few gyms which had eight inch landing mats, hand grips for the horizontal bar were primitive and were not constructed for gymnasts to maintain contact with the bar, there were no three- to four-foot landing pits, but only a few two to three inch mats, which made attempting difficult tricks very dangerous. I have flown off the horizontal bar a number of times and struck my head upon a wooden floor. This poor spotting equipment did not motivate me to try risky exercises so often.

Doing tumbling on wooden or tile floors always limited doing somersaulting movements, mostly to competitive situations, and even the one inch foam pads were not that much better. Thus, in my case at least, fear was a limiting factor. I had no problem in relaxing, energizing, but fear and stress reactions definitely limited my performance, and I applaud those gymnasts at the international level who went through these events with success.

## 16.6 Cognitive skills

### Imagery and mental practice and their implications for learning and performing in gymnastics

The terms imagery and mental practice have been used interchangeably in the sport psychology literature and thus will be discussed together in this section until they are either experimentally or functionally shown to be independent variables. Both concepts are normally practiced outside of the competitive context, while the mental skills of focusing and refocusing are typically executed during training and/or competitions.

Murphy and Jowdy (1993) emphasized the importance of carefully distinguishing between the two terms. Corbin (1972) defined mental practice as the: "Repetition of a task, without observable movement, with the specific intent of learning" (p. 94). On the other hand, Suinn (1993) associated mental practice with techniques which included thinking about a skill by visualizing or *feeling* it. This involved self-talk throughout the steps of a skill, imagining oneself or another individual executing a movement, and incorporating auditory, proprioceptive and emotional elements while visualizing the perfect way of performing. With this definition, it was specified that mentally practicing does not imply the engagement in imagery or mental rehearsal. In the research and interventions using the OMSAT-3 (Durand-Bush, Salmela, & Green-Demers, 2001), the two variables were separated, since it was believed that the generation of images was different than their integration into practice.

Cogan (2006) also indicated that imagery could be used as an energizing or activating force when a gymnast reached a mental block for performing a standing back tuck somersault on the beam: "Instead of focusing on her legs being bolted to the beam, she should imagine her legs feeling like pistons that could propel her off the beam and through the air" (p. 652).

Not all gymnastic movements are of equal difficulty, and imagery when combined with emotions should be directed to the most demanding elements just prior to mounting an apparatus. In this way, each routine is *segmented* into its key difficulties and mental practice, and should be directed with emotion regarding each of these key skills. More comprehensive imaging would probably be appropriate for the pommel horse or the balance beam because of their complex and continuous nature.

Suinn (1993) pointed out that imagery and mental practice techniques can be used to achieve a variety of goals, such as enhancing correct responses, simulating competitive environments, eliminating anxiety or negative thoughts. Mental practice is reported to have beneficial effects on the learning of new skills. Certain factors were identified to mediate the effectiveness of mental practice at different skill levels (Murphy & Jowdy, 1993). Individuals who were better imagers, that is, those who can create clear, real, controlled images, were benefitted more from mental practice than their less able counterparts. Moreover, it was suggested that experienced athletes may benefit more from mental practice than novices (Suinn, 1993).

Mahoney and Avener (1977) found imagery perspective to be an important factor having a possible influence on the effectiveness of mental practice. Imagery outcomes were identified as another mediating factor of mental practice. They suggested that negative imagery, that is individuals rehearsing a task with a negative outcome, has a debilitating effect on performance. One explanation was that "Negative mental practice affects performance through its impact on dynamic properties of the subjects such as confidence, concentration or motivation" (Murphy & Jowdy, 1993, p. 230).

Some have viewed imagery rehearsal as a procedure individuals use to optimally arouse or physiologically activate them for a given performance. However, results of studies attempting to test these ideas have been inconclusive. In some studies, the introduction of arousal in imagery rehearsal did not enhance performance. One explanation for this has been that the arousal included in the imagery increased the activation levels beyond optimal levels (Murphy, Woolfolk, & Budney, 1988).

One popular skill that has often been cited in sport psychology is that of *self-talk*. From our perspective, this is a sub-component of imagery, mental practice, focusing and refocusing and is the sub-vocalizing of key words or phrases requiring special attention during the performance of a skill in a routine. For example, during a somersaulting vault, the imagery or mental practice may include a short verbal cue or trigger like: "Attack the horse and drive back your heels." In more complex events with multiple skills, the performance should be segmented into automated, or well learned, components with an emphasis on the more difficult or dangerous elements, where self-talk during practice or performance is most critical. For example, for a skilled gymnast, a glide kip on the parallel bars requires little attention nor self-talk, while a twisting double somersault requires full attention and vigorous self-talk for greater mental practice with the sub-verbal cues of "attack now," "wrap tightly" or "stick the landing."

The most important thing about imagery and mental practice is making the images clear and precise, and then putting them into practice in the most efficient way which may include the anticipated actions, thoughts, emotions and results. Sylvie Bernier, a Canadian who won the 1984 Los Angeles Olympic gold medal in springboard diving, reported that she felt foolish when she first attempted mental imagery and practice which had been suggested to her by a sport psychologist. But she began to do it anyway, hidden in her bedroom.

After a few days, when she was on the board, she had a strange revelation that she had already practiced this difficult dive in her mind, but now it was much simpler. As she approached the Olympics, she was often doing 80% mental practice and only 20% physical training, which of course was much easier on her body. She expanded her mental framework and imagined herself winning the gold medal, then walking up to the medal podium and hearing the playing of "O Canada," her national anthem, when she was awarded the gold medal! It is truly powerful what the mind can accomplish.

But you must first establish what the most effective form of imagery and mental practice is for you. Is it best to see yourself, as if on a video, or to see the routine as you would actually perceive it internally in real time? I know that I used the latter technique before going to sleep in high school and my palms would be sweating as I went through each routine.

Self-perceptions of one's gymnastic movements are not always the most accurate measures of what your body is actually doing. Most often, the coaches provide critical feedback of the demands of the key elements of your gymnastics performance. However, with the advances in video technology, it is very simple for gymnasts to observe themselves on their cell phone videos for an accurate performance assessment, which either can be assessed alone or with their coach. This form of external imagery must then be transformed and internalized into personal imagery, so that the feelings of the movements and the emotions can be rehearsed during mental practice sessions.

The first time that I ever saw a gymnast use this technique was in 1963 at the University of Michigan with twice National Collegiate Athletic Association parallel bar champion Arno Lascari. Lascari was way ahead of his time. He had someone film every one of his routines over the season with a Super-8 film camera and he would develop the film every week and study his technique on each event to help both his mental and physical practice.

Since not all the elements in a gymnastic routine are normally of equal difficulty, it is essential that there are sequencing and priorities established, for it is essential that there is a plan for the next day for the successful execution of the full gymnastic routines. The imagery and mental practice processes should be employed mentally in the same way that we function in our professional or academic lives or sport activities.

For example, when performing on a concrete or wooden floor in the free exercises, I began with a balance Y scale, which required no attention, and then I did a front handspring front somersault, which required my full attention, given the predictable heel bruises which would occur on these hard surfaces. In the middle of my routine, I included even simpler tumbling, balance and flexibility sequences and then my attention turned to the now simple, round-off, flic-flac, back somersault. It is clear with the increased complexity of current routines accomplished on sprung floors that the imagery sequencing processes are much more detailed and the complex skills are increasingly difficult, but the imagery concepts remain the same.

Another related concept in terms of training, imagery and mental practice is that un-supervised normal athletes spend more practice time on the easy elements that they can easily do. I spent a lot of time doing forward and lateral splits, since I was naturally flexible, but should have devoted more time to the dangerous tricks which terrified me, because of the number and the thinness of the mats and the primitive state of our hand grips, compared to the current ones. While I still love and appreciated my loving, high school gymnastics coach, Don Cochrane, and my inspirational football coach, Ivan Livingstone, they were not demanding enough on me. In retrospect, being a tough cookie, I would have run through thicker walls if they asked me to. I would have responded more to the coaching philosophy of the great Dallas Cowboys' football coach Tom Landry: "To get men to do what they don't want to do in order to achieve what they want to achieve. That is what coaching is all about" (Irwin, 1993, p. 1).

## Focusing and its implications for learning and performing in gymnastics

Most mental skills and techniques, including goal-setting, relaxation, activation, imagery and mental practice, require excellent attentional control or focusing abilities. In fact, the ability to consistently attend to the most relevant tasks and environmental stimuli is often referred to in the popular literature as focusing, a central aspect of athletic performance (Boutcher, 1993; Nideffer, 1987). Over the years, this construct has been studied from various perspectives, including information-processing and social psychology.

Within the information-processing perspective, attention has been characterized as "The ability to switch focus from one source of information to another and the amount of information that can be attended to at any one time is limited" (Boutcher, 1993, p. 252). Researchers using this perspective have concentrated their efforts mainly on selective attention, capacity and alertness. Selective attention is believed to play a central role in both the learning and performing of sport skills. It occurs when individuals process certain amounts of information at a particular moment, while other information is screened out or ignored. Research has revealed that selective attention can be voluntary or involuntary, and can take place in a wide variety of behavioral situations.

Attentional capacity is another aspect of focusing that has been investigated. Studies have indicated that there is a limited capacity for processing information at any one time, and that this capacity is even more limited when individuals are engaged in controlled rather than automatic processing. Athletes performing multiple tasks or attempting to focus on more than one source of information could thus experience reduced performance. This often occurs when coaches give too many detailed instructions just before a gymnast's performance. Shriffin (1976) revealed that although control processing may be dominant in the early stages of learning, it will eventually be replaced by automatic processing if the skills are to be performed in an effortless and efficient manner.

Arousal is a third aspect of attention that has been examined through an information-processing perspective. Studies have shown that when emotional arousal is increased, focusing fields are reduced and their ability to respond to peripheral stimuli may be decreased (Easterbrook, 1959). Boutcher reported that this attentional field-narrowing phenomenon may be important to consider in sport performance, since many sport skills are performed in aroused states. Nideffer (1987) reported that when high arousal levels persist, attention may be directed inwardly to dimensions such as fatigue and pain, and the external environment is not considered anymore.

Nideffer demonstrated that individual differences exist in one's ability to use different attentional processes and concluded that individuals possess various styles. It was suggested that the attentional demands of any sport will vary along two dimensions: width (broad or narrow) and direction (internal or external). A broad external focus should be used to focus attention on a wide area of the external environment, such as the tasks of a midfielder in soccer. Whereas a broad, internal focus should be adopted to direct attention internally on various strategies and past experiences, such as for a coach during team games. A narrow, external focus was most useful to focus attention on a single aspect of the external environment, such as in rifle shooting, while a narrow, internal focus was effective for attending to specific internal images or bodily cues, such as balancing in gymnastics. In the sport of rhythmic gymnastics (RG), narrow internal attention must shift between balance and turning moves, to narrow external cues for catching the thrown object into the air. Research findings on the relationships between attention and performance have incited researchers to develop optimal attentional training programs for athletes. To help the future development of such programs, Boutcher (1993) suggested that: "The precursor to successful attentional control during actual performance may be the establishment of a series of behavioral, physiological, and cognitive cues that optimally prime both body and mind for the ensuing skill" (p. 262).

Cues or behaviors that lead to optimal attentional states have interested many scholars. In sport, optimal attentional states have often been termed peak performance or *flow states* (Csikszentmihalyi, 1975). Flow states have been associated with positive emotions, extremely focused attention and total connection or oneness with the task at hand.

It has already been mentioned in the previous mental imagery/practice section that it is important to separate into discrete sequences the competitive performance that gymnasts are about to accomplish. They are no longer lying in bed thinking about what they might be doing, they are now focusing on the competitive battlefield and are about to go to war!

Hardy Fink, now in charge of the FIG Coaching Academies, related the following to me regarding Boris Shaklin, six times gold medalist in the Olympics for the USSR. Historically, Boris was probably the first gymnast to systematically prepare himself mentally in a consistent fashion, even before mental training was even conceived of:

> I could always recognize him as he prepared to perform a routine. He stood facing away from the apparatus with his arms hanging at his side and his head hung forward and upper back rounded – kind of a position of relaxation, but also ideal for focusing on the task ahead. I guess he probably also had his eyes closed because that seemed natural in that position.
>
> *Personal communication, February 23, 2009*

Nideffer (1987), who was the first to coin the term focusing, reported that it was essential to direct one's attention towards something that was important in the sporting environment. It has been demonstrated that gymnasts can either focus externally on the crowd, the judges or other competitors or upon their mental images or on the state of their own emotions.

In gymnastics, the physical stresses are not the same, but the mental ones are. I was at the World Championships in 1987 in Rotterdam with the Canadian men's team trying to qualify for the Seoul Olympics the next year. Canada was currently in ninth place after the compulsories, with the top 12 teams qualifying. Our youngest gymnast went onto the floor for the warm-up, and it was clear to me from 25 rows away that this lad had lost his mental control. He was not focusing on the apparatus and was wandering around looking lost. To put it concisely: "His lights were on, but nobody was home!"

Obviously, the stresses of the moment overcame his mental resources for this situation. Canada fell from ninth to 13th place because of our disastrous team pommel horse performance, and we had not considered a back-up strategy. It was only because Cuba withdrew from the Olympics, for either political or financial reasons, that we eventually qualified. At Seoul, with better technical and mental preparation, we ranked again in ninth place, our best international and focused classification. We had learned to better mentally train the team, how to direct our focus to the most important external and internal cues, and how to avoid the many distractions.

A final example is presented on how our mental training program helped the Canadian team in the 1988 Seoul Olympics. Since we had extensively trained using the relaxation, imagery and mental practice procedures previously outlined, we borrowed an idea from the mental room or bubble concept, and asked them to create a *team bubble* while competing.

All six gymnasts were asked to remain together to be protected from outside influences like other competitors, coaches, the crowd and judges. After their successful performances, many of the gymnasts reported that the team bubble concept really kept the team focused.

## Refocusing and its implications for learning and performing in gymnastics

Researchers have also used distraction theories to try to explain the relationships between attention and sport performance. They have postulated that individuals lose their focus because certain factors attract their attention to task-irrelevant cues. According to Boutcher (1993), processing task-irrelevant information could explain performance decrements in both competitive and less important sport situations. An unlimited number of factors may cause athletes' attention to be directed towards irrelevant stimuli. Some of these identified sources were the presence of worry, self-awareness, family members, teammates, coaches, competitors, scores, officials, media, sponsors, close relationships, unrealistic expectations, and changes in performance levels. Two of the best predictive OMSAT-3 items for the refocusing scale were: "Errors generally lead to other errors when I am competing" or "I think that it is difficult to gain control if I am disturbed during my performance."

Orlick and Partington (1988) reported upon the refocusing skills of the 1984 Canadian Olympic delegation as follows:

> Those athletes who performed at their highest levels consistently had excellent strategies for getting back on track quickly when things didn't go well, or when faced with distractions. Those who were less consistent appeared to need more work in this area to improve the consistency of their high level performance.
>
> *p. 117*

Because distractions in sport and in life are numerous, researchers have emphasized the importance of developing distraction control or refocusing plans. According to Orlick (2008): "Refocusing appropriately before, during and after the competition are some of the least practiced, but most important skills for high-performance athletes" (p. 49). It is believed that to obtain consistent performance in training and in competition, athletes must regularly develop the skill of distraction control. Refocusing is also consistently the weakest of all of the 12 OMSAT-3 scales, probably because it is neither well understood, nor taught.

Anshel and Payne (2006) introduced two interesting strategies regarding the refocusing dimension, i.e., what to do when things go wrong, using both approach and avoidance coping. Approach coping, much like in Hans Selye's (1974) classic dichotomy of the "*fight* or *flight*" syndrome, is to fight. The difficulty with the fight strategy is that it takes time and requires performance analysis of oneself or of the opponent. The avoidance strategy, or flight, means that stressful, instantaneous events, such as lucky scores by the opponent, unscored points or bad calls from the judges, must be ignored or put aside and one's total focus should remain on what lies ahead, especially in gymnastics. Failure to deal with a distraction in gymnastics on one apparatus can sometimes carry on to performance errors on the next event.

The fifth best discriminating OMSAT-3 item on the refocusing scale between the Canadian international and national athletes (Durand-Bush, Salmela, & Green-Demers, 2001) was: "If I start losing, I find it hard to come from behind to win." Elite athletes in this study had better refocusing skills than their less elite counterparts, and more specifically, they had the ability to redirect their attention to the task at hand when faced with important distractions to have success in competition. Orlick and Partington (1988) found similar results in their study. They reported that the ability to refocus after distractions varied considerably among elite athletes, compared to the other skills that were assessed through interviews and questionnaires.

Unlike in team sports where a bad call or spectator reactions may cause athletes to misdirect their attention or focus from the current situation to inner negative thoughts, in gymnastics it is different. When a gymnast falls from an event, they can only see and feel the whole negative panorama of disturbances on this event, and often center upon their emotional reactions, such as anger and frustration. They also can focus upon their thoughts of the future consequences of reducing their final rankings. Also, in gymnastics, it is different from other sports since they only have 30 seconds to deal with the error. The most frequent cause of distractions is falling from the apparatus, which is most often from the pommel horse or from the beam. Lower skilled athletes will just chalk up and while still in their disturbed emotional and cognitive states they immediately remount in a worse mental state and then fall off again.

A disconcerting event occurred halfway through the women's all-around competition during the 2000 Olympics in Sydney. An alert Australian gymnast noticed that the vaulting horse, which was specified to be set at a height of 125 cm, had been set 5 cm too low. The officials immediately raised the horse and allowed any gymnast who had already previously vaulted the opportunity to vault again.

It was too late, however, for the Olympic favorite and leader at that time, Svetlana Khorkina from Russia, who had already vaulted and crashed earlier in the competition. Distraught that she had ruined her chances for the Olympic all-around gold, she went to the next event, the uneven bars, and also fell there. Later, when the height error in the horse was discovered, she was told she could redo her vaults, but with her low score on the bars, her all-around hopes were already dashed. This instance brought to light two cognitive mental skill errors she had committed. The first dealt with competition planning, where either she or her coach should have taken the responsibility of verifying the height of the horse, and the second one was in refocusing, or getting back into a normal competitive state of mind, as has been previously outlined.

Ken Ravizza, who has worked with women gymnasts in California, outlined a *Six R* strategy of dealing with adversity in 30 seconds or less:

1. *React.* First of all, curse or swear, but immediately rid yourself of this emotional state, because this anger will never help when you remount. You can quickly call yourself an idiot and the then forget about it.
2. *Release.* Try to forget about the fall, or as Orlick (2008) termed it, to "park" the incident, as if putting it outside in your parked car to be dealt with later, or "change the channel" from a bad movie to a good one.
3. *Review* (if you have time). Try to figure out what went wrong and what you should do when you remount, i.e., either repeat the same move, or start with the next sequence. Your choice will affect the judges' final evaluation, since correctly repeating the missed element will demonstrate your confidence.
4. *Regroup.* Try to get into your normal pre-competitive state and assure that you have placed your shoulders back and that your chest or sternum is up.
5. *Ready yourself.* Refocus on the movement cues that you will shortly perform, check your hand grips and take a hard look at the apparatus that you are about to conquer. An extra five seconds is rarely deducted if it looks like you have a plan.

6. *Relax and go!* Take a deep breath, and exhale slowly and blow away all of the anger or frustration that you had built up and prepare yourself for a positive performance, like the thousands that you have already accomplished in practice and in other competitions.

## Competition planning and its implications for learning and performing in gymnastics

Researchers have suggested that planning is a very important step in the achievement both of peak performance or flow states. Williams (1986) reported:

> Each athlete must learn how to create consistently at competition time the ideal performance state (thoughts, feelings, bodily responses) typically associated with his or her peak performance. Rarely will this occur if pre-competition preparation and competition behaviors are left to chance or, to good and bad breaks.
>
> *p. 314*

According to Williams, establishing pre-competition and competition routines not only help athletes develop a consistent performance approach, they also help control their arousal levels. It was recommended that athletes organize their internal thoughts, feelings, mental images and the external environment in a way that they can maximize their feelings of control, and cope with unforeseen events. Developing pre-competition and competition plans is a long process which requires constant evaluation and refinement. Williams indicated that trial-and-error experimentation, combined with consultations with a coach or a mental trainer, may be necessary before athletes can establish their most effective pre-competition and competition routines for achieving optimal performance.

One interesting empirical study demonstrated the importance of competition planning and other mental skills in high level sports. Orlick and Partington (1988) assessed 235 Canadian Olympic athletes' mental readiness through questionnaires and individual interviews. It was found that these elite athletes: (a) possessed high levels of commitment; (b) set clear short- and long-term goals; (c) did imagery and simulation training; (d) focused and refocused under distractions; (e) had an established mental training plan that was used and refined throughout the season; and (f) had clearly established mental plans for competition, which included pre-competition and competition mental plans, distraction control plans and post-competitive, constructive, evaluation plans. Orlick and Partington found that between physical, technical and mental preparation, mental preparation was the only variable that significantly predicted the athletes' actual Olympic rankings.

Planning for a given competition can begin months, even years, before the actual event. Gymnasts must plan what are the elements that they must consistently master based upon not only current gymnastic norms but also upon what types of routines will be performed in the next two, three or four years. A dozen years ago, no one could anticipate the multiple, consecutive tumbling runs, both forwards and backwards with various twists, that are currently being performed. If the plan for the next years includes a single tumbling run with a brief pause for the performance of the next series of simpler skills, they need to renew their plan.

In the events that led up to a selected target competition, a list should be made up of necessary equipment for their kit bag. The night before, all of these items should be in the bag, which should be packed by the gymnasts and not by their parents. A sense of responsibility must be developed and personal ownership of the gym bag is essential. For instance, has it been planned on packing an extra pair of broken-in hand grips in case one of the old ones breaks in competition? Are both sets of competition tops or pants packed in case there is an on-site change by the coach? Are there extra wrist wraps and adhesive tape, or hand spray like "Tuff Skin," in case of a minor hand rip?

Cogan (2006) and Cogan and Vidmar (2000) elaborated a series of competition planning activities that used competition *simulations* to ready their gymnasts. This occasionally included "mock meets," where the gymnasts were dressed in their full competition gear, a "one touch" three-minute warm-up with judges, scores and sometimes even awards. Other coaches I have talked to have included playing in practice, recorded crowd sounds on loud speakers during a competition, as well as cheering and booing after the events to help gymnasts develop their refocusing skills. Cogan also suggested videotaping the routines of all gymnasts to be used in non-gymnastic sessions for technique corrections by the team and the coaches.

All of the previous mental skills including goal-setting, relaxation, activation, imagery, mental practice, focusing and refocusing can be integrated into the pre-competition plan. Each mental skill element must be practiced and not left to chance. Most importantly, the plan should include practicing the elements that the gymnasts are having difficulty with, and not only those that they have already mastered.

Gymnasts may have had a specific time of day for mentally rehearsing their sport, but they might not have had a plan of what they were exactly going to rehearse. When I was coaching the men's university team at Laval University, on Thursdays before the Saturday competition I would hold a competition simulation or "mock meet." The gymnasts were dressed in their full competitive uniforms, would march in after having a general warm-up to be greeted by four beautiful women judges who knew nothing about gymnastics. The gymnasts always competed better on Thursdays and the judges gave me sheets of paper that were either blank or had some bogus score on them.

Prior to the 1988 Seoul Olympics, we had a very successful three-week training camp which included some unique contingency planning procedures. The national coach and I prepared a list of things which could occur or go wrong at the Olympics and how to deal with them. For example, "What do you do if a teammate gets injured during the competition?" Some said that they had to help him. *Wrong answer!* Step over the body of this warrior and move on to the next event, because there are plenty of sport medicine doctors and physiotherapists on site, and you cannot lose your focus. This actually did occur. Or another, "What do you do if a rival head judge tries to ignore you to unsettle you and does not give you the hand signal to begin your routine?" Some said wait until he acknowledges you. *Wrong answer!* They were told just to look at the red light at the judging table and when it was lit, they could begin their routine. Many mind games occur in these important events between both gymnasts and judges and you need to have a plan.

## 16.7 Implications of cognitive skills and performing in gymnastics

Cognitive, or thinking, skills are central to exceptional performance in gymnastics. Gymnasts must be able to imagine what the ideal gymnastics movement is – either imaged as watching a perfect video or as from their personal perspective of actually performing the skill. This is usually accomplished after doing relaxation methods. They should be able to see and feel the ideal movement patterns, once learned.

However, the mental training on these skills must be carried out in the gym, while performing, and at home in a rested environment. Gymnasts often tell about vivid images, with accompanied sweating of the palms and an increased heart rate. These cognitive skills go well beyond actual practice situations, and can be effectively used in competitions.

Refocusing is the least taught mental skill and refers to what to do when something goes wrong in competition, usually from falling from an apparatus or receiving a lower than expected score on a given event. There are a number of strategies outlined in the text, which facilitate regaining of the optimal mental state to continue performing well, and recovering from these negative incidents. In the above sections, some strategies are provided which include rapid readaptation to the current event, and how to quickly "get back on track" and resume their normal routine and forget the fall and its consequences on their current performance.

Competitive planning includes all of the above mental skills, including the foundation, psychosomatic and cognitive dimensions. This means all of the states which proceed may occur during competition, and an assessment during post-competition, both positive and negative, must be planned for in order to succeed.

## 16.8 CONCLUSION

Competitive planning relates to all of the above mental skills. How does a gymnast feel that they will devote at least ten years of their lives to excel in their sport? More importantly, how will you invest in achieving excellence, even if it is not in the sport of gymnastics? You cannot go anywhere if you do not have a plan, otherwise you will end up somewhere that you do not want to be!

Gymnasts must plan on what they want to think about, how they want to feel, how they have to prepare themselves, organize their equipment and act during both training and competition, on a daily basis. One of the most effective ways is by maintaining a daily list in which are included their goals, intended emotional states and how to control them, and their necessary thinking states and how to react when something goes wrong. A small notebook will suffice, and notes should be taken on how to improve for the next practice, or competition. These simple steps can also be used for the rest of your career, when you start your profession.

# 17

# CLOSING THE CIRCLE OF MENTAL SKILLS TRAINING

## Providing mental skills feedback to gymnasts

*John H. Salmela*

Since it is believed that all of the 12 OMSAT-3 (Ottawa Mental Skills Assessment Tool) scales are interactive, but measure distinct variables (Durand-Bush, Salmela & Green-Demers, 2001), this permits more effective interventions in the mental training process. This learning can also occur by trial and error experiences in pressure situations, but this latter process is more haphazard and lengthy. It was evident from the interventions with the Iranian delegation, both coaches and athletes, at the Asian Games that even former world champions still sought out consultations regarding their perceived mental training short-comings, which indicated the ongoing importance of having the availability of mental training consultants on-site at important competitions. Apparently, the best in the world still want to become even better. A generic overall view of providing mental skill feedback to athletes will now be elaborated with reference to the most recent interactions with the Iranian delegation in the 2006 Asian Games.

Ericsson's (2007) belief is that skill learning is an ongoing life-long process, if sufficient deliberate practice is accumulated. Unfortunately, the data on the number of practice hours was not collected on these Iranian champions, but the number of years of training for the women approached the ten-year limit for expert performance, while the men surpassed the women by three years.

The athletes in each discipline were first given an overview of the advantages of the mental skills orientation, in relation to fixed mental traits or capacities, such as personality or IQ tests, which are commonly used by clinical psychologists in Iran. The total delegation of athletes was administered the adapted Persian version of the OMSAT-3, either in a classroom, or, if there were time constraints, at their residence. All groups received at least one team mental training session along with their coaches from the senior author and his two Iranian associates, which included cognitive–behavioral relaxation, and when deeply relaxed this led to the visualization procedures described in Chapter 16. Additional individual sessions were also scheduled for those athletes who wished further consultations, and who were usually the best performers.

All athletes and coaches were shown their individual and team OMSAT-3 profiles in a graphic form on a laptop and were asked whether they made sense of them. They also received a printed version of their own personal profile for their individual study. Their strengths and weaknesses were pointed out and their perceptions validated the assessments and they were shown that improvements in one scale could affect positive changes in others. For example, by learning that they could take control of their state of stress through the relaxation techniques they were taught, this could positively affect their focusing, refocusing, stress control and confidence levels. They were also shown that competitive planning could have powerful effects on almost all variables. However, since the delegation was so large and I did not speak Farsi, my interventions were more limited. Also, the women on the teams, in a follow-up questionnaire, responded more positively than did the men, since it was my belief that they appreciated more the attention that was given to them (Salmela, Mosayebi & Monfared, 2007). However, possibly due to shyness or cultural norms, I was never asked for personal consultations with them.

Many interventions related to mental skill training were also carried out by my trained associates in 15 disciplines over a three- to five-month period, which included the administration of the OMSAT-3, the interpretations of individual and team profiles, and subsequent individual and team interventions, based mainly upon relaxation and imagery skills training. Individual consultations based upon the profiles of the selected athletes in nine sports were conducted over the month prior to the Games.

Interestingly enough, an elite athlete in Canada in the Durand-Bush, Salmela and Green-Demers (2001) study wrote on his retest questionnaire that:

> After having filled out this questionnaire for the second time, it is very clear to me that the majority of the items were consciously answered in a different way from the first to the second time. I have, therefore, been tremendously influenced by the events (particularly the sporting ones) in which I have participated when answering these questions.

During the initial testing, this subject obtained means of 5.67 and 4.78 on the self-confidence and commitment scales, respectively. When retested, this subject scored higher on the same two scales, and the mean scores had increased to 5.83 and 5.78.

It is important to note that in between the two testing sessions, this subject won the Canadian Championships in his sport, and it was therefore not surprising that his level of self-confidence and commitment had increased. This reflects the fluctuating nature of developing skills in relation to situational outcomes. It also points out that the simple administration of the OMSAT-3 is an *intervention* in itself. The items revealed some aspects of mental preparation that some athletes perhaps had never considered, or in other cases, reinforced some mental skill dimensions that they had discovered from trial and error, but were never taught.

As an example of the latter, in 1992, on my first visit to Iran, I had a university professor translate the items of the original OMSAT-3 to a group of four women rifle shooters. When I asked them for feedback on the test, most of them replied that they could never believe that their inner most thoughts and feelings regarding shooting had already been written down on paper! This again indicated that these mental skills can be learned by trial and error, but it just takes longer, it occurs at a later stage in their careers and that there may be gaps in their knowledge.

For instance, during a graduate level class in Brazil, I had a dozen students complete the OMSAT-3 and a computerized graph of their mental skills profile was generated of what they were doing, thinking and feeling when they were at their best in sport. They then exchanged their profiles with a classmate and were to do a diagnostic report of their colleague for the next week's class.

During the debriefing the following week, one expert indoor soccer (or futsal) coach showed extremely high levels in his foundation skills, but his cognitive skills were extremely low. When questioned about these low values, he said that coaches in Brazil, when he was a developing athlete, never mentioned imagery, mental practice, focusing, refocusing or competitive planning. This fact alone is a good justification for the teaching and practicing of these mental skills at a young age to developing players to increase their chances of winning.

It was thus concluded that the OMSAT-3 is more a state or situation-oriented inventory that will yield different scores when the athletes fill it out depending on the time of the season or the stage of their career in sport. Because it was anticipated that the athletes' responses would vary within the various training and competitive phases that they progressed through in a season, it was suggested that the reliability of the OMSAT-3 scales should not be assessed entirely upon test–retest coefficients. The strengths and weaknesses of athletes' states of mental training should be considered as a reflection of the mental skills and practices over that period of time.

## 17.1 CONCLUSION

It is scientifically evident that high scores on the three foundation skills are prerequisites for exceptional performance in gymnastics and in many other of life's challenges. If a gymnast does not have clear and, sometimes, public expression of their short- and long-term, achievable goals, where do they think that they are going? If they do not have the self-confidence that they can achieve these goals, either from within, from their coaches, family or a mental trainer, chances become even slimmer for success. Finally, if they are not committed to investing over years of thinking and training, by themselves, or with their coach or mental trainer, they should get themselves measured for a shovel and work as a gardener! Fournier et al. (2005) demonstrated that most of the selected OMSAT-3 mental skills, with the exception of stress control, which may be another form of mental toughness, developed in their upbringing, or to genetics, showed improvements after a season-long training with mental skill interventions.

It was concluded in Orlick and Partington's (1988) study that crucial elements of success for the best athletes in the world were: (a) total commitment, (b) quality training which included daily goal-setting and imagery training, and (c) quality mental preparation for competition which entailed developing pre-competition, competition focusing and refocusing, and post-competition evaluation plans. Similarly, it was found by Mahoney, Gabriel and Perkins (1987) that the top level athletes in their study: (a) were more confident, (b) were better able to focus before and during competitions, (c) were less anxious, (d) had better internally focused imagery abilities, and (e) were more committed to excelling in the sport than competitive athletes in the lower ranks. Finally, with a large sample of international Iranian athletes, the total number of OMSAT-3 scale differences was almost identical between Canadian national and international level athletes. However, the scale values were reduced to two, stress control and refocusing, for selected and non-selected international athletes, and then only the one scale of stress control for medalists and non-medalists, which indicated that training and the relative levels of expertise of the athletes showed up in their mental training abilities (Salmela, Monfared & Mosayebi, 2009).

Many of the above findings clearly demonstrate both the scientific and practical reasons for the selection of the OMSAT-3 methodology. First of all, the OMSAT-3 provides quantitative data and interpretations via the internet at a reasonable price. Other instruments are more limited in their scope, and it regroups the theoretical underpinnings of these measurements. From these perspectives, I personally feel that the present comprehensive perspectives from sport psychology research dedicated to the single sport of gymnastics have allowed the collation and integration of new insights for practical applications, neither of which existed when I edited my first book on gymnastics, more than 30 years ago (Salmela, 1976).

The final mental skills interactions required to close this performance circle will now be addressed. Research in Canada (Durand-Bush, Salmela & Green-Demers, 2001), France (Fournier et al., 2005) and Iran (Salmela et al., 2009) has shown a number of relevant elements which contribute to the understanding of mental skills and exceptional performance in sport. The most important elements are their interactivity, or how learning within one domain influences one or many other mental skills.

First of all, the research on the development of expertise in sport began qualitatively with champion swimmers and tennis players (Bloom, 1985), quantitatively with musicians (Ericsson, Krampe & Tesch-Römer, 1993), qualitatively in a variety of sports in different age groups in sport (Côté, Baker & Abernethy, 2003), and with multiple world and Olympic champions (Durand-Bush & Salmela, 2002). Based upon his personal experiences with expert sport performers, Orlick (2008) suggested some key elements of mental skills demonstrated by champions, such as commitment, belief or self-confidence and commitment. However, the links between the interactions of expertise in sport and a variety of mental skills were first demonstrated by Durand-Bush, Salmela and Green-Demers (2001) using quantitative measures with the OMSAT-3 and by Durand-Bush and Salmela (2002) with multiple Olympic or World Champions using interviews.

What was clear from the OMSAT-3 research was that expertise levels were clearly related to almost all of the assessed mental skills between the more experienced international athletes and national level athletes, who were skilled, but who had not yet competed internationally. Based upon Orlick's assumptions, the foundation skills were prerequisites for excellence. In the OMSAT-3, the three foundations skills were somewhat modified from Orlick's intuitions. In the Salmela et al. (2009) study, it was demonstrated that with the international level Iranian athletes, whether they were selected for the Asian Games or not, or were medalists or not, all showed no significant differences between their foundation skills, as was demonstrated with the Canadian sample. The question is: Does success in sport increase the levels of goal-setting, self-confidence and commitment, or the reverse? My guess would be that success improves these levels of mental training, especially given the lack of interventions in sport psychology in Iran.

Within the categories of the nine remaining psychosomatic and cognitive skills, several issues remain unclear. While it has been shown that high levels of anxiety narrow the focus or attentional fields of performers (Easterbrook, 1959; Nideffer, 1987), there is very little research to support other interactions that could close the circle of all of the mental skills in sport performance. One obvious place to begin such investigations would be the consideration of the role of relaxation with the other scales, since it was close to being significant between the selected and non-selected Iranian athletes. Durand-Bush (1995) showed that relaxation was positively correlated with ten of the OMSAT-3 variables, with the exception of fear control, while self-confidence was positively correlated with all of the other scales.

Interestingly enough, goal-setting was positively related to most scales, with the exceptions of stress, fear control and refocusing, while commitment was not positively correlated with stress and fear control. Obviously, stress and fear control, as well as refocusing, require further attention both by researchers and mental trainers.

It also seems obvious that competitive planning, when combined with relaxation and imagery, would also have positive effects upon almost all of the variables, with the exception of fear control (Durand-Bush, 1995). Good coaching from both the conceptual and biomechanical perspectives, given the appropriate audio-visual, material and technical resources, could greatly reduce fear factors in gymnastics.

Much work in research and interventions in gymnastics have been accomplished worldwide using the various sport sciences, but there still remain many questions to be resolved, with perhaps the most in sport psychology and mental training.

# 18

# PERCEPTION OF COMPLEX MOVEMENTS

*Alexandra Pizzera*

## 18.1 Learning outcomes

After reading this chapter, you will be able to:

- Describe the different human senses and relate them to sport specific examples.
- Explain the different types of tasks representing specific gaze control strategies specifically with regard to gymnastics.
- Describe how gymnasts use their visual perception to control aerial skills.
- Select and use appropriate methods involving visual perception to support the learning process as a coach.
- Name advantages of using other senses such as the auditory sense to support the learning process and optimization of complex dynamic skills.

## 18.2 Introduction and objective

While observing movements in sports, I perceive different aspects of the movement, depending on the role I have in the respective sport. As a coach I might try to detect errors and then give feedback and tips for error correction. As a gymnast I might put myself into the movements I'm observing and compare it with my own. As a judge I might evaluate the performance of the skill according to the judgment criteria. As a teacher I might think about an appropriate method to teach such a skill in physical education. The objective of this chapter is to take up these different viewpoints and provide an overview on the perception of complex movements by using theories from psychology, examples from sports and specifically gymnastics. Up-to-date research and practical recommendations with respect to gymnasts, coaches and judges will be provided.

## 18.3 Perceptual senses

Perception is the process of selecting, organizing and interpreting information through our senses (Myers, 1998). Humans gather information through different senses, giving them the opportunity to take up information from the visual, auditory, kinesthetic, olfactory and gustatory senses. To take an example of gymnastics, let's consider a gymnast performing a split leap on the beam. She might be looking at the beam and therefore control where the foot should be located to land (visual sense). In addition, the gymnast might infer from the sound of the landing (auditory sense) how well the skill was performed in terms of height and force. Or she might use the feeling of the knees and hips to control the split in the air and ensure a larger angle between the two hopefully straight legs (kinesthetic sense). The olfactory (smell) and gustatory (taste) senses are scarcely used in sports. However, there is increasing consent that athletes are dependent on continuously receiving accurate and reliable information from the environment before, during and after skill execution. This information is not exclusively restricted to visual information, although this is the primary sense used. To successfully perform a skill, give feedback, manually guide athletes or judge movement quality, athletes, coaches and judges need to make use of all senses to get a holistic picture. Such cross-modal or multimodal effects have been shown to exist for optimal information processing. For instance, experiments showed that perceptual organization in the auditory modality can influence perception in the visual modality (Vroomen & de Gelder, 2000). The integration of multiple senses is therefore considered central to adaptive behavior, enabling the use of increased salience by multisensory redundancy (Calvert, Spence, & Stein, 2004). Or in our example, the gymnast might use the information provided by the kinesthetic sense during the flight phase to improve visual information take-up immediately before landing.

## 18.4 Visual perception in sport

In most everyday actions, people use their visual perception to guide their movements. Visual perception is the "[...] process of picking up environmental information, which instils form (of objects, surfaces, events, patterns) within a perceiver" (Williams, Davids, & Williams, 1999, 6). For instance, if we want to pick up a cup, our eyes look towards the task-relevant object (the cup) and the information which is taken up is used to guide the corresponding action (moving the hand towards the cup). Therefore, for such manual tasks, but also other types of tasks, there is a spatial and functional relation between our eye movements and our whole body or body segment movements. In sport, information processing and the use of information is intended for actions in sports and/or cognition. Depending on the perceiver and his or her role in the sport, visual perception serves different functions (see Figure 18.1). A gymnast may use the visual information of the bar for action, for instance while learning and performing a kip. A coach might also use visual information for action, however not with the intention to perform the skill but to spot the gymnast at the right time and right location to provide manual guidance. Or the coach might use visual information for *cognition* to give feedback (see Myers, 1998, 87, for a definition). Similarly, a judge might use visual information for cognition to evaluate the quality of the skill.

With the knowledge of the perceiver and the role of perception for these different perceivers, how do we know that a person is actually using visual or optical information? Somehow attention seems to be directed at a specific location. Attention refers to three different processes (Williams, Davids, & Williams, 1999), which will be explained by using an example from gymnastics. Imagine a female gymnast standing on the floor, waiting for the beginning of the first tunes to start her routine (*alert attention for action*). She then continues with some split leaps with leg change, focusing specifically on the angle between

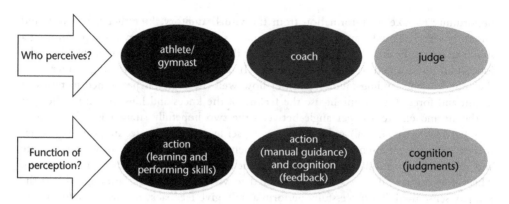

**FIGURE 18.1** Overview of the different roles of a perceiver in sports and the function of perception depending on the role.

the legs (selective attention), while also paying attention to her toes being pointed and moving in accordance to the music (divided attention). The process of selectively guiding attention to specific areas of interest is of major importance in sports, especially in gymnastics, and will therefore be covered in more detail in the following section.

## Visual selective attention

If our attention is directed at a specific location, based on a visual cue, this is termed visual attention and usually gives us a hint whether a person is attending to or perceiving optical information. As we look about a scene, we have periods of stability in our gaze and periods when we move our gaze rapidly between objects and locations. For these two periods, two types of gaze are defined: *fixations* and *saccades*. A fixation is described as the gaze being held on an object or location for 100 ms or longer within 3° of visual angle (Carl & Gellmann, 1987; Carpenter, 1988; Fischer, 1987). One hundred milliseconds is the minimum amount of time necessary to become aware of and recognize stimuli. Saccades are defined as eye movements shifting quickly from one tracked or fixated location to the other. Due to these fast eye movements (between 60 and 100 ms), information cannot be picked up and is therefore suppressed. However, the information gained during fixation or tracking is maintained while performing saccades, allowing us to perceive a stable and coherent scene (Bridgeman, Hendry, & Start, 1975; Irwin & Brockmole, 2004). Saccades therefore are an indicator of a shift in attention, because it is impossible to make an eye movement without a shift in attention.

Considering that in many sports, athletes, coaches and judges need to make decisions in split seconds in very complex and changing environments, focusing their attention on only the most crucial or relevant sources of information is a necessary skill for these people. But how do skilled athletes perceive visual information from these highly dynamic and complex environments to perform such exquisitely timed and consistent actions? Performers usually pick up the relevant information by engaging in the process of visual search. For instance, goalkeepers in soccer usually do not have enough time to initiate their movements in time to catch the ball, after reacting to the predicted flight direction of the ball. For this, they need to make use of advance visual information. But which cues are relevant and should be used? Studies by Savelsbergh and colleagues have shown that expert compared to novice goalkeepers are more accurate in predicting the direction of the penalty kick based on a more efficient search strategy involving fewer fixations of longer duration to less disparate areas of the display (Savelsbergh et al., 2002, 2005). So this process of visually guiding selective attention on relevant environmental information is crucial to prepare own actions or adapt already executed actions.

## Gaze control framework

The example of how a goalkeeper controls his or her gaze and attention for anticipation and movement execution was just one of many examples that exist in sports. Gaze and attention are controlled in distinct ways depending on the task at hand. With respect to the sports environment, rather than each sport affording unique gaze and attention characteristics, there are three main categories of tasks for which gaze and attention show specific control processes (Vickers, 2007):

1. Gaze control in targeting tasks.
2. Gaze control in interceptive timing tasks.
3. Gaze control in tactical tasks.

*Targeting tasks* involve situations in which gaze and attention need to be directed at a target in space (e.g. a basketball hoop) with the aim of controlling an object intended to reach the target area. A very famous example of what an important role gaze and attention play for successful targeting tasks is the quiet eye phenomenon. In numerous tasks, such as basketball free throw, darts, rifle shooting, golf putting and billiards, studies have shown that the final fixation just prior to the first observable movement of the hands into the shooting action is longer in experts compared to novices and is also related to success. This final fixation has been termed the quiet eye and is at least 100 ms up to 1 s long in experts. For an overview of the literature on the quiet eye, see Vine and Klostermann (2017). In gymnastics, such targeting tasks can be found especially in aerial skills in which the target may be a bar or the beam. For instance, a gymnast may perform a Jäger salto on the uneven bars by which the bar needs to be released and regrasped. For this regrasping action, high spatio-temporal demands are placed on the gymnast to reach the target and grasp it successfully. Although this kind of task in which the target is stationary and the body instead of an object is moving towards the target is different from the kind of targeting tasks mentioned in the gaze control framework by Vickers, the spatio-temporal coordination demands are considered similar. This idea was also taken up by one study in horse-jumping (Hall et al., 2014) and one in skateboarding (Klostermann & Küng, 2016). In both sports, the athletes (plus horse) approach the target or obstacle with the aim of jumping over it. The results showed the functional adaptability of the visuomotor system with different gaze strategies used depending on the environmental constraints in highly dynamic situations. Especially the look-ahead gaze strategies were observed, showing the use of anticipatory cues to guide movements (Patla & Vickers, 2003).

If, however, the object travels towards the performer, it needs to be recognized, tracked and controlled as it is delivered, approaching and received by the gaze and attention system of the performer. Such an *interceptive timing task* affords the control of dynamic visual information and involves, for instance, a ball being thrown at the performer with the task to catch it. For objects to be intercepted, visual information needs to be picked up to move the hands to the correct location at the correct time (e.g. when catching a ball with the hands). For this processing of dynamic visual information people use the so called time-to-contact, or tau (Lee, 1980; Lee et al., 1983). As an object approaches a person, its size on the retina of the receiver becomes increasingly larger. However, at some point, an object may not be thrown with enough precision towards a person, so that the receiver needs to run towards the object to catch it. Such processing of dynamic visual information to intercept an object while moving has been termed the angle of elevation (Schmidt & Lee, 2008). The receiver needs to run in a way that the angle between the horizontal and the link between him/her and the ball remains constant. In gymnastics, numerous examples for such interceptive timing tasks can be found from the perspective of a coach. Especially when gymnasts are learning new skills, coaches are continuously asked to provide manual guidance. For these spotting tasks, coaches also need to adapt their location and timing in order to be at the correct spot at the correct time. Specifically during the learning phase of aerial skills, unpredictable motion of the gymnast needs to be taken into account, placing even more demands on the correct use of this dynamic visual information. As with the baseball player attempting to catch the ball, a coach will then use time to contact and angle of elevation information to 'intercept' the flying gymnast and bring him or her safely to the ground.

*Tactical tasks* are often found in both targeting and interceptive timing tasks. For instance, in sport games it is important to read and comprehend complex patterns of moving objects, like players or balls, to make the best decision often under severe time constraints (e.g. a playmaker). Such pattern recognition and tactical awareness is also important in activities such as ski racing or speed skating, which represent the tasks of gaze control during locomotion. Here, while moving one's body through the environment, people fixate stationary objects to take up information on surfaces, obstacles or perceptible landmarks to navigate through, between or over them. The apparatus vaulting table represents such tasks for gymnastics elements. The aim is usually to adjust the strides during the approach run in a way that the springboard is hit at the correct spot, then, to take up visual cues to fly over the table, and to use the table to generate even more height for the second flight phase. This gaze control framework by Vickers (2007) shows that visual attention and the control of gaze and attention is highly task-specific. Therefore, different foci are advantageous; however, these are not restricted to a specific sport, but rather dependent on the task (e.g. targeting or interceptive), situation (e.g. stationary, dynamic, complex) or the person (e.g. athlete, coach, judge or referee).

## Aerial skills

A type of skill that is typical in gymnastics and that hasn't really been considered in the gaze control framework by Vickers are aerial skills. What about the relationship between gaze behavior and movement goals in complex skills incorporating flight phases and rotations about one or more body axes? If you ask one of the most talented gymnasts worldwide, the answer is rather surprising, considering the well-developed gaze control of experts mentioned before in other sports. During her tumbling runs, Simone Biles (several times World champion and Olympic champion 2016) says, "the only thing she sees is the colors of the ceiling and the floor, whizzing past in revolving blurs. When she vaults, she sees nothing at all" (Wiedemann, 2016). So in such complex aerial skills, does the angular velocity of the head surpass the visual system's ability to take up information? A number of studies have looked at this aspect. If the gymnasts' statements were true that they don't see anything while tumbling, there should be no difference in the somersault performance between vision and no vision conditions. However, both Davlin, Sands, and Shultz (2001b) and Luis and Tremblay (2008) found more stable landing positions during vision compared to no vision conditions. In addition to using gaze to control for landing, visual orientation cues were also used to orient the body in space (Rézette & Ablard, 1985). Yet, as the aerial skill became more complex (back tuck somersault with a full twist), visual cues seem to have been used less easily and probably replaced by vestibular cues. Therefore, the visual system primarily provides the athlete with information to control the landing of aerial skills, which in turn is thought to result from controlling body orientation during the flight phase in a prospective manner (Davlin, Sands, & Shultz, 2004; Hondzinski & Darling, 2001). In another study in gymnastics, evidence was provided that gymnasts use optical information during or even before the handspring on the vault, during the run-up (Heinen et al., 2011). The aim of the study was to examine if the position of the springboard is used as an information source in the regulation of the handspring on the vault. Gymnasts performed handsprings on the vault, while the position of the springboard was manipulated. Without letting the gymnasts know, the springboard was placed 10 cm closer (– 10 cm) or 10 cm further away (+ 10 cm) from the vaulting table with respect to the individual springboard position of each gymnast. The results showed that, on average, gymnasts placed their feet in the same position on the springboard irrespective of the manipulation. So even without knowing about the manipulation, the gymnasts visually took up the changed springboard position and adapted their approach run. The analyses revealed that this adaptation occurred during the last three steps of the approach run. In addition, the different springboard distances to the table had effects on different parameters throughout the movement phases (e.g. shorter first flight phase, longer second flight phase). Taken together, the springboard position is one relevant visual information source in gymnastics vaulting, showing large effects on the movement phases following the take-off. Such links between visual information sources in the environment and the resulting regulatory processes of athletes are crucial for performance and therefore important for athletes and coaches in terms of training programs designed to optimize performance.

## Guiding visual perception during the learning process

Besides the function of motor control in aerial skills or vaulting, such focus on visual information also offers a wide range of possibilities for coaches. Manipulating perceptual information in the environment during skill acquisition may support action changes with learning and enhance flexibility and greater motor control. Following Williams, Davids, and Williams (1999), a coach has three possibilities to work with perceptual information:

a. To increase the sensitivity to other potentially useful information sources (auditory, haptic, proprioceptive).
b. To direct attention of learners to key information sources (e.g. by using videos).
c. To direct attention away from visual cues to decrease reliance on central vision.

With regard to (a), a coach may direct the attention of the gymnast to the rhythmic footstep sounds during tumbling with the aim of adapting their tumbling sounds to those of expert tumblers. An example of directing attention away from visual cues (c) would be to tell gymnasts to walk over the beam with their eyes closed. This would increase their focus on tactile information in their feet, which in turn could potentially be more important for learning some skills in which visual cues cannot be used. The predominance of visual information during skill execution sometimes even hinders the use of other potentially useful information sources. Directing attention of learners to key information sources (b) is a method often used by coaches and will therefore be discussed in more detail.

The earlier-mentioned studies have shown that gymnasts generally use visual information for different functions (e.g. orientation, landing). However, making use of such knowledge in a way that learners can benefit from the guidance of their gaze behavior is a different aspect. A study by Heinen et al. (2012b) nicely demonstrated how the gaze behavior of gymnasts can be manipulated during the backward salto dismount from the uneven bars and then influence the movement behavior. Expert gymnasts were asked to fixate a light spot on the landing mat during the downswing phase. From the individual landing distance, the light spot was varied systematically in a way that the light spot was located 15 or 30 cm further away or nearer to the high bar. The results revealed that different movement parameters changed as a function of the light spot location. The gaze direction during the downswing motion seems to serve as a prospective type of control, influencing the swing motion itself and as a result also the take-off conditions. These, in turn, effect the flight phase and consequently lead to a particular landing location (see Figure 18.2). Therefore, directing the gaze towards the landing area serves as an anticipatory type of control when performing dismounts on the uneven bars. So if you want your gymnast to increase landing distance, directing his or her gaze to a landing spot further away might automatically result in a larger landing distance.

Another question one might ask is what exactly 'happens' with the gymnasts' gaze during the execution of the handspring on the vault. More specific, how can coaches direct gaze behavior of their gymnasts during the handspring to enhance performance? A study with 30 gymnasts aimed at addressing this question by directing the visual attention during

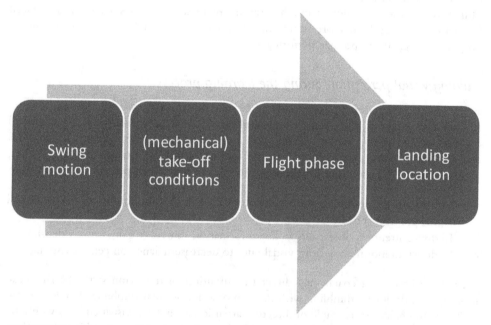

**FIGURE 18.2** Overview of the interrelationship between gaze behavior and the different movement phases during the uneven bars dismount.

the execution of the handspring on the vault (Heinen, Vinken, & Fink, 2011). A perceptual instruction group was verbally guided to fixate specific areas during the different movement phases. For example: "Try to fixate your gaze on the top side of the vaulting table/hands during the repulsion phase." The perceptual instructions + visual cues group was verbally and visually guided with red dots (15 cm) as visual cues attached to the fixation spots. Although all three groups showed a similar performance immediately after the acquisition phase (six training sessions), the two instruction groups were able to maintain their performance level two weeks after the training ended, whereas that of the control group declined. The fixation location areas instructed by the coach were taken from expert gymnasts who had a high movement quality of the handspring. The results showed that teaching learners to use an eye movement strategy employed by skilled athletes can facilitate the acquisition of such a complex motor skill. The fact that performance was retained even after training ended additionally provided evidence that this strategy can be developed through practice.

To sum up, visual attention is used by athletes during motor execution, serving the function to anticipate, orient or control the landing of the body. The coach can support the learning process of the gymnasts by directing their gaze behavior or instructing specific gaze strategies. Especially beginners learning new skills are at first unable to interpret information from velocities or limb displacement on the basis of proprioception and therefore might need additional cues to rely on. In addition they might not be able to know which visual cues in the environment are relevant and important for successful skill execution.

## 18.5 Auditory perception

So far I have focused on visual perception, because this is the most prominent sense we use in our everyday life, but also in sports. However, auditory perception also plays a role, although we are not always aware of it. For instance, while walking, we produce individual footstep sounds. So if you are sitting in your office you may infer from the type of footstep sounds approaching your office in the hall who this person is. Let's now span the bridge to gymnastics. Imagine you are a coach and you are standing with your back to the floor and your gymnasts are practicing their tumbling skills. Would you be able to identify individual gymnasts from their footstep sounds produced during tumbling? Or would you even be able to judge performance quality based on these footstep sounds? And do gymnasts hear their movement sounds during execution and if yes, do they use them for motor control? Similar to visual information perceived in the environment, acoustic information is also picked up in a way that we have different acoustic representations of dynamic movements. These acoustic representations are thought to link to and overlap with our motor representations, which in turn are essential for motor performance (Agostini et al., 2004).

A study in hurdling provided evidence that hurdlers are able to perceive and identify their own performance based on the movement sounds they produced (Kennel, Hohmann, & Raab, 2014). And not only are they able to perceive their sounds, they use them to control their movements. Kennel et al. (2015) asked athletes to clear four hurdles with the typical four-step rhythm. While running, athletes wore headphones which presented either their own running rhythm acoustically, however with a slower tempo than currently running, or they heard white noise, or were prevented from hearing any noise. Running performance did not change in the white noise and control condition, however it decreased significantly in the delayed condition. The fact that the delayed rhythmic sounds disturbed running performance indicates that athletes need and make use of the auditory sense for hurdling. Agostini et al. (2004) used this idea to enhance performance through training with acoustic feedback. They found that hammer throwers who listen to optimized movement sounds immediately before throwing show better throwing performance. Although the use of such natural movement sounds is commonly used in gymnastics gyms, to the best of my knowledge, studies examining the effects on performance do not exist so far.

However, what about movements that do not produce such perceivable natural movement sounds, such as, for instance, the giant swing in gymnastics? Another possibility of using acoustic information for performance enhancement is that of sonification. Here, rather than using natural movement sounds, artificial sounds are developed based on kinematic parameters (Effenberg & Mechling, 2005; Effenberg et al., 2016). For instance, acceleration measures of material, such as the boat moving during rowing or of body parts, are recorded and then related to pitch, based on the mapping to specific tones. Sonification enables access to biomechanical parameters and provides a useful tool to support the learning process (Schaffert, Mattes, & Effenberg, 2011). In gymnastics, this method was used to enhance the circles performance on the pommel horse (Baudry et al., 2006). Gymnasts trained the circles on the pommel horse for two weeks with 300 circles in total. The experimental group received auditory feedback on their body segmental alignment. Specifically, a feedback apparatus linking the upper and lower body part activated a buzzer as soon as the trunk leg flexion increased by more than 20°, because of an increase of the cable tension. As soon as the body of the gymnast was aligned again, the buzzer was switched off. Compared to a control group receiving the typically used verbal feedback, the performance of the experimental group increased by 2.3% from pre- to post-test. Additionally, even after two weeks without training on the pommel horse, performance was maintained. A major advantage of using acoustic information as an athlete for motor execution is that often the eyes are already busy with controlling the movement. The use of auditory perception is therefore helpful for learning and optimizing movements and in addition offers even better information on the temporal structure and rhythm of continuous movements (MacPherson, Collins, & Obhi, 2009).

## 18.6 CONCLUSION

In sports involving complex movements such as gymnastics, perception plays a major role for athletes, coaches and judges. Although the visual sense is considered the most prominent sense, perceptual processing can comprise other senses such as the auditory or kinesthetic sense as well. The chapter has shown how, when and why gymnasts, coaches and judges use their perception in sports in general and specifically in gymnastics. Considering the large complexity, speed and difficulty of gymnastics skills, gymnasts need to make use of their senses for controlling their movements, orienting themselves during aerial skills and preparing the landing of such aerial skills. Coaches, on the other hand, may support the learning and optimizing of skills by making use of the gymnasts' perception as well as making use of their own perception. By guiding and directing the attention of gymnasts to relevant and helpful cues, gymnasts learn to use those cues to help themselves during movement execution and ignore those that are less relevant.

However, coaches also need to attend to information in the environment that will guide their own movements when providing manual guidance to the gymnasts. In addition, they need their perception to take up information regarding the quality of the movement in order to give feedback and provide error correction respectively. Finally, judges share the cognitive part of coaches with respect to the task of identifying and judging errors. Here, visual perception is used to attend to the most relevant cues to judge movement quality. With respect to all three groups of people, the gaze strategies play an important role for their task. Depending on the respective task, situation and person, different gaze strategies may be used to direct attention to the most promising areas of interest and taking up information needed for action, perception and/or cognition. Knowing about these aspects is crucial to develop training programs and/or optimize evaluation and judgmental processes.

## 18.7 Take-home message

- Apart from focusing on well-established methodological approaches when teaching new gymnastics skills and giving feedback on form errors, another important aspect to attend to is that of perceptual information automatically provided by the body.
- Gymnasts are able to use a number of different cues when performing complex actions, with most of them based on our prominent visual sense. However, also auditory or kinesthetic information is used in gymnastics.
- Gymnasts use visual cues in an anticipatory manner, for example during the approach-run to the vaulting table or during the downswing phase of a dismount from the uneven bars.
- Visual cues are also used during aerial skills for body orientation and landing preparation.
- Coaches can make use of such perceptual information by guiding gymnasts to relevant cues during movement execution.
- With the help of time-to-contact information, coaches are able to predict the landing location of gymnasts flying through the air to adapt their movements for optimal manual guidance.

# PART IV REVIEW QUESTIONS

Q1. Discuss which men's and women's events have similar task demands and others which are completely unique.

Q2. Can gymnasts be evaluated by a single sport science method, and why?

Q3. Why is gymnastics an unnatural sport for human beings?

Q4. What is the most appropriate method of evaluation of a gymnast's potential: sport science methods or coaches' observations, and why?

Q5. Discuss the differences between the four models of expertise development in sport.

Q6. Discuss the roles of coaches in a gymnast's development.

Q7. Discuss the roles of parents in a gymnast's development.

Q8. Discuss the most important interactions between the foundation skills.

Q9. Discuss how psychosomatic skills can affect cognitive skills.

Q10. Discuss why you have to plan ahead of training or competition, regarding your goals, emotional states and cognitive activities.

Q11. Why is it worthwhile to keep updating your note book when you are progressing throughout your gymnastic career, or your future lifelong activities?

Q12. Describe which types of gaze control strategies exist and provide an example for each one of them.

Q13. What do we call the process that directs visual attention to locate relevant environmental cues?

    a. Selective auditory attention
    b. Optical information
    c. Using advance cues
    d. Visual search

Q14. How can you support the learning process as a coach, using visual attention?

Q15. If a gymnastics coach attempts to spot a gymnast during a dismount from the balance beam, he or she uses:

a. the angle of elevation
b. the quiet eye
c. the angle of dilation
d. estimation of the moment of inertia.

Q16. Discuss the advantages of using acoustic information during performance enhancement.

# PART V

# Interaction between physiological, biomechanical and psychological aspects of gymnastic performance

# PART V

# Interaction between physiological, biomechanical and psychological aspects of gymnastic performance

## Learning outcomes (*Monèm Jemni*)

At the end of this part you will be able to:

- Identify how physiology, biomechanics and psychology interact with each other to contribute to understanding and enhancement of gymnastics performance.
- Pinpoint the physiologist's actions in the performance enhancement process.
- Pinpoint the biomechanist's actions in the performance enhancement process.
- Pinpoint the psychologist's actions in the performance enhancement process.

## Introduction and objectives (*Monèm Jemni*)

Coaching gymnastics requires crucial skills allowing a progressive transition from "baby gym" until the very high levels of performance. Technical skills are as important as the scientific skills when it comes to coaching children and adults a complex sporting discipline such as gymnastics. This later contains a risk-taking element and a decision-making process in many practice situations. The psychological stress that may apply, the high level of physical preparation required and the biomechanical constraint that the apparatus imposes added to the strict rules of the events make the performance a high-challenging task. This part of the book "works" from the sport to the science and provides views, applications and ways where the science has influenced the practice. Real gymnastic examples and situations are given as support to show how physiology, biomechanics and psychology are interacting and how they influence the quality of performance. This part draws together all three scientific strands to illustrate specific sporting skills and giving an expanded picture on "Applying Sciences to Gymnastics".

The objective of this part is to give an insight into how physiology, biomechanics and psychology interact with each other in order to contribute to a better understanding of gymnastics performance. We aim to address the coaching processes, the pedagogies and performance analysis as an entire picture.

# 19

# THE PHYSIOLOGIST'S POINT OF VIEW

*Monèm Jemni*

As explained in the introduction of Chapter 6, the aim of exercise physiology is to understand how human systems work under different exercise conditions and regimes. Humans react differently under different stress levels and at different arousal states. Such conditions have various impacts on hormonal regulation which affects muscle physiology, force production and changes the neural control of the movement pattern (Mikulas, 1994). Indeed, the impact of a "stress/situation" differs between the context of a gymnastics training and/or competition. The way the gymnast interacts with the external environment (coach, team mates, spectators and even the apparatus) changes according to the situation and subsequently influences the quality of his/her performance.

The "environment" in which gymnasts perform is very special because of the specificities of the apparatus. Safety is a major concern in this sport where high risk of injuries is associated with the high acrobatic elements. Equipment engineers have contributed to the immense gymnastics evolution from the 1970s to nowadays. Coaches, gymnasts, medical staff, physiologists, biomechanists and psychologists interacted with manufacturers, each from their respective point of view in order to improve not only the safety of the practitioners but also to ensure a parallel evolution of the equipment that equals the increasing gymnastic difficulties.

The following paragraphs give more evidence of these interactions.

## 19.1 Body composition versus physiology, biomechanics and psychology

Bale and Goodway (1987) analysed the performance variables associated with the competitive gymnasts. They showed that male gymnasts generally reach their peak of performance in their early twenties, whereas female gymnasts tend to reach their peak in their mid-to-late teens. Working with different age groups is one of the biggest challenges for a coach. Each age group has different "group psychology", "group personality" and also "group fitness" which might be totally different to the respective individual components. In fact, the "individual personality" merges within the group's personality to create a "trend". In the meantime, some individuals might have stronger roles/status than others in each of the above "groupings". The coach has to understand all these "group components/ psychology" while considering the individual variables. There are many coaches who succeeded in working with younger age groups but completely failed with older gymnasts and vice versa. Shall we end up by believing in common sense that says "Oh, this coach is born to work with children!"? In addition, how many times have you heard about conflicts between a coach and a gymnast? How many times have you heard about a very talented gymnast who made a huge success at junior level but disappeared/burnt-out after a few years? How many times have you heard about severe injuries in gymnasts? Have you ever tried to understand why?

Working with younger gymnasts has a crucial advantage from a technical versus body composition point of view. Young gymnasts are typically lighter and shorter than older gymnasts. This slender physique has biomechanical advantages in performing high risk acrobatic skills common to contemporary gymnastics. It is widely accepted that the short stature and reduced weight, typically observed in elite gymnasts, offers less inertia to the angular momentum dominant in gymnastics performance (Faria & Faria, 1989). Coaches consider the pre-puberty phase as an important period of action; in fact gymnasts are very receptive to "technical learning" and strength and power gain during this period. It is also well known that puberty is associated not only with morphological and hormonal transformations but also with an increase of the maximal power (Bedu et al., 1991; Falgairette et al., 1991).

It is very common that high level gymnasts practise for more than 20 hours per week during this period. However, it has been proven that the high level of stress imposed by training and conditioning has an impact on menarche, bone health and growth and development (Courteix et al., 2007; Jemni et al., 2000; Sands, Hofman, & Nattiv, 2002; Theodoropoulou et al., 2005) (see Chapter 3, section 3.4). Thus, young children are potentially at greater risk of different types of injuries. It is, therefore, very important to understand all these issues when working with young gymnasts. Key elements that physiology provides in order to avoid these health risks are: overload and progression, individual response, readiness and periodisation. These are, in fact, some of the most important principles of training that a coach has to fully consider. A very clear and progressive periodisation of the training seasons taking into account all the above is necessary to avoid burn-out. In addition, physiology provides some tools to assess the progress of the gymnasts before raising the load stimulus versus stage and readiness. Some of these tools might be performed as tests in the laboratory or in the "gym".

Chapter 3 has demonstrated that gymnasts (mainly female artistic and rhythmic) are prone to an increased risk of eating disorders and mineral deficiencies with a remaining issue: if they eat enough, what do they eat?! (Filaire & Lac, 2002; Jankauskienė & Kardelis, 2005; Lindholm, Hagenfeldt, & Hagman, 1995). In addition, many gymnasts and coaches are at risk of severe behaviour in an attempt to maintain a lean body size; including extreme dieting, dehydration by the use of diuretics, punishment and food deprivation by the coach if gymnasts put some weight on, etc. Evidence showed that inappropriate diet leads to a decreased performance and an increased risk of injuries and some severe health issues (Benardot, 1999). In order to prevent such issues, physiology provides tools to monitor gymnasts' health and body composition via standardised tests and blood/saliva analysis. Regular medical/physiological exams are indeed recommended, particularly at high level. It has also been proven that gymnasts are more open to talk to the medical staff than to their coaches; this, in fact, might improve prevention, help diagnosis of certain conditions in order to get treated appropriately and ensure healthy training and practice. Moreover, authors such as Borgen and Corbin (1987) and Rosen and Hough (1988) suggested making the role models, such as parents and coaches, more sensitive to the pressure the gymnasts have towards exaggerated slenderness and give sensible advice on nutrition. The development of the sport psychologist's role in this context can be of great support in many difficult cases.

## 19.2 Skills design

One of the main areas where biomechanists, coaches and physiologists interact effectively is the "invention/design of new skills". It is thanks to the collaborative work of this trio that gymnastics has seen an extraordinary expansion of the technical repertoire. There is evidence of work that has been previously performed using computer simulators in order to address the required biomechanical variables, in particular for high aerial acrobatic elements on the high bar, parallel and uneven bars and vault (Hars et al., 2008; Holvoet, Lacouture, & Duboy, 2002; Mkaouer et al., 2008; Mkaouer et al., 2005; Sands et al., 2005; Sands et al., 2006c; Sands et al., 2006d; Yeadon, King, & Hiley, 2005). Some of these studies have been published but others kept "secret". Also, some of these simulations have been successful but some others did not show any applications (Know, Fortney, & Shin, 1990; Milev, 1994; Petrov, 1994a, 1994b). A close link with the physiologists is a key point in order to guarantee the success of the "new designed element". Body composition modelling versus gravity and motion laws is indeed very important to consider in order to guarantee a successful new skill.

## 19.3 Growth and development versus personality

Heredity, maturity, gender, nutrition, rest, sleep, level of fitness, illness/injury, motivation and environmental conditions influence the response of a gymnast to training stimulus. Obviously, each gymnast's response is different to another. For these reasons, one of the principles of training to fastidiously apply is individualisation. A wise coach should detect individual responses and formulate appropriate reactions for each athlete. However, as explained in Chapter 5, section 5.4, there are many cases when coaching a team makes individualising a practical impossibility. The history of gymnastics provides several examples of clashes between gymnasts and their coaches. In order to protect privacy, this section will not mention any names. However, it has been shown that most of these issues occur around adolescence and could have been avoided if coaches had a wider range of knowledge and manoeuvres about this sensitive period. Indeed, teenagers undergo various physiological changes which directly affect their mood and psychology. Consulting sport psychologists can resolve major clashes and retrieve situations. Few high level gymnasts have indeed seen their career ending because of the lack of collaboration between coaches, psychologists and physiologists.

# 20

# THE BIOMECHANIST'S POINT OF VIEW

*Patrice Holvoet*

The number of biomechanical studies is continually increasing. Even if many skills can be grouped together relative to common biomechanical principles (Bruggemann, 1994), it is unfortunately impossible to study all the gymnastics movements performed on the different apparatuses.

A primary purpose of biomechanical studies is to contribute to a better understanding of complex human movements. Therefore, descriptive motion analysis of existing skills is a necessary step in the development of meaningful principles that explain gymnastic techniques. Biomechanical investigations of gymnastic skills reveal that movements can be performed in many different ways. Due to the multitude and redundant degrees of freedom of the body's articulations, the segmental movements need to be controlled by perceptual-motor processes. For a better understanding of these strategies, mechanical principles have to be combined with psychological knowledge that can explain the spatio-temporal organization and regulation of the movement of the body's segments.

A second purpose is to develop specific principles applicable to the creation of new skills. For example, computer simulations attempt skill creation by trying a skill under different conditions to answer questions such as: what is required to execute complex new airborne dismount figures? But are these computer simulations realistic? Initial computer models are then checked to determine physiological values of a joints' torque and muscular activities to ensure that the simulations represent what can be performed by real gymnasts.

The third purpose consists of the contribution of biomechanical studies for increasing safety via equipment development. For that purpose, examinations of contacting-conditions with the apparatuses (landing on mats, bar gripping, springboard takeoffs, etc.) are conducted to guarantee better absorption of shocks, reduction of force and impulse peaks and greater stability.

Let us look now at some findings of biomechanical gymnastics studies and talk about how biomechanics must interact with other scientific disciplines to contribute to a better understanding of gymnastics performance.

## 20.1 Floor exercise, trampoline and tumbling

Numerous studies investigate the takeoff requirements for performing acrobatic skills such as double somersaults with or without twists (King & Yeadon, 2004; Yeadon, 1993a,b,c,d; Yeadon & Mikulcik, 1996). Takeoff velocities, linear and angular momentum, body position and contributions of the body segments to these kinematic characteristics define the main factors that affect performance. The build up and the control of body configuration changes are examined by comparisons between techniques actually used by gymnasts and computer simulations. These analyses indicate what body orientations to give performance focus and what the segment motions are which adjust and control rotation during flight.

Even if it is difficult to trickle the results of these studies down to the coaches and gymnasts, they are an interesting starting point for investigations on the function of the perceptual systems (such as visual and vestibular systems) that control body orientation and angular velocities during complex acrobatic skills.

For example, the regulation of the body's moment of inertia during aerial skills is of crucial importance for a safe landing. Indeed the gymnast must first modulate the angular momentum to execute the required degree of body rotation and must secondly reach a body configuration that permits absorption of landing shock while maintaining balance. One of the main characteristics of the gymnast's expertise is the stability of aerial movements. When performing complex aerial skills, high level gymnasts are able to reproduce appropriate body configurations with a great consistency.

Recent psychological studies explain this stability as the result of a control strategy which requires that the gymnast pick up information concerning future body orientations relative to the ground, distance to the ground, and time remaining before ground contact (Bardy & Laurent, 1998). This information is used to initiate body extension for decelerating the angular velocity and regulating the angular displacement of the body at landing. These studies based on protocols of trials "with and without vision" executed by novices and experts indicate that the body configuration is visually controlled during flight by the experts but not by the novices (Davlin, Sands, & Shultz, 2001a,b, 2002, 2004).

## 20.2 Vault

Many studies examine the correlation between kinematic springboard parameters, parameters while in contact with the horse and landing parameters with the scores given by judges to vaults such as the handspring, handspring with twist or somersault, and Yurchencho vault (King & Yeadon, 2005; Koh & Jennings, 2007; Takei, 1998).

It has been demonstrated that the running horizontal velocity and the takeoff linear and angular parameters are very important factors to optimize in order to realize an effective performance. Simulation models confirm that having appropriate initial kinematics at touchdown is essential and that the use of shoulder torques during horse contact plays a minor role in vaulting performance. Consequently, the meaningful information for coaches and gymnasts to consider is that it is very important for enhancing vault performance to focus on generating high body running velocity and a high level of angular momentum during the takeoff from the springboard.

Even if these studies get more realistic, some important physiological factors are needed to improve the biomechanical models. For example, individual values of stiffness and damping governing the visco-elastic characteristics of joints' elasticity and realistic activation time histories of joints' torques are required to determine the ability of joints to resist the compression action during springboard and horse contact.

## 20.3 High bar and uneven bars

Mechanics of giant swings and takeoff requirements of dismounts and release–regrasp techniques are the most commonly examined performance factors (Bruggemann, 1994; Holvoet, Lacouture, & Duboy, 2002; Holvoet et al., 2002). Kinematic, kinetic, energetic and electromyographic data are reported in order to identify differences, to establish profiles for different techniques and to classify techniques relative to common principles (Arampatzis & Bruggemann, 1998). Even if the construction and the design of these apparatuses are different, similarities between mechanics of the horizontal bar and uneven bars are established in takeoff conditions for dismounts. Mechanical differences between these two apparatuses result more from the "beat" or "tap" action through giant swings (Sheets, 2008). Differences in the anthropometric and physiological characteristics between female and male gymnasts should be taken into consideration when explaining the specific techniques used to perform similar skills on the two apparatuses.

Actually, biomechanical studies develop optimization processes using computer simulation models (Hiley & Yeadon, 2008). These models become more and more complex. The bar is often represented as a spring and the gymnast is considered as a 3D articulated system. The shoulder stiffness and damping are also taken into consideration. Comparisons between experimental and simulation performances are conducted to verify how the model reproduces the movement and what parameters can be optimized to minimize joint torques to maximize performance (Begon, Wieber, & Yeadon, 2008). As has been seen in the preceding section relative to the vault, there is a need for the addition of individual physiological values for the strengths of joints and muscular activities in validating these optimization processes.

## 20.4 Rings and parallel bars

Biomechanical studies on these apparatuses are less extensive than those on the horizontal bar. Kinematic analysis and computer simulations have been conducted on feldge, backward long swing and dismounts on rings (Yeadon & Brewin, 2003). On parallel bars, 2D models have been developed for making predictions on the swing to handstand, dismounts and for analyzing the dynamics of the bar acting at the hand grip (Linge, Halllingstad, & Solberg, 2006). Forces exerted during static movements such as the handstand were also analyzed on these two apparatuses (Prassas, 1988).

Because of the important part that the strength elements play in the performance on these apparatuses, development of the upper body muscular strength and organization of progressions in strength training programs are essential aspects to maximize technique changes.

## 20.5 Other apparatuses

The pommel horse and balance beam are often considered very difficult apparatuses because loss of balance and falls are the primary cause of large score deductions during competitions. Biomechanical studies on the pommel horse are limited to some case studies. Because circling is an essential prerequisite of performance on this apparatus, the control of the angular velocity is analyzed when performing basic circles, Thomas and Magyar elements

(Baudry, Leroy, & Chollet, 2005). Dismounts performed by women gymnasts on the balance beam are the main skills that are examined by the biomechanical studies (Brown et al., 1996). There is a lack of information concerning the nervous control mechanisms that can regulate the posture and the balance involved in the execution of both static and dynamic skills performed on these apparatuses. Biomechanical analyses should be combined with neurological and psychological information in order to better understand a gymnast's balance ability.

## 20.6 Studies on safety and equipment development

Because gymnastic activities involve repeated large weight bearing impacts on the hands and feet, biomechanical procedures can estimate the internal loads exerted at joints during critical phases of movement (Davidson et al., 2005; Mills, Pain, & Yeadon, 2006). For example, three times body weight tensile force is exerted at the wrist and shoulder joints when performing the "beat" swing during a backward giant swing before a release–regrasp skill on the horizontal bar. When performing a simple forward somersault on the floor, 19 times bodyweight compressive force is applied at the ankles. Studies of these types provide useful data for improving the gymnastic training process and for protecting gymnasts from excessive muscular overload. These studies must be associated with medical knowledge about sports injuries in order to be effective in injury prevention.

We have seen that an important source of force used in performing many gymnastic skills is the elasticity of apparatuses such as springboards, trampolines, horizontal bars and asymmetrical or parallel bars. The elasticity of a material is a characteristic which represents a measure of how this material will reform after it has been stretched, bent or compressed. Large efforts have been made to develop models to understand the dynamics of springboard, trampoline and apparatuses. A mass-spring system is often used to represent the interaction between the gymnast and apparatus. Stiffness and damping properties of apparatuses are analyzed in bouncing or stretching experiments in which given loads are exerted on the materials. In this way, foot contact and hand grasp conditions can be examined during critical phases such as landing. These biomechanical studies are used in several other disciplines that include industrial engineering and manufacturing, research in material and equipment, and innovation in design. For example, many innovations are conducted in the manufacture of anti-slide textures that guarantee large safety contact zones covering the gymnastic apparatus surfaces.

So the improvements in the area of gymnastics depend not only on a better understanding of the skills execution but also on the technical progress and innovation in the manufacture of material and equipment.

# 21

# THE PSYCHOLOGIST'S POINT OF VIEW

*John H. Salmela*

## 21.1 Foundation skills and exercise physiology

There is no question that the physiological and morphological status of gymnasts from 9–14 have a strong, determining influence on the performance of young gymnasts (Régnier & Salmela, 1987). If you are strong, fast and flexible, this will enhance your goal-setting, self-confidence and commitment to gymnastics. There have been some exceptions to the rule, such as Eberhart Gienger who was quite tall compared to his teammates, but obviously if you are powerful, fit and strong compared to your cohorts, your foundation skills will be high.

Usually, in the western world, most gymnastics coaches do not have specialized degrees in exercise physiology, but have degrees either in physical education or in the sport sciences and know basic principles for strength, speed, power and flexibility training. They are also able to quickly evaluate whether their young gymnasts have set their training goals, are committed to them, and have developed the necessary levels of self-confidence. This usually results in greater gymnastic skill levels, because of these physiological endeavors on the above physiological dimensions.

However, there are also downsides to commitment from a physiological perspective, and which result in breakdowns, especially in young girls. Dr Monèm Jemni, editor of this book and a researcher at the University of Greenwich in the UK, has extensive data on this subject, since girls need to perform earlier in their life cycle than boys, principally because they have to be able to perform before puberty to aid in the spotting of complex skills, while men must wait until post-puberty to develop sufficient strength to perform, for example the strength moves required on the rings (Jemni et al., 2001; Jemni & Robin, 2005; Sands et al., 2000).

For example, as an athlete, both in gymnastics and Canadian football, I remember well my coach having the boys' team before practice do five sets of 10, 20 and 50 yard sprints, back and forth. It killed me. Then I looked down the field and saw two of my idols, Don Clark and George Dixon, who played for the Montreal Alouettes, coming to watch my coach Ivan Livingstone's practice. Ivan called out to me: "John, you are the only guy who tries hard in every practice, and is in shape. But, I forgot to bring the bag of footballs! Can you run down to the club house and bring them back?" I was physiologically exhausted, but sprinted again across two 100 yard football fields with my professional heroes watching me. I refused to fail! Once I got there, I lay down on the club house floor for two minutes to partially recover, and then sprinted back another 200 plus yards with a bag of more than ten balls. Ivan reminded me of this 20 years later, but I would rather die in front of my idols!

## 21.2 Foundation skills and biomechanics

Bill Sands, a biomechanist at the United States Olympic Committee, was reported on in a recent article on how to become a champion, based upon physiological and physiological training. I had the privilege to work with him for a short time, both in Canada and in Qatar, during seminars. He has a profound knowledge of sport science and gymnastics, and was named at the team coach for the US women's team, which unfortunately, like Canada, boycotted the Moscow Games because of the Russian attack on Afghanistan, which is now quite ironic. He has keen insights on the inter-relationships between the sport sciences and the sport of gymnastics.

I remember, when I was doing my doctorate at the University of Alberta, that the gymnastics coach Tanaka told me that my handstand was too "archy," or curved, even though I was 25 at that time! He made me lie down on the floor in a totally stretched and contracted position while two other gymnasts picked up my stiff boy and placed me in the perfect handstand position seen today. Upon graduating, I used the same biomechanical techniques in my university teaching and, of course, the students' goal-setting, self-confidence and commitment increased, and they were then able to learn more complex skills by maintaining this position in both static and dynamic skills, and were thrilled with their mechanical control of their bodies.

## 21.3 Psychosomatic skills and exercise physiology

The psychosomatic skills are probably the most affected by exercise physiology, especially those regarding relaxation and activation. Exercise has profound effects, especially once it has terminated. There is a great feeling of relaxation and relief after an extensive period of intensive effort. This is the moment for deep relaxation, which often enhances mental training and rehearsal upon the day's activities and can be combined with imagery and mental training for the next day's training or competition. The combination of the recovery from extreme exertion and its positive effects are ideal for mental training, such as stress control.

On the other side of the coin is that activation activities, such as brief sessions of exertion, even screaming out your key words, usually in your mind, may pump you up for better performance for certain events, such as floor or vault. Normally, gymnasts do not need to be psychologically activated, especially young girls, since they have proud parents who have invested a great deal of money, have driven them across the country, or even abroad.

In Sabae, Japan, in 1995, at the World Championships for the 1996 Olympics, the whole team and the technical team delegation lived on the same floor in a hotel. Although I was forbidden to work with the women's team, many of the girls would come up to me and talk, without their coaches. The men were six to ten years older, and the girls often wanted to know why they were so relaxed and joked around, while they felt so stressed. The answer seemed so obvious: you are much younger, you worked rather than played at gymnastics, and the men were friends with the coaches, rather than being dictated to by them.

## 21.4 Psychosomatic skills and biomechanics

From my experience, biomechanical knowledge for gymnasts is most useful for the controlling of stress and fear. I have previously described how, at a training camp with teenage gymnasts, who were potential Olympians for the Canadian team in five years, our coaches explained the relatively new move on the horizontal bar, the Tkatchev. For this the gymnast had to perform a backward giant swing, and before the three-quarter point in the giant, had to arch, blindly counter-rotate over the steel bar and regrasp it again. The biomechanically trained coaches went through how the proper execution of "timing techniques," the use of spotting mats to cover the bar to prevent back injuries, and the use of deep foam landing pits would minimize any possibility of injuries. I previously reported how one courageous gymnast went for it, hit the mat on the bar and landed in the deep foam pit and came up laughing. After that, all 14 gymnasts attempted this element many times within the next 20 minutes. As described earlier, I spent several hours talking to the young gymnasts about real and irrational fears, as outlined by Feigley (1987), and their gymnasts found out that this was, in part, an irrational fear, when using good coaching and some biomechanical and appropriate spotting principles

## 21.5 Cognitive skills and exercise physiology

In relation to the OMSAT-3, the exercise physiology effects on cognitive skills have already been discussed in the foundation and psychosomatic skills sections regarding the gymnasts' general perceptions of their potential to succeed in the sport in terms of their goals, self-confidence and commitment. Obviously, a non-muscular, overweight, slow and inflexible person will not succeed in gymnastics.

However, if they can learn to relax or physiologically activate themselves, either physically or through mental processes, they may have a chance of greater achievements in gymnastics. However, the psychological dimensions drive gymnasts to work hard, do strength and flexibility training, set goals, keep self-confidence and remain committed, usually through the influences of family, peers and good coaches.

## 21.6 Cognitive skills and biomechanics

The science of biomechanics, from my perspective, has provided me with the greatest insights on why I was neither a great gymnast nor a university coach. Much of this I have learned in my later years while participating as a spectator with revolutionary coaches such as the national coach of Canada, Yuri Iarov, at international FIG courses in different parts of the world. Yuri was the first coach to have a gymnast, Valeri Luikin, perform a triple back somersault on the floor exercises, and it has since been done only by seven other gymnasts. He could do a double back after just a back handspring! These techniques for tumbling must be mastered within mental practice and imagery skills, since they appear to be much quicker than those traditionally taught.

I also observed unbelievable learning curves within men's support skills on the pommel horse and on the parallel bars. From the periods before the 1950s until a decade ago, male gymnasts were taught to swing with the shoulder structures contracted, which elevated and contracted these muscles, and thus limited their degrees of freedom regarding lateral movements in the transverse planes. The upper torso would remain stiffer and thus limit flexible movements. Iarov showed me and a team of Moroccan gymnasts how the relaxation of their shoulder structures freed up their torsos and allowed them to improve their circles on the pommel horse by 50% within a couple of days.

I also learned from Iarov how within one to two days to teach the complex skills of the Diamadov skill, which required a 180-degree turn on one arm to a handstand, a skill that usually takes months to learn. But the gymnasts were also instructed to imagine the execution of the movement in the evening, and to keep in mind the coach's innovative methods.

Much work in research and interventions in gymnastics have been accomplished worldwide using the various sport sciences, such as mental training, exercise physiology and biomechanics, but there still remain many questions to be resolved, with perhaps the most in mental training.

# PART V CONCLUSION

As a conclusion for Part V of this book, it is clear there is no single cornered way to emphasize the contribution of physiologists, biomechanists and psychologists in the enhancement of the gymnasts' performance. Multiple examples have been shown throughout this part; scientists alone or together with other contributors have been and remain committed to the development of many related aspects of practices that have directly and/or indirectly affected gymnasts' performance. Equipment engineering, model designing, medical instrumentations, testing theoretical to artificial 3D models, better physiological understanding together with the psychological factors and many more examples have marked this discipline in particular during the last 30 years. The proofs are obvious; just take a look at an international competition video from the late 1980s compared to one from the present day. You would not only witness the performance development but also the technology that accompanies all the show.

# PART V REVIEW QUESTIONS

Q1.  Summarise the point of view of the physiologists on how they interact with the biomechanists and the psychologists in order to enhance gymnastics performance.

Q2.  Summarise the point of view of the biomechanists on how they interact with the physiologists and the psychologists in order to enhance gymnastics performance.

Q3.  Summarise the point of view of the psychologists on how they interact with the physiologists and the biomechanists in order to enhance gymnastics performance.

Q4.  Analyse a real example taken from gymnastics performance where physiology, biomechanics and psychology interact with each other.

Q5.  As a coach, how could you guarantee a long-lasting career for your gymnasts, taking into consideration physiology, biomechanics and psychology?

Q6.  Analyse the interaction of physiology, biomechanics and psychology in the context of skills design.

# PART VI

# Motor learning in gymnastics

# PART VI

# Motor learning in gymnastics

## Learning outcomes (*Monèm Jemni*)

The reader will be able to identify, understand and appreciate:

- The difference between academically oriented knowledge and practical application of such knowledge to motor learning and pedagogy.
- The stages of motor learning and coaching competence.
- The roles of practice and skill transfer in effective coaching.
- The different types of feedback and their application to gymnastics teaching and learning.
- How to examine the motor control concepts that a gymnast applies during the running approach to vaulting.
- The scientific basis underpinning rotations and spinning/twisting in gymnastics.
- How to evaluate the importance of elastic training in modern gymnastics via a real case study from Brazil.

## Introduction and objectives (*Monèm Jemni*)

This part will provide a framework on the theoretical and practical aspects of teaching and learning gymnastics skills.

It provides real examples of simple to advanced skills where motor control concepts could be identified, such as the running approach to vaulting, the rotation and twisting and the elastic training in gymnastics.

Gymnastics is all about skills. Gymnasts perform skills for the sake of themselves, not to gain ground in an opponent's territory, increase the distance of a projectile, move an opponent, send a projectile precisely to a target, beat an opponent senseless, or simply try to cover some arbitrary distance faster than anyone else. As such, gymnastics shares performance objectives with diving, figure skating, aerial skiing, synchronised swimming, dance and others. Gymnastics preparation is somewhat unusual because there is no opponent such as a team or combative rival. Gymnastics does not seek to exceed some personal record (e.g. a personal best), nor does the gymnastics athlete attempt to compete by something the athlete has never done before. The gymnast does not face rapidly changing terrain or environmental conditions and thus does not prepare for rapid environmental changes. Gymnasts are measured by their ability to perform a skill or series of skills that they have performed hundreds of times in the past, and they must perform them with excellence in the decisive moment. They push their performance envelope to increase the spectacle of their work, higher levels of difficulty, elegance and greater consistency. Unlike the circus performer, it is the job of the gymnast to make skills look easy, as an aspect of elegance, not threatening or death defying. Many athletes and coaches lay claim to their sport as art. Gymnastics can embrace both the visual elegance of art in motion, and the Zen-like aspects of art as described by discussions of the spiritual aspects of gymnastics.

Many sports have a handful of skills that are repeated over and over. Previously working with mentally challenged students in physical education showed me that they could play a version of most sport games relatively early in their learning. For example, passing, dribbling, shooting and so forth can be used to create a basketball game. Basketball skills do not have to be performed perfectly, there are no judges, and a pass is effective if it finds its way from the passer to the intended receiver. Lebron James performs dribbling, passing and shooting. His skills are differentiated, not so much by the skills, but his level of perfection within the skills. Gymnastics pursues the problem in the opposite direction. The high demands on perfection are in place immediately. A beginner learns and performs beginner skills, but the demands for execution perfection are already assumed, taught and emphasised. An advanced gymnast continues the idea of perfection but applies it to more difficult skills. Bragging rights in gymnastics come mostly from the skills performed.

The general hierarchy of gymnastics preparation (Figure VI.1) begins with physical preparation. The gymnast needs strength, speed, stamina, flexibility and power before the other aspects (Bompa, 1990; Sands, 1990a). If the gymnast cannot produce the forces and positions required by the skills, then the gymnast needs to work primarily on fitness. If the gymnast's fitness meets or exceeds the fitness requirements for the skill, then he or she can begin learning the skill with expectations of success. If the gymnast attempts a different route, such as trying to "trick" themselves into the fitness requirements for a skill (work on the skill while expecting fitness transfer to occur), then problems usually ensue (Sands, 1984, 1990b). Skill learning readiness is assessed by a gauntlet of decisions starting with physical readiness and followed closely by technical, tactical, psychological and theoretical aspects of readiness (Sands, 1990b). Of course, the specificity of working the skill to gain skill fitness is laudable, but while learning the skill the athlete will make many errors, some uncomfortable and some dangerous. Learning errors that have their roots in fitness are not reversed and unlearned easily after the habituation of hundreds of repetitions. Moreover, there are numerous benefits to learning basic or "root" skills that can be easily added to and modified leading to other skills (Irwin, Williams, & Kerwin, 2014).

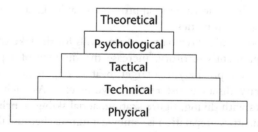

**FIGURE VI.1** Preparation hierarchy.

Note: "Technical" (i.e. skill) ranks second in importance and is built from a foundation of fitness.

# 22

# MOTOR LEARNING VERSUS PEDAGOGY

*William A. Sands*

Gymnastics coaches pride themselves on detailed knowledge of their craft. Coaches have a great deal to know, and the best coaches have a large assortment of gymnastics-specific knowledge and coaching skills. Knowledge is most commonly thought of as academic-type knowledge or the "know-that" of something. For example, a coach may "know-that" force is the product of mass and acceleration, he or she may "know-that" gymnastics skill performance is dominated by anaerobic energy metabolism, and the coach "knows-that" landing forces dissipated over a greater distance or over a greater impact area result in lower pressures. Coaches also "know-that" the sun rises in the east and sets in the west, if you drop something it will fall, and water is wet. Perhaps more important for the coach is another kind of knowledge that is suggested by a one-word hyphenated concept, "know-how" or practical knowledge. Philosophy and artificial intelligence have attempted to codify knowledge in the hopes of clarification and to develop expert systems (Jones, 1988; Sands, 1991a, 1991b; Sherald, 1989). This concept also includes know-what (facts), know-why (science or theoretical understanding) or know-who (social discourse or politics). Certainly, application without know-that is too often foolish and dangerous; knowledge without know-how is sterile.

Intelligence (know-that or know-why) is not sufficient for great coaching. In fact, past experience indicates that most coaches would argue that the most important knowledge orientation for successful coaches comes from "know-how". Coaches must know how to teach a skill, know how a particular execution mistake may be harmless or severely damage future learning, and know how to paint a rich word picture, or design just the right drill to help an athlete learn (Sands, 1994b, 1999a). Of course, know-that forms a basis for further knowledge in all of the types listed above, but it is not uncommon to find excellent gymnastics coaches who were never gymnasts themselves, never studied sport or exercise science, and never coached athletes in another sport beyond minimal levels. Clearly, there is a lot to be studied in regard to the development, progression and long-term education of coaches (Martens et al., 1981).

# 23

# FUNDAMENTAL CONCEPTS AND DEFINITIONS IN MOTOR LEARNING

*William A. Sands*

The term "motor" refers to "movement," and the reader should feel free to replace "motor" with "movement" throughout this chapter. The coach should understand that first and foremost, coaching is teaching (Sands, 1984). "It is in the process of technical preparation that the educational ability of the coach and his professional qualification as a gymnastics teacher especially come out vividly" (Arkaev & Suchilin, 2004, p. 125). As such, the coach and athlete are in a cooperative situation where the coach attempts to teach the fitness, skills, tactics, mental orientations and philosophies of gymnastics while the athlete attempts to learn these concepts. Fundamental definitions of crucial concepts are as follows:

- Motor/movement learning – a change in behaviour, associated with practice, that signifies the relatively permanent acquisition of a new or novel motion.
- Motor/movement control – a process by which the central and peripheral nervous systems combine to produce motion through the application of muscle tension.
- Motor/movement performance – the outcome of a motion-related behaviour.
- Motor/movement development – the age- and maturation-related changes in movement behaviours.
- Learning – a change in behaviour that is relatively permanent and involves a new capacity.
- Practice – an exercise or task behaviour that is repeated to gain proficiency in the exercise or task behaviour.
- Skill – "A skill is learned, and it is distinguished from capacity and ability because an individual may have the capacity and ability to perform a skill but cannot do it because it has not been learned" (Adams, 1987, p. 42).
- Technique (technical actions) – "... are actions connected with the implementation of a certain technique of performing an exercise" and "The sum total of technical actions determines the technique of performing gymnastics exercise and their technical structure" (Arkaev & Suchilin, 2004, pp. 128–129).

Learning is somewhat of a "black box", because learning is not observable. In order to identify and verify learning one will observe an "input", desire, motive or demand that goes through a transformation (the black box), and later an observable behaviour change results. The observable parts of learning are connected by a link that is not observable and can only be inferred from the initial behaviour and the outcome behaviour. Using the definitions described, motor learning consists of some, usually repeated, behaviour (i.e. practice) that is observable and transforms via the neurophysiology of motor control to finally emerge as a changed and observable motor performance.

# 24

# STAGES OF MOTOR LEARNING

*William A. Sands*

Motor skills acquisition tends to pass through three stages. These stages do not have sharply defined borders, but rather a blending or merging of behaviours, much like the colours transition in a rainbow. The three stages of motor learning are: cognitive, associative and autonomous (Schmidt, 1988) or initial, intermediate and advanced (Halsband & Lange, 2006). Although the naming conventions of the stages of motor learning vary, nearly all are tri-phasic. Motor learning is also cyclic. Skill learning begins with a goal, proceeds through learning and practising, and ultimately achieves success or failure (Arkaev & Suchilin, 2004).

The cognitive stage of learning is first. The athlete encounters a new skill, he or she will watch a demonstration of the skill, listen to a description from the coach or other athletes, and attempt to assimilate the basic structure and dynamics of the skill into movement patterns he or she already possesses. The athlete will engage in active thinking about the skill; attempt to see, feel and otherwise sense the skill; and finally attempt the skill with the jerky awkwardness of a beginner's movements. Self-talk is often used by athletes to describe salient features of the skill as mental reminders of the movements to perform. The cognitive stage tends to show rapid progress.

The associative stage of learning is second. This stage involves the consolidation of the separate and serial movements that were attempted in the cognitive stage. The athlete has assimilated enough of the skill that he or she can perform most or all of the skill without needing conscious thought engaged at each movement or sub-movement. Smaller details now become apparent and important. The athlete goes through a sort of "debugging" process as faults are recognized and corrected leading to a more refined skill. Performance is more controlled and consistent. There is less reliance on self-talk and the need to translate a movement "script" in cognitive steps during skill execution. Progress is relatively rapid, but not as fast as the cognitive stage.

The associative stage is third. The athlete now has complete or nearly complete control of the skill. Skill performance is automatic, requiring no cognitive intervention. Progress is much slower during this stage as the subtler aspects of learning and performance are addressed. For example, the skill is now stable enough to move to new circumstances (Harre, 1982). As learning proceeds, there is a need for a rich flow of information.

Information must flow to and from various structures in the central and peripheral motor and sensory nervous systems to produce movement.

A stage-like concept describing the information flow of motor performance consists of: motivation, ideation, programming and execution (Halsband et al., 1993).

- Motivation comes from the limbic or "emotional motor system" that works on needs and wants:

  o   Biological drives (hunger, thirst, etc.) often serve as motivations. Biological drives are coupled with emotions to initiate motor performance.
  o   The limbic system can override pain sensations during stressful situations.
  o   The hypothalamus is the focal point of this system, while also being hierarchical including the forebrain and midbrain.
  o   Central nervous system input from the limbic system is integrated with higher centres to form "ideas".

- Ideation consists of the linkages and selections of appropriate motor programmes to act on the idea.
- Programming takes an idea and converts it to a motor action:

  o   Programming takes an "idea" and "converts" it to action.
  o   The brain structures that perform this conversion are: premotor cortex, motor cortex, basal ganglia and cerebellum.
  o   The result of the conversion is called a "central command".
  o   Central command works simultaneously in two primary directions, from the brain to lower centres such as the spinal cord and periphery, and to higher centres in the brain.

- Execution proceeds from a central command:

  o   A motor programme is either constructed at the moment or accessed from memory.
  o   Movement commands in the form of the motor programme are sent to appropriate areas of the central and peripheral nervous systems for processing.
  o   Feedback from sensory receptors returns to the central nervous system by several afferent pathways.
  o   Integration of the motor commands and sensory feedback occurs in the sensorimotor cortex.

Information flows through the various stages listed, and nested within the three stages of motor learning. Information is a single term that belies an underlying complexity of processing that incorporates everything from perception to action, recursive, reliant on memory, and various reflexes and feedback loops.

# 25

# FEEDBACK

*William A. Sands*

Feedback is defined as any perceived sensory information regarding a movement. Feedback can be intrinsic (i.e. coming from within) such as the information derived from the feeling of a movement, noting the position and motion of limbs, and so forth. Intrinsic feedback occurs as a natural result of the movement. Feedback can be extrinsic, such as information obtained from coaching comments, video images and watching oneself in a mirror. Extrinsic feedback is basically coaching – feedback that arises above and beyond that of the natural result of a movement.

Intrinsic and extrinsic feedback are themselves composed of two further types of information: 1) knowledge of performance (KP), and 2) knowledge of results (KR). Intrinsic KP includes noting the motion and positions of limbs (i.e. the process of the movement) while intrinsic KR is the self-talk based on information obtained from the environment such as the sound of a take-off or landing and is usually outcome-related (i.e. the "success" of the movement). Extrinsic KP is information obtained from coaches, video, film and mirrors while extrinsic KR is information such as scores, lap or finish times, and a measured distance. Generally, KP is the most useful, especially in gymnastics when the nature of the outcome is not obvious. Gymnasts usually cannot see themselves perform and rely heavily on a coach's interpretation of their performance to gauge the need for change and their proximity to success (KR). If the outcome was obvious, KR is not helpful. While the gymnast is learning, there are more than ample opportunities for a coach to provide KP in the form of comments about techniques, body positions, speeds and a myriad of other factors (McClements & Sanderson, 1998).

Extrinsic feedback usually serves as a motivator. The purpose of the motivation is to encourage the athlete to try again and thereby accumulate more practice trials in order to move closer to the ideal skill model. The coach must make every effort to augment the athlete's access to feedback by encouraging more practice, and providing information that is relevant and timely. Motivational extrinsic feedback serves to fulfil basic needs, such as described above in regard to the limbic system. More specifically,

> Psychological well-being and optimal functioning and learning in a broad range of domains appear to depend on support for, or satisfaction of, basic needs: competence, autonomy, and social relatedness ... The need for competence refers to the need to

experience oneself as capable and competent, whereas autonomy is related to the need to control or actively participate in determining one's own actions and behavior. Social relatedness describes the need to feel connected with others or to experience satisfaction in one's involvement with the social world.

*Lewthwaite & Wulf, 2012, p. 175*

Self-controlled practice (i.e. autonomy) and the ability to control the timing and richness of observational practice enhances learning when compared to "yoked" practice in which the learner has no input into the learning trials (Chiviacowsky, Wulf, & Lewthwaite, 2012; Wulf, Chiviacowsky, & Drews, 2015; Wulf, Shea, & Lewthwaite, 2010). External feedback involving comparisons and language indicating that the learner is performing well, and as good or better than other athletes, also serves to enhance learning (Wulf, Shea, & Lewthwaite, 2010).

Motor learning is linked to perceptions of competence in that motivational extrinsic positive feedback has been shown to enhance learning better than negative feedback. In fact, even false positive feedback tends to enhance motivation for learning better than negative feedback (Lewthwaite & Wulf, 2012; Wulf, Chiviacowsky, & Lewthwaite, 2010; Wulf, Shea, & Lewthwaite, 2010).

Feedback, while considered essential for skill acquisition, can become a barrier to success when the gymnast becomes dependent on the feedback for skill performance. As such, feedback is an optimization problem – not too much, not too little, just right (Sands, 1984). For example, continuous *concurrent* feedback has been shown to be detrimental to learning (Schmidt & Wulf, 1997). Feedback is also a "richness" problem. The gymnastics coach must learn to paint rich, vibrant and detailed word pictures (Sands, 1984). In addition, the coach should include language describing the skill as learnable and that the athlete's competence is sufficient to acquire the skill (Wulf & Lewthwaite, 2009).

Feedback usually involves identification of errors, but certainly all errors are not equal, and the athlete's ability to identify an error improves with experience. Some errors are more serious, some trivial and some may disappear simply by further repetition. When feedback frequency and magnitude gradually fade, commensurate with performance competence as the athlete improves; the feedback is called "bandwidth feedback". Bandwidth feedback is scheduled and elastic in the sense that it punctuates the time course of learning and practice, and expands and contracts based on the performance characteristic being addressed and the error identification capabilities of the athlete (Badets & Blandin, 2010; Sadowski, Mastalerz, & Niznikowski, 2013). The reduction and eventual elimination of feedback are associated with increasing competency. Thus, the application of feedback is complex, depending on content, magnitude, timing, intent, and can be a learning enhancer – even when external feedback is removed (Badets & Blandin, 2010; Wulf, Shea, & Lewthwaite, 2010).

Stages of learning should be used as an indicator of the appropriateness, type and timing of feedback. During the cognitive stage, concurrent and terminal feedback are crucial and should be applied generously. The associative stage involves a reduction of concurrent feedback and the schedule of feedback follows a bandwidth approach, a reduction of both frequency and magnitude of feedback as learning progresses. External feedback during the associative stage is more of a summary than an ongoing captioning of performance as it occurs. External feedback is absent, or nearly so, during the autonomous stage. External feedback follows a path that involves rich description and information, given often, and lots of it. As learning progresses, the feedback is provided more selectively, more summarily and shifts emphasis from external to internal. As the athlete gains control of the skill, the necessity of feedback declines to a period in the autonomous stage when feedback is not only unnecessary but may harm performance by overwhelming the athlete with too much information. In addition to fact-based content to aid the learner, external feedback can also serve as a powerful motivator (Wulf, Shea, & Lewthwaite, 2010).

An important aspect of external feedback involves using an external focus. Several studies have demonstrated that an external focus for gaining feedback information is better than an internal focus. For example, looking for feedback information by focusing attention on muscles during a vertical jump is much less effective than focusing on an external place to touch the fingertips (Lohse, Wulf, & Lewthwaite, 2012; Wulf, Chiviacowsky, & Drews, 2015; Wulf, Shea, & Lewthwaite, 2010). Language and external performance feedback should seek to direct the athlete's attention to an external or environmental objective (Wulf et al., 2002).

# 26

# TRANSFER OF LEARNING TASKS TO A FINAL SKILL

*William A. Sands*

Transfer is at the heart of coaching (Adams, 1987; Rosalie & Müller, 2012). Transfer refers to the idea of doing one thing and then shifting it to augment a different thing. Motor learning uses drills that attempt to transfer aspects of a sub-skill to the goal skill. Learned skills are often transferred to skill series, combinations and routines. The idea of skill transfer is also involved in the concept of generalizability. A skill learned in one way should be generalizable or transferable to a different circumstance. If skills did not transfer, then we would be forced to learn a specific individual skill for every possible movement task – clearly inefficient. However, research has shown that transfer can be complex. For example, simple skills may not transfer to complex skills in spite of performance similarities (Wulf & Shea, 2002). Moreover, while simple skills may not transfer, increased feedback and physical assistance devices may enhance complex skill learning (Wulf, Shea, & Matschiner, 1998; Wulf & Toole, 1999).

The more two tasks resemble each other, the greater the transfer from the first task to the second (Baker et al., 1988; Carnahan & Lee, 1989). Positive transfer tends to occur between skill drills and the skills used in a game. Negative transfer is less common and involves a situation where practice on one skill damages the performance of a second skill. Positive transfer may occur in moving from inline roller skating to speed skating on ice. Positive transfer may occur from a cartwheel to a round off, from a forward entry vault take-off to a front salto for tumbling, from a basket-swing drill on parallel bars to a stoop-in on horizontal bar, and from a split leap on floor exercise to a split leap on balance beam. Negative transfer may occur from a backward roll on tumbling to a backward salto on tumbling, negative transfer may occur from a barani motion on trampoline to a front salto half-twist on a parallel bar dismount, and from performing a left twisting salto on floor exercise to a right twisting salto on an uneven bar dismount. Learning and skill transfer is particularly important in the economy of learning skills. The most desirable skills to learn are those with the greatest potential for transfer.

The acquiring and refinement of technical skill in gymnastics is, in our view, a matter of the forming in the cerebral cortex of a sort of database whose elements are a multitude of functional systems. This is a meta-programme consisting of control of concrete movements programmes.

*Arkaev & Suchilin, 2004, p. 127*

An example of such a skill is a back extension roll. The back extension roll is a fundamental skill in beginning tumbling, but can be used to learn about aspects of a clear hip circle and a toe on and off on uneven bars. If a pirouette is added to the up-phase, then the skill can be used to learn about a blind change. "One should adapt the mastered movement to growth in complexity through purposive refinement of the technique of its performance with the aid of special tasks" (Arkaev & Suchilin, 2004, p. 131).

Positive and negative transfer in motor skills finds an important role to play in gymnastics pedagogy and efficient learning. Historically, gymnastics had little regard for the preferred direction of a gymnast's twist. Generally, multiple twists were performed only in single saltos, and twists in multiple somersaults rarely exceeded a single full twist. When gymnastics progressed to incorporating multiple twists within multiple somersaults, coaches and gymnasts often found that the twist direction they learned on one skill transferred negatively to twisting performances in the second somersault of a double salto. It rapidly became clear that gymnasts might save themselves considerable frustration by learning to perform all twisting in the same direction (Sands, 2000b). This idea also applied to the direction in which one learned a round off (Sands & McNeal, 1999a).

The transfer of skills can occur in what is described as "near" and "far" transfer (Rosalie & Müller, 2012; Vuillerme et al., 2001). Near transfer refers to the use of drills and skills that are similar to the skills that will be used in competition. Again, the closer the similarity, the greater the transfer. A near transfer example would be handstand drills on the floor transferring to balance beam and parallel bars. Far transfer involves the shift of a known, usually basic, skill or drill to tasks that are not considered similar to the original skill or drill. Far transfer occurs when one skill is practised and later contributes to the learning and performance of a relatively wider variety of skills. Jumping and landing are examples of basic skills that can transfer to dozens of other skills, sports and circumstances.

# 27

# TRANSFER STRATEGIES

*William A. Sands*

How can a coach take advantage of learning transfer most effectively? Some of the most common approaches include the use of simulators and simulations, whole-part practice, lead-ups and drills, and mental rehearsal. The basic premise behind these strategies is to engage the learner in associated learning activities by reducing danger and expense, reducing the size of the serial skill load, increasing experience with sub-skills, and using the mind to practise skills with and without physical practice. The goal of transfer strategies is a positive transfer from one activity or task to a larger, longer, or more valuable activity or task.

Simulators and simulations are involved in many other activities that require coaching and learning. Perhaps the most visible and complex are flight simulators where pilots practise flying in a complete cockpit while a computer renders images of the outside and controls instrument readings. The pilot shows his or her abilities by skilfully manipulating the aircraft through virtual air space and dealing with all of the various problems that could arise during flight. There are automobile, submarine, tank, space craft, ship and many other well-designed computer-based machines that allow the operator to practise the tasks involved with the full device without incurring damage, crashes, injuries and so on while learning from mistakes. There are only a few simulators that have been used in gymnastics, primarily as research tools (Bernasconi et al., 2009; Calmels, Pichon, & Grezes, 2014; Krug, Reiss, & Knoll, 2000; Mills, Yeadon, & Pain, 2010). Malmberg (1978) wrote in 1978 that Soviet Union scientists had simulated over 1,000 skills that had not yet been performed, and that the Tkatchev or reverse hecht was one of these skills. Arkaev and Suchilin (2004) describe a number of training aids or simulators involving body position training using a suspension system, the overhead spotting rig with a twisting belt and various pneumatic apparatus additions for controlling and enhancing the elasticity of the apparatus. The suspended plastic "bucket" for pommel horse circle instruction can be considered a simulator, in spite of being rather low-tech (Fujihara & Gervais, 2012). Virtual reality simulation approaches could be of great benefit to gymnasts. For example, visual simulations of landings may help the gymnast learn optimal landing techniques without the risks associated with impact (Yeadon & Knight, 2012). Investigations to determine which exercises and drills best measure and simulate the muscle recruitment of gymnastics skills have been performed for still rings (Bernasconi et al. 2004; Dunlavy et al., 2007; Sands et al., 2006a).

**FIGURE 27.1**    Simulator and training aid for enhancing the gymnast's ability to perform a Maltese Cross on still rings.

Practice methods in gymnastics are designed to promote skill transfer and nearly always involve a whole-part, part-whole or whole-part-whole methodology along with many skill drills (Federation, 1986; Sands, 1981a; Sands & McNeal, 1995a). Complete gymnastics skills are often too complex or too dangerous to simply attempt without some preparation or safety measures. Gymnastics coaches pride themselves on their knowledge and creativity in designing clever drills that help simulate all or part of a skill in a safe environment. Numerous texts have been written which show dozens of skill drills, lead-ups, parts of skills and whole skills: Portuguese (Araujo, 2002), French (Magakian, 1978), Spanish (Estape, Lopez, & Grande, 1999; Estape Tous, 2002), English (Hacker, Malmberg, & Nance, 1996; Sands & McNeal, 1995a; Werner, 1994).

Mental rehearsal consists of practising a skill, without actually physically practising the skill (Schack et al., 2014). Studies of the brain while viewing skills on video and through mental imagery have shown transfer to physical practice (Munzert et al., 2008). Mental rehearsal skills cannot be observed directly, however the ability of the gymnast to report the duration of skills and routines can be measured and provides a means of determining validity of the mental activity (Calmels & Fournier, 2001; Guillot & Collet, 2005). In fact, it is often surprising to note just how close the athlete's mental imagery duration is to his or her actual routine or skill duration (Lee, Young, & Rewt, 1992).

# 28

# PRACTICE

*William A. Sands*

Practice involves repetition. In Latin, *repetitio est mater studiorum*, "repetition is the mother of study" fits as well now as it did during Roman times. However, Bernstein describes an updated concept of practice in his phrase, "repetition without repetition" (Latash, 2012, p. 262). Implied by this phrase is that in spite of performing repetitions, the exact nature of each repetition varies somewhat so that no two repetitions are exactly the same (Latash, 2012). When confronted with the question of "what is repeated in a repetition", the answer is most likely that the learner is repeatedly posing and solving a movement problem to his or her central and peripheral nervous systems (Lee, Swanson, & Hall, 1991). These ideas become apparent in the fact that practice is not simply repetition of a skill or movement. When speed of a movement, environment, timing, sequencing and other aspects of performance are changed, the skill is "learned" again. Some skills progress rapidly as the new learning is acquired quickly; others can appear to force the athlete to nearly start over from the beginning.

Several different types of practice systems are used in sport. Part practice refers to the performance of smaller sub-skills or drills of a goal skill. Part practice is the most common approach in gymnastics. Whole practice involves attempting the entire skill, and relying on practising the whole skill for learning. Part practice proceeds from the assumption that the target skill can be broken down into smaller segments. Most skills can be subdivided into components, but the learning may or may not be enhanced by doing so. Some skills which are not easily broken down, or are executed as a single unit, may be best approached by the whole method. Continuous or cyclic movements are often difficult to break down into sub-skills. Skills which are performed as one unit or have segments that interact are also difficult to break down and may be better practised as a single unit. Such skills might include a single backward hip circle, running or jumping. A flic flac can be broken down into phases: sit and rearward arm swing, jump upward and backward with an explosive arm swing around the shoulder to a position overhead, the jump and arm swing result in backward rotation during the flight, the hands contact the mat with the body in a hyperextended spine and hip, the body then passes through a handstand position, and finally the gymnast pushes from the hands while the torso and hips produce a rapid flexion motion lifting the body from the floor to finally land upright on the feet. There are numerous drills available for each of these skill parts or segments. Figure 28.1 shows several examples of flic flac (back handspring) drills.

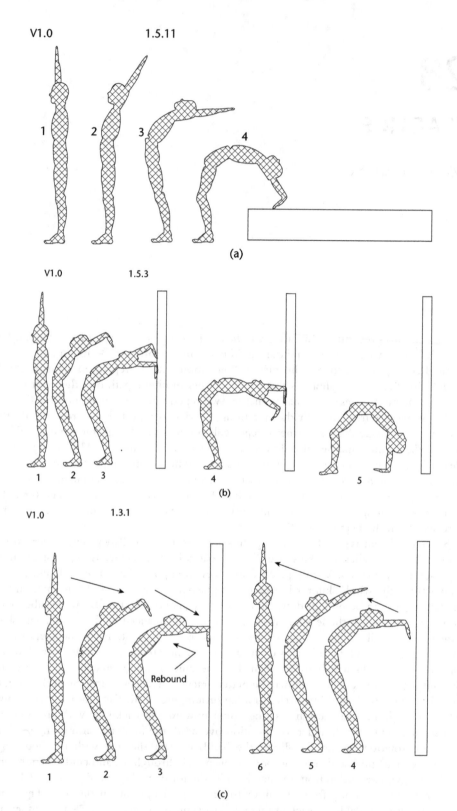

**FIGURE 28.1** Several images of flic flac or back handspring drills.

© Copyright 1995, Wm A. Sands, Ph.D., Jeni R. McNeal, C.S.C.S.

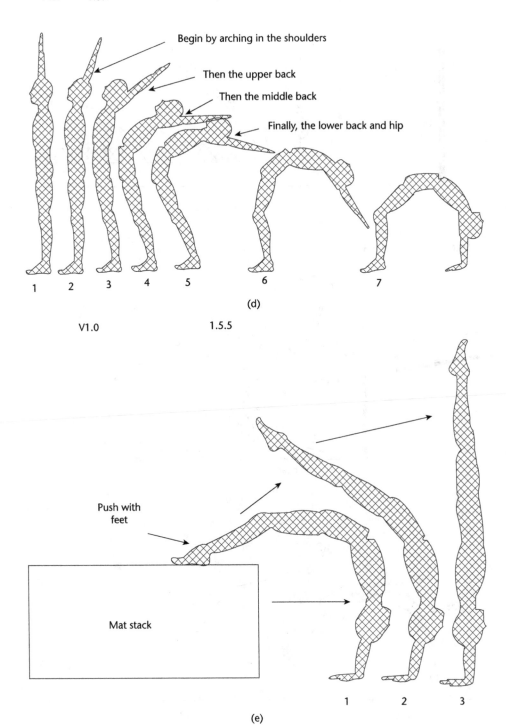

Begin by arching in the shoulders

Then the upper back

Then the middle back

Finally, the lower back and hip

1  2  3  4  5  6  7

(d)

V1.0                1.5.5

Push with
feet

Mat stack

1    2    3

(e)

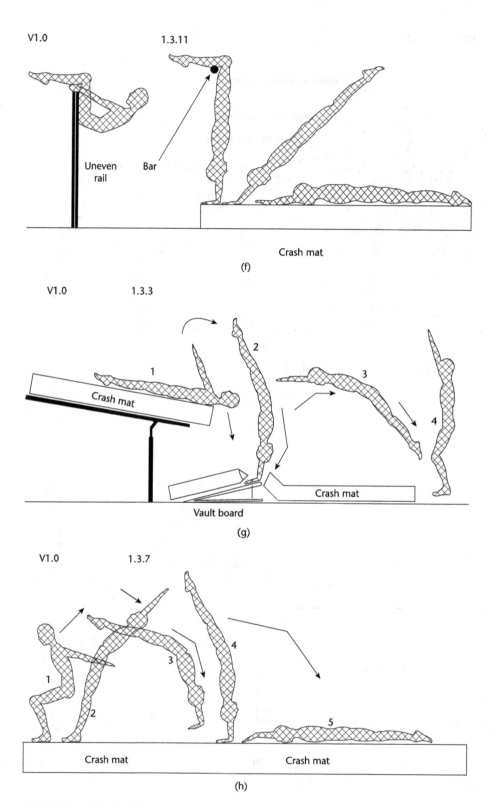

(f)

(g)

(h)

**FIGURE 28.1** *(continued)*

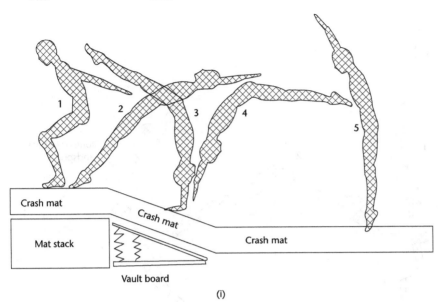

Crash mat

Mat stack

Crash mat

Crash mat

Vault board

(i)

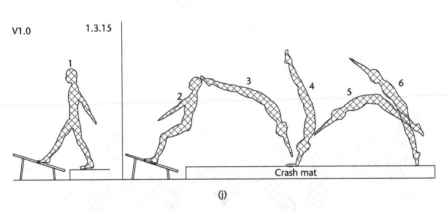

Crash mat

(j)

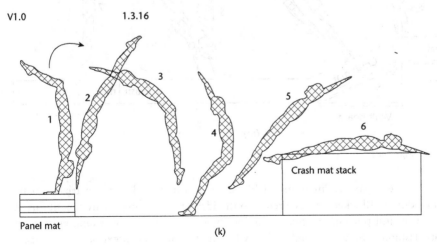

Crash mat stack

Panel mat

(k)

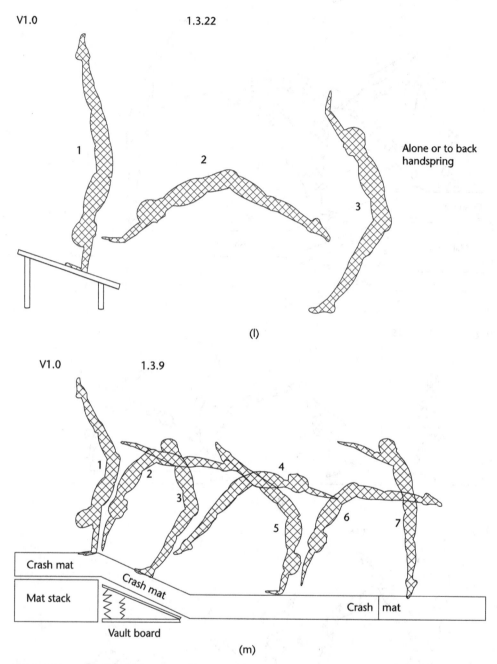

FIGURE 28.1 *(continued)*

Practice can be organized in terms of blocks of learning trials, or the trials can be prescribed randomly. Blocked practice refers to the idea of performing many repetitions of the same skill or drill pursuing mastery. Random practice refers to prescribing tasks in an unsystematic fashion. The term "random" may be misused in this context because it is rare that skills are presented in a totally random way with neither the coach nor the athlete knowing which skill comes next. The term is more meaningful if one considers that intent of the term is to convey that the skills are presented with less attention to continuous skill repetitions, but rather to intermixing skills such that one skill may be practised for a period and then another skill is presented and practised followed by a third yet different skill. Laboratory-based skills used in experiments investigating blocked and random practice were the origin of the ideas and the terms. It is unlikely, and sometimes dangerous, to present and practise skills that are totally unrelated and proceed with no lead up or warm up efforts.

Moreover, simple skills may not transfer well to complex skills, particularly in a blocked practice approach (Wulf, Horger, & Shea, 1999). However, the lessons from the laboratory-based experiments with skill learning still provide important information to the athlete and coach regarding the sequencing of the serial nature of skill practice, but should not be implemented blindly.

The major finding of experiments involving blocked and random practice is that blocked practice (practice by repeating the same skill many times) enhances initial performance and stabilizes skill repetitions quickly. Random practice (practice by performing two or more skills in a non-systematic order) does not enhance initial performances as well as blocked practice. However, when retention trials are later used to evaluate whether the practice resulted in stable skill acquisition, random practice is shown to be considerably better. These experiments provide important information for skill-dominant sports like gymnastics. Common practice schedules and methods usually demand that the gymnast performs any number of skill repetitions in one setting from a few to over 100. The typical sequence involves preparing the skill with a general warm up, then a specific warm up, and then many repetitions of the skill under the watchful eye of the coach. There seems to be a cultural tradition of winning respect from other coaches by boasting of the number of repetitions athletes are forced to complete. The typical sequence described here is blocked practice, repeating the same skill many times. Random practice would have the athlete perform one or a few skills of one type, and then switch to performing one or a few skills of a different type. By using random instead of blocked practice, the coach and athlete will commonly experience a rapid improvement in the single skill, but an inability to maintain and sustain the improvements when the athlete attempts the skill in a few hours to a few days. The initial improvements, which are very exciting, are ephemeral and the athlete often has to start over in the skill learning sequence following a lay-off from the skill. In random practice, the early success of blocked practice is replaced by poorer initial performance. Random practice methods can be considered an investment in skill learning rather than a purchase. Random practice may result in slower initial learning with more stable learning and performance later. While random practice may be better at retention of skills, the skills used in random practice must be developed to a point that they are not dangerous when practised in a non-ordered sequence (Harre, 1982). Past experience has shown that the idea of leaving a skill after only one or a few repetitions to practise another skill is not popular among gymnastics coaches.

The blocked versus random practice approaches can be illustrated with the common learning of multiplication tables in arithmetic. Multiplication table practice using blocked practice methods would have the student repeating one equation over and over, such as, 3 times 5 equals 15, 3 times 5 equals 15, 3 times 5 equals 15, again and again. If the reader tries this exercise one will usually find that disengagement from the task occurs early, and one simply blindly and thoughtlessly repeats the equation. If one is trying to learn the multiplication tables by a random practice approach, then the student might practise by saying or writing, 3 times 5 equals 15, 4 times 6 equals 24, 7 times 5 equals 35, 4 times 4 equals 16, and 3 times 5 equals 15. The random method avoids thoughtless repetitions and forces the learner to engage the process of multiplication not simply the outcome, likewise with skills. By encouraging a more mindful practice, the learner's progress is slower, he or she may appear more frustrated, he or she may make more mistakes, but the retention of the learning is notably better. Coaches and athletes should heed this information by practising using a block format during the earliest periods of learning and to ensure that the athlete is less confused by multiple skill presentations. However, as the athlete progresses, he or she should be tasked with practising one skill and then switching to another skill. The timing and sequencing of such learning is reliant on the judgement of the coach in determining when blocked practice can be reduced and random practice can commence.

The role of practice in skill training has recently come under scrutiny. Dr. K. A. Ericsson, perhaps the foremost expert on deliberate practice and the development of expertise in skills, has shown repeatedly that deliberate practice is what ultimately determines the extent of success and the difference between the highly successful and those less so (Ericsson, 1996, 2003, 2008, 2013; Ericsson & Poole, 2016). Ericsson postulated that about 10,000 hours or ten years of deliberate practice was necessary to reach expert performance (Ericsson, 1996, 2003; Ericsson, Krampe, & Tesch-Römer, 1993). However, B. N. Macnamara and colleagues argue that there are many more factors involved in expertise and success than deliberate practice, and that deliberate practice may only account for about 18% of the variance in expertise (Macnamara, Hambrick, & Moreau, 2016; Macnamara, Moreau, & Hambrick, 2016). The debate of the role of deliberate practice continues in the literature with both sides agreeing on the necessity of deliberate practice, but not on the magnitude of its contribution (Ericsson, 2016; Macnamara, Hambrick, & Moreau, 2016; Macnamara, Moreau, & Hambrick, 2016). Of course, genetics appears to play a large role in the athlete's ability to handle training loads, profit from coaching and avoid injury (Calvo et al., 2002; Montgomery et al., 1998; Puthucheary et al., 2011; Yan et al., 2016). Future research may help disentangle the roles of nature and nurture in sport training and performance, but we are still in the first baby steps of such understanding. Given that gene doping will not be available for some time, the only avenue remaining in the practical approach to enhancing performance and learning is via systematic, vigilant and intelligent applications of skill practice. Genetic profiling and selection for gymnastics has not gone unnoticed (Koyama et al., 2012; Massidda, Toselli, & Calo, 2015; Yeowell & Steinmann, 1993).

# 29

# SPECIAL CONSIDERATIONS IN GYMNASTICS LEARNING

*William A. Sands*

Gymnastics is perhaps one of the best "laboratories" for the study of skill acquisition and performance. Spatial orientation, kinesthesis and balance are involved in gymnastics training at all levels and nearly every day. Of course, gymnastics is not alone in the use, abuse and exploitation of these aspects of motor learning and control, but gymnastics is perhaps the richest activity for demonstrating the wide range of these factors in a single sport.

*Spatial orientation* refers to the ability of the athlete to know where he or she is in three-dimensional space. Spatial orientation relies on nearly all of the sensory systems available to the athlete to convey to the central nervous system the body's position (i.e. upright, upside down, leaning, etc.), the body's motion (still, up, down, forward, backward, sideward, rotation, combinations, etc.) and the position and motion of the external environment. Brain structural changes have been noted in gymnasts and attributed to the gymnast's high abilities in spatial orientation (Huang et al., 2015). The gymnast relies on three interacting sensory systems (Gillingham & Wolfe, 1985; King Hogue, 1990; Krejcova et al., 1987):

- visual system
- vestibular system
- somatosensory system.

## 29.1 Visual system

Vision is the dominant spatial orientation system, as anyone trying to navigate around a room in the dark can attest, but the visual system also relies on a concordance of information from other senses (King Hogue, 1990). For example, most people have experienced the sensation of movement in a vehicle when waiting at a stop light and an adjacent vehicle to the side starts to move forward slowly. When this occurs, the visual system is confused and the perception that you are moving backward becomes apparent. This phenomenon of motion vection occurs in a variety of circumstances (Berthoz, Pavard, & Young, 1975). For example, if you place your hand on a slowly rotating horizontal drum with your eyes closed, you will notice after a few seconds that you feel as though your body is moving in the opposite direction to the drum. If you have experienced walking in a tilted room, such as those found at amusement parks, you will have experienced the sensory conflict that occurs when your eyes tell your central nervous system one thing, but the other sensory systems are conveying something else based on some reliance on gravitational vertical.

Vision involves two sub-types, foveal and ambient. Foveal vision is used to focus attention on something in the visual field, such as reading, and lies in the central region of the retina. Ambient vision includes both foveal and peripheral vision and is the most important vision-type for spatial orientation (King Hogue, 1990; Sands, 1991d). It is ambient vision that can be fooled by motion vection and does not require conscious awareness. Both types of visual systems are responsible for determining self-motion, environmental motion and the relationship between them. The visual system includes the optokinetic reflex which helps stabilize the eyes' focus on an external object (Shupert, Lindblad, & Leibowitz, 1983). Vestibular system motion detection (vestibulo-ocular reflex) is complemented by the optokinetic reflex (Shupert, Lindblad, & Leibowitz, 1983). The gymnast's sensory systems must resolve this visual and vestibular information to "make sense" of the motion of the gymnast combined with the apparent motion of the environment. Coordinated and smooth operation of the vestibulo-ocular reflex allows the eyes to remain focused on an object in spite of head motion. For example, coordinated motion occurs when the head turns to keep the eyes on the object of interest by activating the appropriate eye muscles to rotate the eyes in the opposite direction of the movement of the head. Systematic stabilization of vision during motion is exemplified when you view a video or film obtained with a camera that was not held still. Motion of the visual field (i.e. the field of view of the camera) results in jerky motions of an object in the field and considerable difficulty focusing on the object (Gillingham & Wolfe, 1985).

Injury and visual impairment of the eye can lead to serious consequences for the gymnast. For example, a study of retinal damage from school playground equipment (i.e. swinging from the knees on monkey bars, Rabinovitch et al., 1978) found that the high-speed motion of the head and the centrifugal properties of that motion led to extreme forces at the eye resulting in retinal haemorrhage.

## 29.2 Vestibular system

The vestibular system is composed of the bilaterally paired semicircular canals and the otolith organs. Together they provide orientation and motion information. The semicircular canals sense rotation of the head, while the otolith organs sense linear motion and gravitational vertical. Semicircular canals are located orthogonally to each other (roughly at right angles), and there are three of them corresponding to our three-dimensional world (Schone, 1984). The canals' orientations involve one horizontal canal (or lateral), one anterior or superior canal (vertically oriented) and one posterior or inferior canal (also vertically oriented). The anterior and posterior canals are also called the vertical canals. The horizontal semicircular canal provides orientation information about rotational motion and the horizontal plane. Rotational motion in the sagittal plane is detected by the anterior (vertical) canal, and rotation in the frontal plane is detected by the posterior (vertical) canal. Both anterior and posterior canals are oriented at approximately 45° between frontal and sagittal planes.

A structure called the cupula consists of tiny hair cells in another structure called the crista ampullaris which is covered with hair cells that respond to physical bending. The bending is achieved by the inertia of a gelatinous mass called endolymph. When the head rotates, the endolymph resists rotation by virtue of its viscosity and inertia and the hair cells are deflected by being pulled through the endolymph thus causing the hair cells to activate or inhibit neural traffic to the brain. Moreover, a push–pull system is involved in the activation of nerve cells in the canals such that when the canal on the left side of the head is activated, the canal on the right side of the head is inhibited (Schone, 1984). By analogy, when you are riding in a convertible automobile your head motion is caused by the external movement of the vehicle and the hair on your head is deflected by the wind opposite to the direction in which the vehicle is moving. Your hair does not discriminate its neural traffic based on direction, but you can still sense that your hair is moving and you can determine the direction of that movement.

Conserving the idea of using hair cells to detect motion, the otolith organs use the same mechanism to detect linear motion. Two otolith organs, the utricle and saccule, are paired on each side of the head. Both otolith organs have an area rich in hair cells called the macula. The hair cells of the maculae are embedded in a flexible membrane and have a crystalline structure on their ends made of protein-calcium carbonate granules called otoliths. The otoliths help the detection of gravitational vertical, head tilting and linear acceleration by using their inertia to bend the "stalk-like" hair cells away from the direction of head motion. The brain can determine direction, tilt and acceleration by comparing the outputs from these structures against themselves and the contralateral organs (Davlin, Sands, & Shultz, 2004; Hubel, 1988). The neural traffic from the semicircular canals and the otolith organs reaches many parts of the brain, and the function of each pathway is not clear.

## 29.3 Linking the visual and vestibular systems

Gymnasts require a sensitive interaction between visual and vestibular systems. Combining vision and vestibular systems one finds that athletes have preferred perceptual approaches to determining gravitational vertical. An assessment called the rod and frame test seeks to identify if a subject's perception of vertical is dependent on the visual field (i.e. the entire field and background) or not (Sands, 1991d; Schone & Lechner-Steinleitner, 1978). The rod and frame test has as subject view a "frame", usually a visible square, that is tilted away from gravitational vertical, and a rotatable "rod" that pivots from the centre of the frame (Figure 29.1). The rod is oriented in a position away from gravitational vertical and the athlete is asked to provide information about which way and how far to rotate the rod to align the rod with gravitational vertical. The test is performed over multiple trials when the apparatus and the room are lighted, and when both are in complete darkness. In the dark trials, the rod and frame "glow" such that the only things the athlete can see are the rod and frame. Again, the athlete is asked to orient the rod to gravitational vertical and the difference between actual vertical and the athlete's idea of vertical are compared (Asch & Witkin, 1948a, 1948b). The results of this test can be startling (Sands, 1991d). Although E. Dion noted that divers did not appear to be field dependent (Dion, 1985), another study showed that several highly trained Olympic level gymnasts were field dependent. Being field dependent means that the athlete relies heavily on external environmental cues for locating gravitational vertical. Loss of these environmental and visual cues could be devastating and a risk factor for injury (Sands, 1991c, 1991d, 1993a, 2007). Visual system engagement and somersaulting have been addressed in other contexts. T. Heinen and colleagues (Heinen, 2011; Heinen et al., 2014) have shown that novice gymnasts and high level gymnasts do not use vision in the same way during a trampoline backward salto. The same investigators showed that novice performers closed their eyes during the middle portion of a layout backward salto in a study of video images of the eyes during somersaults. The portion of the somersault performed with eyes closed was roughly the period when the eyes could not see the

**FIGURE 29.1** A rod and frame apparatus showing the rotatable rod inside the tilted square frame.

trampoline or floor. Divers are trained from the beginning to "spot" ("spot" refers to seeing a specific "spot" in the environment) and locate things in the environment for cuing on position of the somersault and when to open for the water entry (Hondzinski & Darling, 2001; Sands, 1991c, 2007). Laboratory studies of somersault rotation and spatial orientation have involved the use of rotating chairs or platforms (Figure 29.2) (Dukalsky & Dukalsky, 1977; Sands, 1991c, 1991d), microgravity (Botkin, 1985), unstable platforms (Milosis & Siatras, 2012) and visual occlusion (Davlin, Sands, & Shultz, 2001a, 2001b, 2004; Lee, Young, & Rewt, 1992; von Lassberg et al., 2012; Vuillerme et al., 2001).

## 29.4 Somatosensory system

The somatosensory system is a type of "catch-all" system that accounts for numerous types of sensory information that provides the central nervous system with indications of where the body is, its orientation and how it is moving. For example, a gymnast performing a giant swing can get an idea of his or her speed by sound of the air as the body speeds through it. When a male gymnast wears competition pants, the rustle of the pants created by the legs speeding through the air can provide a tactile indication of speed and position. In the winter, a heated gymnasium usually has markedly different temperature layers of the air within the facility. Personal experience has shown that gymnasts can recognize a change in temperature during a giant swing. The sounds created by creaking apparatus, thumping of mats, rhythms of serial impacts and the loud sounds of vault board take-offs can serve to tell the gymnast (and often the coach) about the speed and effectiveness of a skill. Vibrations of the bars, beams and floor can also serve to inform the gymnast of position, orientation and speed. Whether a gymnast feels her weight on the inside or outside of her foot can provide important information about the dance turn and information about

**FIGURE 29.2**   Helicopter jump seat with four-point harness used to rotate athletes in the sagittal plane to simulate saltos.

balance corrections that may be needed to correct performance errors. Combining the somatosensory system with kinesthesis allows the gymnast the ability to determine when his body is horizontal in a Maltese cross (Figure 29.3).

## 29.5 Kinesthesis

Kinesthesis is a sensory system for assessing body and limb positions without using one of the five traditional senses. Kinesthesis is what is involved when someone with their eyes closed can touch their fingertip to their nose, or the ability to perform a straight body position without looking (Figure 29.4). Kinesthesis, as joint position sense, is vitally important for avoiding and recovering from injury (Aydin et al., 2002; Mulloy Forkin et al., 1996). Saltos are performed in three basic body positions, tuck, pike and layout, with modified tuck positions called a puck (open tuck usually used with twisting somersaults), and a "cowboy" or "circus" tuck (legs are straddled while flexed at the knee to reduce the moment of inertia in saltos). The athlete's ability to distinguish a tuck from a pike with flexed knees is paramount for showing a clear pike position. Athletes also need to distinguish between an arched or hyperextended body position in an attempted layout salto from a slightly flexed torso position in what is called a "hollow" body shape, at least in the States. The athlete's ability to control body position while somersaulting is paramount to effective skill execution and safety (Bardy & Laurent, 1998; Lee, Young, & Rewt, 1992).

## 29.6 Balance – static and dynamic

Balance is one of the characteristics for which gymnasts have extraordinary skill (Asseman & Gahery, 2005; Bressel et al., 2007; Kerwin & Trewartha, 2001). Balance is divided into two basic categories, static and dynamic (Bressel et al., 2007; Gautier, Thouvarecq, & Larue, 2008; Hrysomallis, 2007). Static balance is observed during hold parts, standing for a period

**FIGURE 29.3** Maltese cross position. The gymnast must be able to determine if his body is level without looking with his eyes. Body position corrections are achieved by use of video that is displayed with video glasses.

on one foot, and dance poses. These elements are an important part of gymnastics and rules dictate the type of balance skills required. Females tend to have better static balance than males, often attributed to their standing posture's lower centre of mass (Schembri, 1983). The handstand is probably the quintessential example of a balance position commonly associated with gymnastics (Asseman & Gahery, 2005; Gautier, Thouvarecq, & Chollet, 2007; Yeadon & Trewartha, 2003). In spite of the "static" term that is used to describe these positions, the gymnast is constantly using muscle tension to maintain the position. You can see this by watching a gymnast's fingers, wrists, elbows and shoulders while holding a handstand. The joints will often show subtle but forceful movements used to maintain the position. At nearly any age and ability level, one can simply stand on one foot while trying to maintain balance to feel the foot and ankle working through muscle tension to maintain the position. Static balance tests such as the "stork stand" are often included in motor skill assessments and talent identification test batteries (Sands, 1993b, 2011b).

Dynamic balance is the ability to maintain postural and positional control while moving. Gymnastics displays these qualities most graphically on floor exercise and balance beam by virtue of held balances and stuck landings, pommel horse and parallel bars by maintaining control of the body in constant motion or stopping the body position changes in midflight and landing in a still position. Still rings for men involves large swings of the entire body using a multiple pendulum system and bringing all pendulum motions to a complete stop for held elements. Of course, the balance beam is perhaps the most obvious example of both

**FIGURE 29.4** Athletes should be able to control their body positions with extreme precision and without looking.

static and dynamic balance. The base of support is constrained to 10 cm, and the gymnast's centre of mass must be maintained over this base throughout slow and fast movements held still and with flight. Assessments of dynamic balance usually involve a simple count of errors that result in severe imbalances and falls or failures to continue the dynamic motion. To more fully comprehend the idea of dynamic balance in gymnastics, one can play a simple game using a drink bottle half-filled with fluid. The task of the game is to count how many times one can flip the bottle resulting in the bottle landing on its bottom without tipping over. Although this is more a hand-eye coordination test, one rapidly finds how difficult such as task can be. Consider that the gymnast must stick landings in a similar fashion after rotating and twisting the body at great heights. Landing strategies used by gymnasts have been studied to find that the height of the fall, the position of the body, previous training, the material properties of the landing surface, vision and other factors influence the gymnast's skill (Arampatzis et al., 2004; Bruggemann, 1987; Cuk & Marinsek, 2013; Gittoes, Irwin, & Kerwin, 2013; McNitt-Gray et al., 2004; McNitt-Gray, Yokio, & Millward, 1993; Sands, 1991d).

At the junction of balance, spatial orientation and kinesthesis, gymnastics training has offered some laudable but sometimes counterintuitive results. A study of stance perturbation (standing on one or both legs, with and without vision) showed that gymnasts' response latencies to being suddenly threatened with imbalance were greater (Debu & Woollacott, 1988). In other words, gymnasts tended to tolerate a balance threat and responded more slowly than age-matched controls. There is speculation as to why this might occur, but perhaps gymnasts are more accustomed to being "off-balance" and are willing to tolerate such actions. One study showed a decreased dependency on attentional processes involved in postural sway and unipedal (one-leg) stance in gymnasts versus non-gymnasts (Vuillerme & Nougier, 2004). In support of these findings a study of gymnasts and non-gymnasts showed better postural control and balance among the gymnasts (Garcia et al., 2011). However, a study of basketball players, gymnasts and soccer players found no difference in dynamic or static balance between the soccer players and gymnasts, but basketball players were statistically worse (Bressel et al., 2007).

Fatigue may play a significant role in the gymnast's ability to balance by decreasing balance ability. This was shown in fatigued trials of standing on a force platform and measuring postural sway (van Dieen, Luger, & van der Eb, 2012), and in the reduction of balance beam falls following fatigue and supplementation with a carbohydrate (20% maltodextrin solution) (Batatinha et al., 2013). Gymnasts were better than controls in maintaining balance in spite of occluded vision (Vuillerme et al., 2001).

## 29.7 CONCLUSION

Motor learning and motor control research are still in their early stages. Investigating the literature, I am struck by the lack of overarching covering laws or principles under which to organize an understanding of motor learning and control. The literature reads too much like a smorgasbord of different ideas that while interesting by themselves are difficult to unite in a coherent knowledge structure. In summary:

- Technique learning, motor learning, rely on an already fit athlete.
- Gymnastics is highly dependent on efficient and effective skill learning.
- Learning tends to occur in stages, for athletes and coaches.
- Intelligent and well-timed feedback is crucial for successful learning.
- Transfer is at the heart of coaching skills via the use of smaller and easier skills shifting to become more difficult skills or sequences of skills.
- Practice is a part of skill learning, and the intelligent conduct, content, sequencing and timing of skill teaching and learning are highly related to learning success.
- Gymnastics has a number of special needs, including spatial orientation, kinesthesis, and balance, among others.

# 30

# ADVANCED APPLIED EXAMPLE OF MOTOR CONTROL 1

## The running approach when vaulting

*Elizabeth J. Bradshaw*

### 30.1 Introduction and objective

The approach, board and table contact phases are critical to performance on the vault apparatus in artistic gymnastics. This chapter will provide insight on the motor control of the running approach when vaulting in artistic gymnastics. After completing this chapter, you will be able to:

- Explain the speed–accuracy trade-off as it applies to vaulting.
- Explain the role of vision in controlling the approach action.
- Expain how the motor control of the approach can be enhanced through training drills.

### 30.2 Role of the speed/accuracy trade-off

Design improvements to artistic gymnastics apparatus in the 1990s and early 2000s resulted in an increased "sweet spot" take-off area on the beat board and a much larger contact surface on the vault. Whilst these design improvements increased safety, they likewise improved the motor control conditions for the gymnast. This resulted in faster approach speeds, particularly for the handspring and Tsukahara vaults (Naundorf et al., 2008). By increasing the 'target' area for take-off from the board and contact area on the table, it reduced the speed–accuracy trade-off constraints (Bradshaw and Sparrow, 2000). Similar to long jumping in athletics, the gymnast must balance the speed requirements with an accurate take-off. The larger the size of the target and the less restrictive the contact conditions (position of the foot/feet/hands with respect to the target), the easier it is to negotiate (Bradshaw and Sparrow, 2001). Therefore, the gymnast can run faster when faced with larger targets.

Complexity of the pre-table movement also affects the speed of the approach, and for that reason the Yurchenko vaults have slower approach speeds whilst the handsprings have the fastest. The Tsukahara is only marginally slower than the handspring due to the differing pre-flight mechanics in order to contact the table in a roughly side-on position. In any vault, the hurdle starts the transition between the run-up and take-off for the vault itself. Because of this transition phase, the gymnast may lose some speed (~0.5 to 1.5 m/s) in order to control their momentum (Van der Eb et al., 2012), especially for the Yurchenko vaults. Greater speed can also be lost simply due to poor technique during the transition phase (e.g. uneven foot contact on the board, or the gymnast landing with the feet not square with respect to the board and runway after the round-off), and poor approach mechanics due to the gymnast not "targeting" the board and table.

## 30.3 Role of visual control

The visual control of the approach has often been an overlooked component of performance. Gymnasts are strongly discouraged from baulking (pulling out of) a vault during training and competition; however, this can be a sign of poor targeting (visual control) skills. Other indications are when a gymnast's run-up suddenly appears to change. This can occur simply as a result of growth or training techniques that have led to increased sprinting speed, or when muscle soreness or injury may decrease their running speed. It can also occur with minor or major changes in the composition of the runway such as between the training and competition podium surface.

Visual control of running towards targets is reasonably well understood in athletics (e.g. long jumping; Lee, Lishman, and Thomson, 1982; Hay, 1988) but has only been examined in Yurchenko vaults in gymnastics (Bradshaw, 2004). It was shown that the gymnast who accelerated earlier and faster, and commenced targeting the board as early as possible, had the fastest approach phase and could perform more advanced vaults. When compared to gymnasts who targeted either during the step preceding or during the hurdle, that gymnast lost less speed during the transition phase. Research from long jumping (e.g. Bradshaw and Aisbett, 2006) has shown that athletes who commence targeting earlier in the run-up are able to make smaller adjustments to their steps, therefore losing less speed. This differs to making adjustments during the final one to two steps which usually need to be much larger in order to successfully transition towards contacting the targets.

## 30.4 Applications to training

Coaching a gymnast to target the board and table is challenging, especially if it hasn't been instilled in them from when they first began vaulting. The quickest way to find out if a gymnast has limited targeting skills is to shorten or lengthen their starting position for their run-up by a small amount (20–50 cm) during training drills for vault. Observations of athletics coaching can provide many ideas for how targeting training drills can be incorporated into vaulting training. Checkmarks traditionally used in long jumping, however, should be avoided, as they are seen to encourage a fixed run-up pattern that reduces the requirement for the athlete to visually control their approach (Bradshaw and Aisbett, 2006). Changing surfaces, intention (e.g. 75% effort) and approach tempo (e.g. slower quarter, accelerate quarter, then maintain speed) are methods of facilitating targeting drills. Additional targets using floor or strapping tape across the runway can also be used for some additional explicit targets.

## 30.5 SUMMARY

The approach phase in gymnastics vaulting involves complicated motor control (speed–accuracy trade-off, visual control) in addition to a solid starting technique and sprinting speed. Even for a strong vaulter, helping them to control their approach better by addressing their motor control may unlock further performance potential.

# 31

# ADVANCED APPLIED EXAMPLE OF MOTOR CONTROL 2

## Rotations and twisting in gymnastics, is there a universal rotational scheme?

*Flavio Bessi*

## 31.1 Introduction and objective

In order to develop the highest possible level gymnasts, coaches and all parties involved have to optimize all aspects that have an impact on the performance. Bessi (2016a) divides the influencing factors into factors regarding the gymnast, the practice, the coach, the infrastructure, the environment and other factors (see Figure 31.1). Looking at the practice, the strategy for developing the skills may have an important role. For example, among other things is the decision of the correct and consistent direction while rotating both in upright stance and upside down around the longitudinal axis. Given that children choose a direction to rotate naturally very early, the question about how to implement the skills so that they respect a systematic and consistent structure may be crucial.

The all-around men finalists at the Rio 2016 Olympic Games will be analyzed with respect to their rotational system and categorized according to the classification system presented in this chapter.

## 31.2 Learning outcomes

After completing this chapter, you will be able to:

- Analyze the different types of turning gymnasts.
- Classify gymnasts according to the rotation matrix scheme.
- Discriminate few unambiguous rules of turning preferences.

## 31.3 The phenomenon of laterality

The phenomenon of rotational preference (which in our opinion is an important part of laterality) has not been researched with the same tenacity as the handedness, the footedness, the eyedness and the earedness. We consider the analysis of the rotational preference as an imperative issue in a sports context, and as fundamental within gymnastics. Figure 31.2 shows the different aspects of laterality even if we still don't know exactly how they interact.

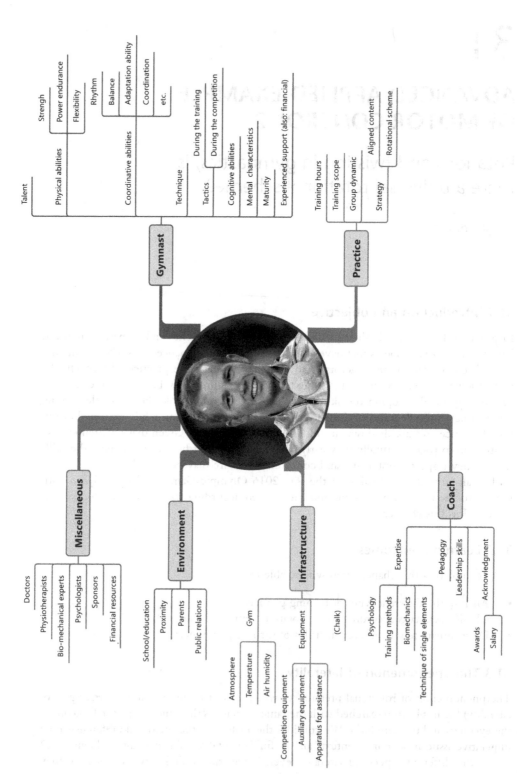

**FIGURE 31.1** Some aspects that determine the final performance of a gymnast.

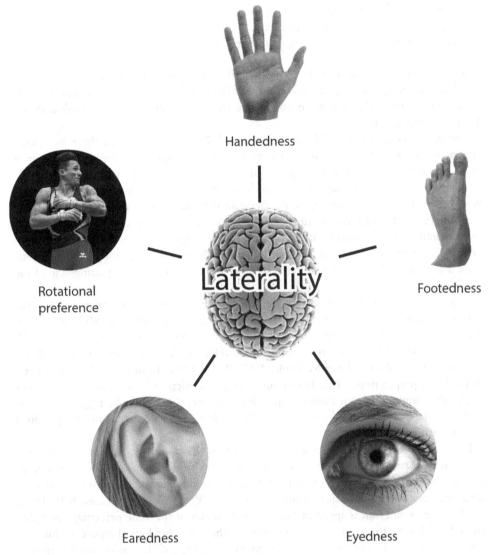

**FIGURE 31.2** Laterality and its different aspects.

## 31.4 Current state of research

We know that there is a tendency among humans to rotate to the left side (Coren, 1993; Dargent-Paré et al., 1992; Iteya and Gabbard, 1996; Mohr et al., 2004). Gymnasts are not an exception to this bias and therefore the majority performs rotations to the left (Koscielny, 2009; Bessi, 2015; Schindler, 2016).

There are different models that try to explain why people chose a certain direction rather than the other. Heinen, Bermeitinger, and Laßberg (2016) collected possible scientific explanations. They presented three hypotheses which could explain the tendency to choose one side instead of the other:

1. Hypothesis of the hemispherical asymmetries in the dopamine system.
2. Hypothesis of the vestibular asymmetry.
3. Hypothesis of the biomechanical or sensorimotor asymmetry.

Some other explanatory attempts focus more on the environmental influences (Figure 31.3).

While observing babies and very young children, it is possible to speculate that the setting of the rotational preference occurs very early, perhaps already in the womb as Previc (1991) also suggested.

The open question is still "Nature or nurture?" We can imagine that there are several influencing factors of the environment (Figure 31.3). For example, what the learning gymnast sees – if to demonstrate some turning movements, a coach always uses the same direction, it is likely that the gymnast may try to adopt the same side to rotate. Also, the already chosen direction of the training partners could facilitate the selection of a certain (perhaps not ideal) direction of rotation. And finally the coach's usual and preferred spotting side may have an influence. A capable coach should be able to spot the elements in both directions. Depending on the skill, he/she must often change the side of spotting. However, that is not always the case, so gymnasts often have to adapt to the favorite spotting side of the coaches.

In past years, there have been some recordings about the rotational preference in gymnastics (Crumley, 1998; Sands, 2000d; Bessi, 2006, Wüstemann and Milbradt, 2008, Schweizer, 2008). However, in more recent times, we can observe a greater interest in the subject as reflected in the current publications (Heinen, Vinken, and Velentzas, 2010; Heinen et al., 2012a, Bessi and Milbradt, 2015; Bessi, 2016b; Bessi et al., 2016; Bessi and Milbradt, in preparation). This is not surprising, considering how important rotations around the longitudinal axis have become. For instance, the current best gymnast in the world, the Japanese Kohei Uchimura, has 40 elements or movements with longitudinal rotation in his six routines.

For many years some coaches have assumed that the dominant hand determines the direction of rotation. The assumption was that right-handed gymnasts should rotate to the left. Contrary to what is often claimed it seems that there is not a relationship between the handedness and the direction of rotation (Heinen, Vinken, and Velentzas, 2010; Bessi, 2016b). The assumption is apparent because there is almost the same percentage of right-handed and left rotating gymnasts. Furthermore, there is some logical aspect in this supposition when analyzing throwing movements: because of the cross-coordination, a right-handed person will perform a kind of leftward rotation with the upper body while throwing. However, the results of both Heinen, Vinken, and Velentzas (2010) and Bessi (2016b) suggest that there may be a slight relationship between footedness and the rotational preference.

### 31.5 How do the best gymnasts in the world rotate?

We analyzed all routines of the men's individual all-around finalists during the Olympic Games Rio 2016 who finished the whole competition determining in which direction the gymnasts rotate around the longitudinal axis when performing elements with this kind of rotation.

The elements with rotation around the longitudinal axis of the 22 gymnasts were registered as was the direction of turning as well. With these data, we determined which

**FIGURE 31.3** Different and complementing hypotheses to explain the rotational preference.

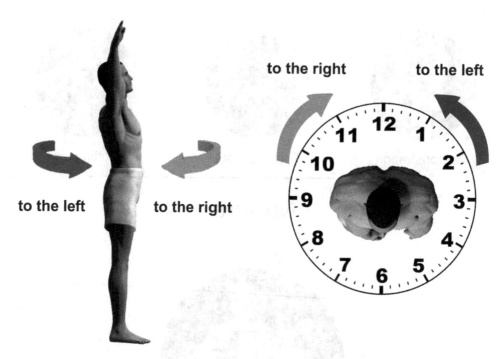

**FIGURE 31.4** Definition of the rotation direction around the longitudinal axis.

rotational scheme they follow based on the new classification system, which we present for the first time in this chapter.

The direction of rotation about the longitudinal axis was defined from the perspective of the gymnast. A rotation to the left in upright stance corresponds to a backward rotation of the left shoulder and forward rotation of the right shoulder. When observing from above the gymnast performs a counterclockwise rotation. A rotation to the right in upright stance was defined vice versa (see Figure 31.4).

## 31.6 Classification system of rotational schemes

The first author who proposed a system for classifying the rotational types was Laßberg (2008). This proposal was first slightly modified by Bessi (2015). After analyzing a large number of gymnasts, we want to propose a new classification system, which in our opinion matches better the reality (Figure 31.5). This new system has two basic pure categories:

*Bilateral consistent rotation scheme (BC)*: A pure bilateral consistent rotating gymnast always rotates in the opposite direction around the longitudinal axis when being in an upside down position, as compared to when being in an upright position. The best way to identify the type of rotation scheme is to start observing the round-off and the back somersault with turn. Such a gymnast performs the round-off left (i.e., rotating right, as shown in Figure 31.6), and the twist to the left.

*Unilateral rotation scheme (U)*: A pure unilateral rotating gymnast always rotates in the same direction, independent of the element or the body orientation in space. Such

Left BC                    Bilateral Consistent                    Right BC

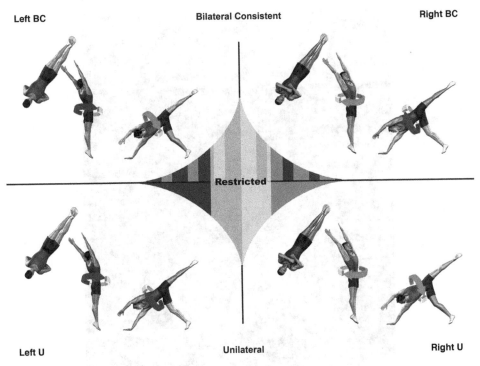

Left U                          Unilateral                          Right U

**FIGURE 31.5**   Classification matrix of rotation schemes.

a gymnast performs, for example, the round-off left (i.e., rotating right, as shown in Figure 31.6), and a somersault backward with turn to the right as well.

Certainly, there are considerations that may lead to the decision that a gymnast has to give up the preferred rotation scheme partly depending on the situation. For example, a gymnast on high bar must make sure that he performs in the middle of the bar. This could require him, for example, to turn around the "wrong" arm. This can also happen on the parallel bar, especially when the gymnast is at the end of the bars. To take account of these eventualities, we counted elements that do not fit into the scheme up to a maximum of two during the whole all-around. In this case, we weaken the pure basic rotational type by identifying it with the word *restricted*.

*Restricted bilateral consistent rotation scheme (BCr)*: A restricted bilateral consistent rotating gymnast is basically a BC gymnast. However, he shows up to a maximum of two elements during the all-around competition that do not fit the pure BC scheme.

*Restricted unilateral rotation scheme (Ur)*: A restricted unilateral rotating gymnast is basically a U gymnast. However, he shows up to a maximum of two elements during the all-around competition that do not fit the pure U scheme.

The terminology of pure and restricted rotation schemes should be preceded by the direction of turning in upright stance. Former analyses indicate that it is best to determine this by the direction of the backward twist. Therefore, the classification system has eight theoretically possible categories: lBC, lBCr, lU, lUr, rBC, rBCr, rU and rUr (Figure 31.5). However, this is only theoretical as we will show in the following.

All gymnasts who do not fit into the preceding categories are labeled with *no distinguishable rotation scheme (ND)*. At this point, it seems appropriate to mention that ND is not an

**FIGURE 31.6**   A round-off left is indeed a rotation to the right around the longitudinal axis.

evaluative category. It only indicates that the conditions of the aforementioned categories are not fulfilled during the analysis of the corresponding elements of the given gymnasts.

That means that a gymnast who turns according to a rotational scheme but has two turn elements in the opposite to the expected direction is assigned to the corresponding category preceded by the weakening *restricted*, while another gymnast doing so but with three unexpected elements passes to the category *no distinguishable rotation scheme*. Even if this limit is based on our experience while analyzing a large number of gymnasts, the definition is arbitrary and serves exclusively for differentiation.

The majority of the gymnasts of our selected sample exhibit a leftward turning preference (Table 31.1 and Figure 31.7). Thus, the results of Sands (2000d), Schweizer (2008), Koscielny (2009), Bessi (2016b) and Schindler (2016) as well as observations by Bessi (2006), Wüstemann and Milbradt (2008) and Hofmann (2015) were verified. The results revealed that 82% of the men's individual all-around finalists rotate to the left when in an upright position, while 18% prefer to rotate to the right. This coincides quite exactly with

**TABLE 31.1** Results of the men's individual all-around final at the Olympic Games in Rio 2016 and their rotational schemes

| Place | Gymnast | Country | Rotational type |
|---|---|---|---|
| 1 | Uchimura, Kohei | JAP | IBC |
| 2 | Vernaiev, Oleg | UKR | IBC |
| 3 | Whitlock, Max | GBR | ND |
| 4 | Belyavskiy, David | RUS | IBC |
| 5 | Lin, Chaopan | CHN | IBCr |
| 6 | Deng, Schudi | CHN | IBCr |
| 7 | Mikulak, Samuel | USA | IBCr |
| 8 | Wilson, Nile | GBR | IBCr |
| 9 | Sasaki, Sergio | BRA | ND |
| 10 | Calvo Moreno, Sossimar | COL | IBC |
| 11 | Kato, Ryohei | JAP | IBC |
| 12 | Yusof, Eddy | SUI | IBC |
| 13 | Kuksenkov, Nikolai | RUS | ND |
| 14 | Brooks, Christopher | USA | rBCr |
| 15 | Deurloo, Bart | NED | ND |
| 16 | Braegger, Pablo | SUI | IBC |
| 17 | Mariano, Arthur | BRA | rBCr |
| 18 | Likhovitskiy, Andrey | BLR | IBCr |
| 19 | Nguyen, Marcel | GER | IBCr |
| 20 | Bretschneider, Andreas | GER | IBCr |
| 21 | Augis, Axel | FRA | IBCr |
| 22 | Stepko, Oleg | AZE | ND |

the results of Bessi (2015) who analyzed the direction of turning in an international sample of 161 coaches and former gymnasts from 14 countries with different levels of expertise and different types of employment based on a questionnaire (Bessi, 2007a, 2007b).

Four (40%) of the best ten all-arounders and seven (32%) of all finalists are lBT. A total of eight (80%) left turning gymnasts are among the ten best and 15 of all finalists (68%). The question may arise if the rotational scheme could have an influence on the performance to determine who takes the medals. Due to the fact that the distribution of the direction corresponds in large part to previous results, it is likely that this is not the case, but more data are needed to answer this question. Very surprising was the finding that among the best participant field of the world are five gymnasts with no distinguishable scheme, one of them (Max Whitlock) even won the bronze medal in the men's all-around gymnastics at the Rio 2016 Olympics. No lU, lUr, rU or rUr gymnasts were among the finalists.

There were eight countries which were able to qualify two gymnasts to the final (BRA, CHN, GBR, GER, JAP, RUS, SUI and USA). If we look at them, we can state confidently that some nations seem to have a preferred rotational scheme. With the current data and without having interviewed the national responsible person we cannot say if this finding is casual or a desired development. However, we can determine that both the Japanese and

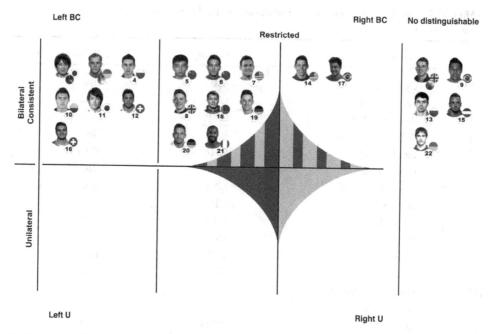

**FIGURE 31.7** The gymnasts of the men's individual all-around at the Olympic Games 2016 with their respective rankings sorted by their rotational scheme.

the Swiss gymnasts are lBC. The gymnasts from China and Germany are lBCr. The elements of the Chinese gymnasts which do not fit into the pure rotation lBC scheme are exactly the same so that it can be speculated that perhaps a national strategy leads to the finding as well. Both lBCr German gymnasts show different non-matching elements so that we cannot assume a national strategy. By the Brazilian, British, Russian and US-American gymnasts no coincidences can be determined in the respective rotation schemes of the two finalists.

There are some other interesting findings. Only two gymnasts, Uchimura (JAP) and Deurloo (NED), perform the circles on pommel horse in a clockwise direction. Especially the Japanese showed something that is very rare for left rotating gymnasts. Several gymnasts (50%) perform Moznik (Tkatchev stretched with half turn to mix el-grip into back upraise to handstand) in the opposite direction than expected. Since the rotation around the longitudinal axis occurs in upright stance, it should be in the normal turn direction in order to match to the corresponding scheme. Very uncommon was Sasaki (Brazil), who performs not only twists to the right and the left but also double Arabian to the right with half turn to the left, which means both directions of rotation during the same acrobatic element. H. Araújo, one of the Brazilian national coaches, confirms that this kind of turn combination was not specifically targeted, but arose spontaneously (personal communication, December 13, 2016).

## 31.7 Unambiguous findings so far

The aim of setting a "logical" scheme of rotation is still unachieved. The unique rule that we can state so far is that the same elements on different apparatus are performed without exception in the same direction (i.e., handstand with half turn forward on floor and parallel bars, round-off on floor and vault, circles on floor and pommel horse, dismounts on rings and high bar, etc.). The dismounts with rotation through the handstand on pommel horse always follow the direction of rotation of the circles; that means that counterclockwise circles turn to the right and vice versa).

Furthermore, we can state that the majority of the gymnasts rotate to the left. Therefore, we can also state that, obviously, gymnasts cannot escape the worldwide bias to rotate to the left. Additionally, the "normal" scheme of rotation is the bilateral consistent or restricted bilateral consistent.

## 31.8 SUMMARY AND CLOSING REMARKS

A consistent scheme has limitations and cannot be applied to all elements. For example, a gymnast performing Makuts on parallel bars has to disregard the "rule" either with the first supporting arm or with the second one.

For quite some time it has become increasingly apparent that gymnasts perform twists in both directions in acrobatic series. This forces us to rethink the classification. Is a gymnast who does not follow a rotation scheme quite exactly, immediately to be deemed inconsistent, only because he deliberately performs a twist in the unexpected direction? Not likely. The further development of the classification matrix should take this kind of previously considered decision into account in the future.

## 31.9 Acknowledgments

I would like to thank Mr. Hardy Fink for his proofreading and advice.

# 32

# ADVANCED APPLIED EXAMPLE OF MOTOR CONTROL 3

## The impact of Elastic Technologies on Artistic Gymnastics – a special case study from Brazil

*Marco Antonio Coelho Bortoleto*

### 32.1 Introduction and objective

Technological development has revolutionized human life, mainly from the second half of the 20th century, drastically modifying sport practices, including Artistic Gymnastics (AG). The goal of this chapter is to present an analysis of the fundamental role of the Elastic Technologies (ET) in AG training, from the elementary to high performance levels, and to discuss its usage within a case study from Brazil.

In our findings we've noticed that coaches and researchers recognize the importance of the ET, however its use is approached as a secondary tool, barely addressed in the literature, and often faces a slight resistance to be systematically incorporated in the AG training process.

### 32.2 Learning outcomes

At the end of this chapter you will be able to:

- Recognize the relevance of the sports technology, especially for AG, and its consequences for training and competitive performances.
- Understand how the ET has become one of the most important aspects for AG apparatus development.
- Discuss the contribution of ET to increase the level of acrobatics difficulties in AG.
- Find out how Brazil is dealing with the adoption of ET for AG training.

### 32.3 The technological revolution in sports: how AG has been affected?

Over the last two decades we have witnessed an exponential increase in acrobatics in a variety of sports (Diving, Parkour, Wushu, Capoeira, Snowboard, etc.) and performing arts (Circus, Dance, etc.). This tendency has become even more significant in the past few years, elevating sports to the level of a "spectacle" that has never been seen until now, as was predicted by Bordieu (1973). It is in this context that we have analyzed AG; a sport that has always appealed to science and technology in the pursuit of new limits, creating a paradoxical reality between an increase in risk-taking and a greater capacity to control such risk (Le Breton, 1995).

We have witnessed, concurrently and dialectically, extraordinary technological advances in this time period, with a great impact on many sports, as has been mentioned by the French historian Georges Vigarello (1988). This impact has been especially seen in sports that feature acrobatics as their core element, among which AG is in the forefront. In these specialized sporting activities, the "technological revolution" has generated profound material, technical, conceptual and aesthetic modifications (Konstantin, Subic, & Mehta, 2008). To summarize: there has been a change of *paradigm* proportions not only in training routines but also in AG competitions themselves.

These technological advances have generated great impacts mostly for physiological, biomechanical, medical treatment (injuries and traumas) issues and, above all, to the development of new equipment and methods for controlling the training process (statistical-computational) (Jemni, 2011a; Bortoleto & Peixoto, 2014; Sands et al., 2016).

However, what particularly interests us is the development of the surfaces and equipment made by "elastic components" that enable the kinetic energy amplification. The innovation of AG equipment has helped to propel higher and increase the speed of the gymnasts, which visually enhances the spectacle and helps to meet the media expectations for such an aesthetic sport. This phenomenon explains precisely what has inspired us to spark a discussion about the impact of the ET on AG.

## 32.4 The impact of ET on AG

Recent scientific studies have underlined the importance of ET to the technical development and improvement of AG, and its consequences for the training process, as an essential requirement for the establishment of safer training (Mills, Pain, & Yeadon, 2006). The relevance of ET lies in its capacity to diminish the impact on the body, featured in the acrobatic training process. The FIG Apparatus Commission member, Ludwig Schweizer, also suggests:

> The support during take-off actions and the reduction of force peaks during landing and impact situations by the use of more elastic and damping components in the equipment allows for a more intensive level of training of difficult skills. Athletes would suffer greater, longer lasting injuries without the new equipment technology in AG.

In fact, since the 1980s (Karacsony & Cuk, 2005), a significant change in material and technology has emerged in AG, with the substitution of old-fashioned rigid apparatus for flexible ones (ET). This represents a paradigm change, that has somehow influenced the Code of Points (CoP) and vice versa, when we think of the hypervaluation of acrobatics and flight elements, and how complex and dynamic the training and also the competitive routines have become. One of the consequences of this upward trend is the higher impact in landings (Thomas et al., 1997), and consequently greater foot and ankle injuries (Chilvers et al., 2007).

As a consequence of these apparatus changes, a remarkable transformation has been seen in AG training practices and competitive routines, introducing significant modifications to the technical and physical preparation of gymnasts (Bortoleto, 2004). The new generation of "Flexible Equipment" (FE) and "Elastic Surfaces" (ES) has led to a number of modifications with regard to the execution time and energy spent on acrobatic elements (Smoleuskiy & Gaverdouskiy, 1996; Sands et al., 2013).

Considering that repetition is a fundamental characteristic of AG training (both men and women artistic gymnastics (MAG and WAG)), either for learning or mastering specific skills, the correct use of ET has considerably contributed to the reduction of impact/stress on joints (Arkaev & Suchilin, 2009), to improvements in safety (Turoff, 1991), and, therefore, to the reduction of a number of injuries, especially to knees, shoulders and ankles (Leglise, 1985; Fink, 1985; Sands et al., 2013).

To exemplify, we observed that in the first decades of the 20th century, Floor Exercises (FX) were performed outdoors, on natural grass, beaten earth or on thin layers of sawdust. There was a gradual substitution of those to artificial surfaces, at first made of "tatami" (straw plates covered with a canvas), then to cotton-filled fabric mats and, finally, to synthetic foam mats (sophisticated materials, densities, forms and sizes) (Karacsony & Cuk, 2005). It was in the 1980s that the "Spring Floor" system arrived, with its springs, wooden boards, an agglomerated polyurethane foam covered with a thick carpet, which modified FX routines significantly (Oliveira & Bortoleto, 2011; Sands et al., 2013). The actual FIG apparatus normative (FIG, 2015, p. 14) reinforces this point of view:

Functional Properties – Performance area and borders:

- Elasticity equality on the surface and dampening;
- When in use, it should not have any hindering motion energy;
- Elasticity and dampening must be balanced in such a way to guarantee the gymnast stability and freedom of movement.

The Horizontal Bar (HB) developed by Nichols-Ketchum in 1998,[1] for example, brought a new steel composition and a dynamic anchoring system to the development of aerial phase elements (or "flight elements") such as Yamawaki/Walstrom, Markelov, Kovacs, Kolman, Gaylord/Pegan, and also allowed the dismounts elements with double straight somersault with a double twist or even triple somersault (Kerwin, Yeadon, & Harwood, 1993).

In rings, after the "the pivoting mechanism" regulation which included the "elastic dampening device," different transversal rotation elements in competitive routines such as Guczoghy/O'Neill and Yamawaki/Jonasson have become more frequent since they are less harmful to the shoulder joints of gymnasts.

In Vault (VT), a technological improvement of the springboard (Richard REUTHER model) in the 1980s allowed more external forces (Sands, 2011a). The "horse" (rigid surface) substitution by the "Pegasus" vault table model (with elastic/flexible components) in 2002 (Oliveira & Bortoleto, 2011) also allowed a considerable repulsion by upper limbs resulting in higher elevation of the gymnast's Center of Mass (CM) during the second phase of the vault (post flight), therefore gaining extra time to accomplish more rotations during a vault.

The Yurchenko (EG III) vault development (e.g. Melinassanidis), incorporated into the CoP in 1983, reinforces this development trend (Seeley & Bressel, 2005; Nakasone, 2015). Forward handspring vaults with multiple rotations on the second flight (post flight) have appeared, among them: Roche, Dragulescu, Zimmerman, Blanik, RI Se Gwang, and more recently in the Rio Summer Olympic Games 2016, the Radivilov (Figure 32.1) performed by the Ukrainian gymnast Igor Radivilov.

Considering the Parallel Bar (PB) routines, the flight elements, such as Belle, Morisue, Lee Chul Hon/Sasaki and Suarez, have been more frequently performed, as more flexible

**FIGURE 32.1**  Radivilov (drawing by Koichi Endo).[3]

bars, made from glass fiber/carbon coating with wooden rails, have allowed the development of techniques that project the gymnast's CM right over the bars. It is important to emphasize that the technical standards highlight that "the bars must have elasticity" (FIG, 2016).

There are evident apparatus differences in architecture and material from the 1960s/1970s to the present moment, observed on components and flexible indexes as exposed by a number of the patents.[2] Recent technological advances reflect the search for AG surfaces and equipment with greater reactive capacity, which means a responsive apparatus to the pressure exerted on them and a greater energy impact absorption (Pérez, Llana, & Alcántara, 2008). At the same time, other complementary equipment used in training has undergone major modifications in architecture and components, enhancing their elastic capacity in order to generate impulsion and dampening. For example, many acrobatic tracks (Fast-track, Tumble-track, Air-track) started to be used more often than the official apparatus (floor) during training sessions (Bortoleto, 2004).

In all AG apparatus, the new technology of mat used for landing, in addition to improving the surface stability, also offers better absorption of the impact energy (dampening), generating significant modifications in landing techniques and permitting dismount elements with more rotations (McNitt-Gray, Yokio, & Millward, 1993; Mills, Pain, & Yeadon, 2006). The training process was certainly facilitated by the use of different types of landing mats, and by the use of technologies which combine different foam types and densities with other materials. In fact, the relevant technological advance of mat manufacturing, supported by recent investigations (Table 32.1), was important to better understand the dynamics of landing impact absorption and to the formulation of new techniques for dismounts, for example.

The patents displayed in Table 32.2, registered from the 1970s onwards, represent examples of ET development, having practical consequences starting from 1980 (Piard, 1982; Pozzo & Studeny, 1987; Turoff, 1991). It is worth noting that this process required tests by FIG for all apparatus regulation and certification for its conformity with the CoP, General Regulations and, mainly, by the Apparatus Norms (FIG, 2015).

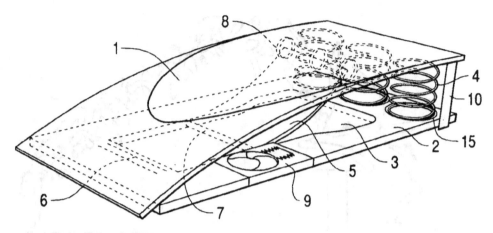

**FIGURE 32.2** Springboard patented in 2003 (European Patent 7175567).

Analyzing these technological devices, we can clearly see a search for a greater elastic capacity that better attenuates impact during the landing phase (damping) and, at the same time, generates more energy for acrobatic elements (an example of springboard design is shown in Figure 32.2). In general, the patent descriptions reinforce how safe the equipment is.

As a consequence of these advancements, certain acrobatic elements that were unexpected a short while ago have become common nowadays. The regular use of the ET allowed the execution of higher complexity acrobatic sequences more precisely and safely when used correctly. Paradoxically, we have observed an increase in risk, a dilemma therefore rising a constant modification in the CoP, reinforcing the need to keep gymnasts safe, as highly recommended by Dr. Schweizer.

**TABLE 32.1** Recent investigations on landing mats in artistic gymnastics (AG)

| Date | Article |
| --- | --- |
| 2005 | Video analysis of the deformation and effective mass (density) of gymnastics landing mats (Pain, Mills, & Yeadon, 2005). |
| 2006 | Lower extremity biomechanics during the landing of a stop-jump task (Yu, Lin, & Garrett, 2006). |
| 2006 | Modeling a viscoelastic gymnastic landing mat during impact (Mills, Pain, & Yeadon, 2006). |
| 2009 | Reducing ground reaction forces in gymnastics landing may increase internal loading (Mills, Pain, & Yeadon, 2009). |
| 2010 | Effects of mat characteristics on planter pressure patterns and perceived mat properties during landing in gymnastics (Pérez-Soriano et al., 2010). |
| 2010 | Modifying landing mat material properties may decrease peak contact forces but increase forefoot forces in gymnastics landing (Mills, Yeadon, & Pain, 2010). |
| 2013 | Different modes of feedback and peak vertical ground reaction force during jump landing: a systematic review (Ericksen, Gribble, Pfile, & Pietrosimone, 2013). |
| 2013 | A systematic review of different jump landing variables in relation to injuries (Aerts et al., 2013). |

**TABLE 32.2** Chronological ET patents for AG

| Technology denomination | Patent | Date |
| --- | --- | --- |
| Roll-fold floor mat for gymnastic and athletic purposes | United States Patent 3636576 | 1972 |
| Resilient floor, especially for gymnasiums | United States Patent 3828503 A | 1973 |
| Arrangement for Floor Gymnastics/Floor Panel System | United States Patent 4135755 | 1977 |
| Tumbling floor | United States Patent 4316297 A | 1980 |
| Aluminum balance beam | United States Patent 4272073 A | 1981 |
| Gymnastics springboard (*Tremplin de gymnastique / Reuther Turn-Und Sportgerate*) | European Patent 0086274 B1 | 1982 |
| Gymnastics mat and floor mat | Germany Patent 3416644 A1 | 1984 |
| Gymnastics floor structure having vertical elasticity | United States Patent 4648592 | 1985 |
| Gymnastics apparatus for vaulting exercises performance | European Patent 0885634A2 | 1997 |
| Gymnastics springboard with adjustable elasticity designed for training and competition | European Patent 7175567 Germany Patent 21033115 A1 (Figure 32.2) | 2003 |
| Gymnastics springboard with adjustable elasticity for training and competition | European & France Patent 1314454 B1/Germany Patent 60212789 D1 | 2006 |
| Gymnastic floor structure | United States Patent 7849646 B2 / US20090139172 A1 / Japan Patent 2007-240445 | 2007 |
| Air-cushion floor | China Patent 201099969 Y | 2007 |
| Impact absorbing gymnastic mat | United States Patent 20070173379 A1/Japan Patent 2004/009824 | 2007 |
| Flexible mat with multiple foam layers | United States Patent 20130017372 A1 | 2011 |
| Foam pit assembly | United States Patent 20150343249 A1 | 2015 |

Over a decade of research has shown us, unequivocally, that currently in most high-performance training centers, the incorporation and systematic use of ET has become imperative in the optimization and maximization of acrobatic improvements. In this context, we will present an example of how ET impacts the training process in one of the main high-performance gymnasiums in Brazil.

## 32.5 The use of ET in Brazil

Brazil has only eight high-level AG training centers in operation. Recently, we conducted an ethnographic investigation in one of them, located in the state of São Paulo, used by some members of the national team who are World and Olympic medalists. According to the national competition results, this gymnasium has been classified over the last ten years as the best nationwide in MAG (Bortoleto & Coelho, 2016). Although it is one of the leading gyms in Brazil, equipped with all official AG apparatus (all of them imported) this gym has only one "in-ground" trampoline (I-TRA), an old fashion Tumble-track (TT), and some additional official apparatus installed on the pit foam.

It is a clear example of the insufficient ET availability when compared to those renowned international gyms.

In general, the use of ET (ES, FE) represents a *non-significant* part of the time dedicated to training, around 40 minutes daily, taking both training sessions (morning and afternoon, 6–7 hours in total) into account. This timeline may vary depending on the training period and that amount may be even further reduced during pre-competitive periods. In the first three age groups, the ET are used more frequently and regularly when compared to *seniors*, given that *senior* gymnasts practice more on the official apparatus.

Gymnasts who are treating injuries, especially on the upper limbs, or are under restriction to train in the official apparatus often use the ES in order to keep themselves active (as suggested by Chase, Magyar, & Drake, 2005), integrate with the group and connect with the training program.

The pit foam, in particular, is frequently utilized during technical training, mainly because there are some apparatus over it which permits greater safety, as affirmed by the Brazilian coaches. The acrobatic elements from the MAG CoP that most frequently used ET for teaching and/or improvement are shown in Table 32.3.

Although coaches recognize the importance of using ET in the gymnast's training process, their regular utilization becomes more complicated due to the scarcity of equipment, and/or the lowering quality of them, and mainly the lack of habit of using them by the coaches. This contradictory situation can be noted in the discourse of the head coach, Mr. Marcos Goto:

> I believe that using the ES helps to increase the gymnast's longevity, to reduce the incidence of injuries and to increase the number of repetitions with less effort. The ES are surely beneficial but we use them less than we should![4]

The coach's opinion reinforces the use of the use of the ET and contributes to the learning *transfer* (Turoff, 1991; Arkaev & Suchilin, 2009) of different skills, as the back somersaults performed on the I-TRA, and the Kovacs elements and its variations (among them, the Kolman) are currently very common in the HB routines. In Goto's own words:

**TABLE 32.3** Acrobatic elements trained on ES

| Elements | Apparatus | ES |
|---|---|---|
| Cross, unlock, dismounts (Somersaults) | Rings (SR) | I-TRA |
| Sequence of backward Somersaults; Tempo; Sequences of forward Somersaults; Cartwheel, Round off, Flic-Flac/ Backward Handspring | Floor (FX) | TT |
| Pirouette (rotations in longitudinal axis); Somersaults (Simple; Double) | All apparatus (except Pommel Horse) | I-TRA, TT |
| Front Handspring Vault, Round off, Flic-Flac/Backward Handspring, Yurchenko, etc. | Vault (VT) | TT |
| Suarez, Diamidov, Carballo, Belle; Double Somersaults (dismount) | Parallel Bars (PB) | I-TRA |
| Yamawaki, Tkatchdev, Kovacs, Giant, Jäger, Gaylord, Endo/ Stalder (upswing); Double Somersaults (dismount) | Horizontal Bar (HB) | I-TRA |

Many exercises, such as the Diamidov for the Parallel Bars, are trained on ET. It is common to train the Tsukahara or Kovacs (both on the Horizontal Bar) in the I-TRA. The Tkatchev on the HB is also taught with exercises on the TT or on the I-TRA.

The TT is used almost daily for training acrobatic tumbling sequences which significantly increase in the number of repetitions. When comparing the FX acrobatic sequences with those performed on the TT, one clearly notices that on TT a greater height is achieved with less impact, less demand of power from the lower limbs, making the learning and performance improvement easier. It is worth underscoring that the coaches complained about the TT's inferior quality, which has been manufactured by a Brazilian company, with lower quality standards than those approved by FIG. Moreover, the fact it is not to be installed in-ground increases the risk of accidents. In Brazil, only two training centers have the in-ground TT, in contrast with what can be observed in the best international gymnasiums, as emphasized by Mr. Goto.

On the other hand, for training high difficulty acrobatic elements, such as the HB or PB dismounts, the I-TRA is used on a more frequent basis, corroborating Turoff's position (1991, pp. 238–239). An example that was observed repeatedly was the PB *Suarez* element, executed at first on the I-TRA in order to master the activity and, subsequently, reproduced on the official apparatus itself. The aforementioned coach confirmed that "it is possible to work many of the Parallel Bars elements and almost all flight elements of the HB on I-TRA, and then transfer to the official apparatus."

## 32.6 CONCLUSIONS

AG has shown a tendency to increase acrobatic and flight elements and to make its routines more "spectacular," adjusting the performances to the concept of "sport as spectacle." A move towards more dynamic exercises can be observed, with the exception of the pommel horse, according to the sporting development logic reported by Vigarello (1988) in the 1980s.

Technological development, in particular of the ET, can be viewed as responsive to this trend, noticeably modifying AG's training system for decades. As mentioned earlier, different ET are already a part of the day-to-day routine of the main AG high-performance training centers. Nevertheless, the reality of the gym analyzed shows the challenge for Brazilian clubs to follow this tendency, very likely due to the difficulty of importing the equipment, given that Brazilian-made materials have not been approved by FIG, at present, as they do not offer the same quality. This delay inhibits Brazilian clubs from following the international tendencies.

It seems clear that the difficulties coaches face in incorporating the use of ET in training can also be attributed to the nonexistence of the continuing education program in Brazil, showing minor chance to change the situation observed previously.

Although we mentioned a specific case study, a number of other investigations suggest that it is a nationwide reality related to the difficulty of acquiring ET and, consequently, with the lack of expertise for its use (Schiavon, 2009; Oliveira, 2014).

It seems evident that the use of ET is considered vital by the experts, although their inclusion depends as much on the financial conditions of the institutions – therefore, on access to these technologies – as on the capacity and the motivation of coaches to incorporate them into the training process, which, in a conservative and traditional practice such as AG, is not a trivial matter (Foster, 1973; Barker-Ruchti, 2007; Bortoleto & Schiavon, 2016).

Finally, we were very surprised to find so few scientific references about the incorporation of the ET into the daily training of artistic gymnastics.

## 32.7 Summary

- The technology has strongly influenced sports and has promoted, especially in Artistic Gymnastics (AG), a particular revolution in training and competitive performances.
- The apparatus development for AG training and competition has been looking for more elasticity of the materials, creating a new generation of Elastic Technologies (ET).
- The ET has become the main ally for the training of AG athletes, being useful to decrease injuries incidence and increase the acrobatics difficulty (height and number of rotations).
- The access to ET is still restricted in Brazil since there are no national providers recognized by FIG.

## 32.8 Acknowledgments

We would like to thank CNPQ (Brazilian National Research Council) and CAPES (Coordination for the Improvement of Higher Education Personnel), Mr. Ludwig Schweizer (University of Freiburg/Germany), Professor Dr. Keith Russel (FIG Scientific Comission; University of Saskatchewan/Canada) and Mr. Marcos Suzarte Goto (Brazilian MAG Senior National Team Head Coach).

## 32.9 Notes

1 Nichols-Ketchum, M. Gymnastics bar and method of making the same. American Sports International, Ltd. (Jefferson, IA) (US Patent 6,475,118. / Registered in 1988).
2 The patent register requested by Nissen Corporation in 1962 at the EUA ("Parallel bar gymnastic apparatus" US 3232609 A), and the 1994 registered by Ted Winkel also in EUA ("Parallel bars" US 5720697 A), represent good examples.
3 As mentioned by FIG "two elements not successfully performed in the men's competition will be published in the men's Code of Points, though they will not be named for the gymnast who attempted them. If done well by their originators in a future competition, they may be named retroactively" (https://live.fig-gymnastics.com/news.php?idevent=6405).
4 Interview with Brazilian Senior National Team head coach, leading the squad in several international competitions, having won two World Championship medals and three at the Summer Olympic Games.

# PART VI REVIEW QUESTIONS

Q1.  Explain the differences between "Know-that" and "Know-how." How do the science and art of coaching work together?

Q2.  What are the differences between learning and practice?

Q3.  List the stages of motor learning.

Q4.  List the stages of coach learning.

Q5.  Explain the differences between intrinsic and extrinsic feedback.

Q6.  Physical and mental practice are thought to work together to enhance overall learning. Which is the most important and why?

Q7.  Balance is a particularly impressive skill-set in gymnastics, such as the handstand. The handstand is held statically, but does the handstand involve all static tensions from muscles? How is the handstand position maintained?

Q8.  What are the main technological advancements in equipment used for vault exercises?

Q9.  Why did changes in equipment design for the vault enable gymnasts to approach faster in the forward entry vaults?

Q10.  What is the difference between an implicit and an explicit target?

Q11.  You measure your gymnasts' approach speed from the start of their run-up until they touch down on the beat board for handspring vaulting. You notice that your gymnasts lose speed (~1 m/s) in the last few metres before the board. Should you shorten their run-up?

Q12.  What is a sign of poor visual control of a vault run-up?

   a.  Poor running technique
   b.  Baulking
   c.  Runs well in our training gym but not in competition on the raised podium
   d.  Both B and C

Q13. Which of these training drills will NOT encourage your gymnast to visually control (target) their run-up?

  a. Completing some vault running drills on an athletics track or grass
  b. Fixed start position
  c. Changing intention between 50, 75 and 100% effort
  d. Variable start position

Q14. Which of the following hypotheses explains "laterality"?

  a. Hypothesis of cultural influences
  b. Hypothesis of the Coriolis effect influencing the behaviour of gymnasts in the southern and northern hemispheres
  c. Hypothesis of the hemispherical asymmetries
  d. Hypothesis of the vestibular asymmetry
  e. Hypothesis of the biomechanical or sensorimotor asymmetry

Q15. A left consistent turning gymnast performs a right turn when upside down?

  a. True
  b. False

Q16. Which statement is true?

  a. The rotational preference correlates with the handedness
  b. The rotational preference does not correlate with the handedness

Q17. Relate both columns:

| | Column A | Column B |
| --- | --- | --- |
| Line 1 | A bilateral consistent turning gymnast | always rotates in the opposite direction around the longitudinal axis when being in an upside down position, as compared to when being in an upright position |
| Line 2 | A unilateral turning gymnast | always rotates in the same direction, independent of the element or the body orientation in space |
| Line 3 | A gymnast with no distinguishable rotation scheme | is a gymnast who shows more than two elements during the all-around competition that do not fit the other scheme |

Q18. The majority of the men's individual all-around gymnastics competition at the Olympic Games 2016 shows a ......... rotation.

Q19. The most common rotational scheme in the world is .........., independent of the gymnast's turn to the left or the right.

Q20. Please mention two reasons that confirm the importance of the Elastic Technologies for the Artistic Gymnastics training.

Q21. Which flight elements (aerial) are common to the Horizontal Bar routine (choose three of the following elements)?

a.   Markelov ( )
b.   Suarez ( )
c.   Belle ( )
d.   Kovacs ( )
e.   Yamawaki/Jonasso ( )
f.   Kolman ( )

Q22.   The access to Electronic Technologies is the same in all FIG member federations? How many countries have the approved equipment by FIG?

Q23.   Knowing the importance of Electronic Technologies in the contemporary Artistic Gymnastics, all coaches had incorporated it into their training? Why?

# PART VII

# Injuries in gymnastics

# PART VII

# Injuries in gymnastics

## Learning outcomes (*Monèm Jemni*)

On completion of this part, the reader will be able to:

- Understand the depth and breadth of injury factors in gymnastics.
- Distinguish the causes and effects of an injury.
- Recognise both chronic and acute injuries to seek appropriate medical care.
- Integrate preventive strategies in different contexts.
- Identify Emergency Action Plan in case of injuries.
- Select subjective and objective monitoring methods for specific groups useful to prevent and to rehabilitate.

## Introduction and objectives (*Monèm Jemni*)

This part familiarises gymnasts, their parents and coaches with the effect, the role and the mechanisms of injuries in gymnastics and a perspective on the information coaches should know when training and competing young gymnasts. The overarching purpose of this part is to help coaches prevent injury by application of science and medicine to training.

This part of the book provides the reader with not only an overview of the possible risk of injuries in different age groups and gender practising artistic gymnastics but also methods, techniques and technologies to prevent them. The authors tackle the topic from different perspectives:

- The science-based theories underpinning the causes of injuries
- Injuries from medical perspectives
- Injuries from coaching perspectives
- The monitoring perspectives

Injuries and accidents are integral parts of the training process. Gymnasts and coaches should be aware of the risks that exist and consider them as a normal event that could happen. Anyone who does not exercise has indeed less risk of injury compared to someone who practises a sport on a daily basis; however, the first example could develop other health conditions and risks, such as obesity and cardiovascular diseases. The more you exercise the more you are prone to different types of accident indeed (acute, chronic, etc.). This part of the book provides examples of those accidents whilst highlighting methods to reduce them and/or to avoid them.

# 33

# INJURY IS A SERIOUS ISSUE

*William A. Sands*

## 33.1 Objectives and learning outcomes

The goal of Chapters 33 to 37 is to provide an overview of the role of injury in gymnastics, and a perspective on the information coaches should know when training and competing young gymnasts. The overarching purpose of these chapters is to help coaches prevent injury by the application of science and medicine to training.

On completion of Chapters 33 to 37, the reader should be able to:

- Understand the depth and breadth of injury factors in gymnastics.
- Identify some of the reasons that make injury the most serious problem facing gymnastics.
- Distinguish the events, sexes, levels of competition, rules and many other factors that interact to create a rich tapestry of linked causes and effects of an injury.
- Identify some of these interactions and the potential causes of gymnastics injury.

## 33.2 Introduction

Modern gymnastics has ancient roots in acrobatics that arose from the sheer joy of jumping, somersaulting, rolling, climbing, swinging and other natural activities (Cousineau, 2003; Gardiner, 1930). Playground observations demonstrate that children enjoy all of these activities spontaneously with only their judgemental constraints on what movements are safely within their abilities. As a naturally competitive species, people began comparing their abilities, which led to an organization, and rules that continue today via officials, rulebooks, assignment of difficulty to skills and scoring methods designed to differentiate between performances. As long as children use acrobatics for play, there is little danger and injuries are rare. As the demands for more challenging competitive exercises increase, the performance envelope is stretched and skill errors become more prevalent. A consequence of the escalation of skill complexity is more severe exposure to injury.

Artistic gymnastics, hereafter referred to simply as "gymnastics", consists of training and competition on four women's events (i.e. vault (WVT), uneven bars (UB), balance beam (BB) and floor exercise (WFX)) and six men's events (i.e. vault (MVT), pommel horse

(PH), still rings (SR), parallel bars (PB) and floor exercise (MFX)). Each event apparatus evolved from equipment used historically in military preparation, school curricula and ethnic and community physical fitness. The apparatuses have evolved over long periods (Davis, 1974; Joseph, 1949a, 1949b; Smith, 1870). Each apparatus places different demands on the gymnast's body, skills, artistic expression and courage. Apparatus skill needs are mirrored in the incidences and rates of injuries based on age, sex, anatomy, event, experience, competitive format, environment and many other factors (Sands, 2000a). As such, the discourse on gymnastics injury is highly context dependent. It matters whether one is describing injuries that occur to men versus women, training versus competition, hanging versus support events, and competitive levels among many others.

Gymnastics changes its rules, sometimes fundamentally, every four years (Federation, 2013; Örsel et al., 2011; Sands, 2000a). Gymnastics rule changes are analogous to basketball changing its rim height, court length, rules of what constitutes a foul, number of players on a team and the types of skills that are encouraged or allowed (Nassar & Sands, 2008). Even the apparatus in gymnastics can change suddenly and with little warning, such as the vault table (International Gymnastics Federation, 2000; Irwin & Kerwin, 2009; Sands & McNeal, 2002), SR (Federation_Internationale_de_Gymnastique, 2009), FX characteristics (Arampatzis & Bruggemann, 1999; Janssen, 2007; Paine, 1998; Sands, 2010b; Sands & George, 1988) and others. Particular skill learning and relearning, enhancement of specific fitness characteristics and the necessary trial-and-error exploration of the new demands resulting from new rules can often find gymnasts and coaches in a serious quandary in training and competition choices. Changing the rules so frequently, even *within* the four-year cycle, places the athlete at unnecessary risk of injury (Nassar & Sands, 2008).

## 33.3 How serious are injuries?

Gymnastics injuries are too common and too severe across gymnastics (Sands, 2000a). In spite of numerous countermeasures such as foam pits, inflatable structures, thick mats, spotting belts, hand spotting and other techniques, gymnastics still finds itself among the more hazardous activities in sport (Backx, 1996; Caine, 2002; Caine et al., 1996; Caine & Nassar, 2005; Kerr, 1990; Sands, 2002a; Westermann et al., 2014). The lay press has also identified gymnastics and acrobatics as significant causal activities for athletic injuries (Associated Press, 2011; Becker, 1998; Daly et al., 1998; Meyers, 2016; Reinhard, 1998; Scott, 2012). It is important to keep in mind that there is a need for risk assessment and risk management in nearly all physical activities (Bernstein, 1996; Clarke, 1998; Gerstein, 2008; Sands, 2002a).

Safety, as a part of injury prevention, relies on risk management from the governing body level at the broadest, all the way to the gymnast and his or her family. The development of rules, the design and manufacture of gymnastics equipment and the training and education of coaches all play a role (Sands, 2000a; Sands et al., 2011). Gymnastics, like all sports, poses many risks. Managing these risks often falls to the coach (Sands, 2002a). An injury arises not simply from the mechanical disruption of tissues, but more broadly from the culture, attitudes and philosophies of the participants (Sands, 2002a). Injury prevention depends largely on a commitment to safety, and careful application of sound coaching methods and encouragement of healthy training behaviours.

Injury and injury prevention are equally complex. Among the countermeasures to injury are mechanical aids, supervisory practices, fitness thresholds, teaching methods and injury prevention practices themselves. One of the problems with injury prevention is that the safer people may be the more likely it is that people will exhibit risky behaviour, "Improving safety also encourages risk-taking" (Gerstein, 2008, p. 105). This aspect of the interaction of people and risk has been called "revenge effects" (Tenner, 1996). As a result of these and other factors, perhaps the most critical approach to injury prevention must come from a system of social and cultural redundancy (Gerstein, 2008; Sands, 2000a; Sands et al., 2011). The idea has been presented metaphorically as a block of Swiss cheese. Each slice of cheese is analogous to an injury countermeasure. If there is a hole in the slice, the countermeasure failed, and injury prevention falls to the next lower slice. If the next lowerst slice has one or more holes that do not line up with the previous hole, then the injury is prevented. The goal is to avoid enough holes from lining up that the injury circumstance or mechanism cannot pass through all of the available countermeasures (Gerstein, 2008; Sands et al., 2011). The "Swiss cheese model" was proposed by Reason to characterize human error (Reason, 2000). Gymnastics skill learning proceeds with a wide range of countermeasures to injury (i.e. cheese slices) such as specific physical conditioning, use of soft landing pits, then thick mats with spotting, then thick mats alone, then spotting alone, then thinner and thinner mats with and without spotting while attending to optimum learning environments and the amelioration of fatigue. Such "systems" of countermeasures help ensure the safest approach to skill learning and performance.

Consistent planning of training and competition can provide a sound foundation for injury prevention, treatment, communication, medical intervention and others. Failure to do so too often leads to injuries that were preventable had someone simply bothered to invoke their knowledge, vigilance, imagination and courage.

# 34

# WHAT IS AN INJURY?

*William A. Sands*

Whether one is injured or not may seem obvious, but some contextual factors have resulted in inconsistent definitions (Finch, 1997; Meeuwisse & Love, 1997; Sands, Shultz, & Newman, 1993; van Mechelen, 1997). Inconsistent definitions have plagued some scientific efforts to ascertain the cultural and practical milieu that surround gymnastics injury (Bradshaw & Hume, 2012). Injury definitions include several common concepts from medical trauma, while gymnastics may provide a separate and relatively new category of sports injury (Caine, Caine, & Lindner, 1996).

Perhaps the most common definition of an injury is some harm to a body part that results in time lost from training and competition (Noyes, Lindenfeld, & Marshall, 1988). A second definition involves diagnosis by a medical professional such as a physician, therapist or athletic trainer (Noyes, Lindenfeld, & Marshall, 1988). A third definition includes an anatomical location and grade or severity score (Noyes, Lindenfeld, & Marshall, 1988).

Gymnastics and some other sports may require another explanation. Gymnastics injuries can be defined as *any damaged body part that would interfere with training* (Noyes, Lindenfeld, & Marshall, 1988; Sands, 2000a; Sands, Shultz, & Newman, 1993). The reason for a particular injury definition for gymnastics is that most gymnastics injuries are skill-specific. For example, although a sprained ankle will usually sideline an athlete in other sports, a gymnast simply braces or tapes the injury and works on hanging events or uses a foam pit or thick mat for landings on the back or seat thus avoiding ankle impact exposures.

Injuries are usually studied via methods of epidemiology. Epidemiology comes in two flavours: descriptive and analytic. Descriptive epidemiology is most interested in incidence and prevalence (Caine et al., 1996). Incidence refers to a simple count of the number of new injuries or the probability that someone in a group will be injured over a particular defined period (Caine et al., 1996). Prevalence includes the number of new and old injuries present in a specific group, or the probability that someone is injured during a given period. Injuries can be reported based on the number of athletes within the total group of athletes who were injured (i.e. athlete rate, the total number of injuries divided by the total number of athletes), or the number of injuries reported based on the number of athletes and the number of exposures to injury. For example, an athlete is not likely to be injured if they are not included in a class or competition. Expressing injuries relative to athlete exposures is the current "best practice" for such reporting (Caine et al., 1996).

While descriptive epidemiology deals with the who, what, where, when and how of injury, analytic epidemiology deals with the why of injury. Analytic epidemiology uses the concept of "risk factors" in an attempt to characterize both the nature of the causes of injury and the probability of these causes being present and their magnitude (Caine et al., 1996). Risk factors are further divided into intrinsic and extrinsic factors. Intrinsic factors involve those aspects of an injury that the injured athlete brings with him or her. Extrinsic factors are those that the athlete encounters or confronts that lead to an injury (Caine et al., 1996).

These concepts are important for the coach in that a profile or conceptual structure can be developed that helps identify how and why an athlete might be injured. Defining the concepts and structure of these ideas is a continuous process that can map a path around injury based on attention to the details provided by the epidemiological information.

# 35

# CHARACTERISTICS OF GYMNASTICS INJURIES

*William A. Sands*

Gymnastics nearly automatically classifies injuries with regard to sex based on the events in which the injury occurs (vaulting and floor exercises involve both males and females). However, there are great deficiencies of injury reports involving male gymnasts, countries other than the U.S. and non-competitive artistic gymnasts. As such, while the following will provide information about gymnastics injuries, the reader should acknowledge that the entire universe of gymnastics and gymnastics-related injuries is simply unknown (Caine & Maffulli, 2005).

## 35.1 Gymnastics participation

According to USA Gymnastics (https://usagym.org/pages/aboutus/pages/demographics. html?prog=pb), the National Governing Body for gymnastics in the U.S., there are 5,273,000 gymnastics participants over the age of six. As many as 902,000 are considered active members engaging in over 100 hours of activity per year. Seventy-six per cent of the participants were female, many taking part in over 4,000 gymnastics schools or clubs. Approximately 80% are under the age of 18.

A 2014 report by the Sports and Fitness Industry Association (www.sfia.org/reports/52_ Gymnastics-Participation-Report-2014) placed gymnastics participation in the U.S. at 1,763,000 in 2013. The International Gymnastics Federation (FIG) (www.fig-gymnastics. com/site/about/federation/population) estimates worldwide gymnastics participation at approximately 50 million at all levels. School gymnastics programmes, at least in the U.S., appear to be declining in numbers. While programmes still exist, very few are being started, and most have a difficult time finding athletes among high school students. According to data from the National Federation of State High School Associations, a U.S. organization that provides rules and governance for many sports, the number of high school gymnastics programmes dropped 75% from 1977 to 2003, often because of an inability to find a qualified coach. However, some states show flourishing programmes, such as Texas, where boys' programmes are strong and growing (www.athleticbusiness.com/high-school/ balancing-act.html). The preceding serves to emphasize the fact that characterizing gymnastics participation worldwide can be wrought with contextual problems. The most common and often necessary approach to study injury is to reduce the problem to a specific group, context or situation. In studies of National Collegiate Athletic Association gymnastics, athletes present at a specific competition, athletes from a specific geographic area or those from a specific programme will likely continue to be the primary targets of injury investigations. Culling information from this mosaic of studies is difficult, given the various issues described earlier. The gymnastics coach should be aware of the study population, duration and level of the athletes when interpreting injury-related publications, and that information is often incomplete.

## 35.2 Injuries depend on time of the season and duration

The timing of injuries is of both theoretical and practical interest. Do gymnasts get hurt more or less often at the beginning or end of practice? Is the early season riskier than the latter period? Is one age group more susceptible than another? Do injuries have an acute onset or appear to develop gradually over time? Only a few studies have described the timing of injury relative to the competition or practice and showed that the injuries tended to occur early rather than late (Caine et al., 1989; Caine et al., 2003; Lindner & Caine, 1990). The roles of fatigue and lack of concentration have also been discussed with increasing risk shown when the athlete had been on the apparatus for a longer period (Lindner & Caine, 1990). A study of error distributions among national level female gymnasts revealed that the majority of significant errors in routines occurred early in the routine and at the dismount (Sands et al., 1992). This finding was later supported by conversations with judges who indicated that the majority of their large deductions occurred at the beginning rather than later in the routine, except for the dismount (personal communication). A non-typical means of training was proposed by these authors that recommended routine training begin at the dismount and add skills moving toward the mount rather than from the mount to the dismount. In this way, the majority of routine repetitions would result in the most practice on those skills likely to be performed in a fatigued state (Caine & Maffulli, 2005; Caine et al., 1996; Sands et al., 1992).

Regarding seasonal variations of injuries, three patterns tend to emerge. First, injuries tend to occur more frequently during the early part of the season when skills are still unstable (Caine & Maffulli, 2005; Caine et al., 1996; Sands, 1981b). Second, injuries tend to increase immediately before competition, during early routine preparation and after time off from vacation or an existing injury, and when performing new skills (Kerr, 1990; Kerr & Minden, 1988; Marshall et al., 2007; Sands, 1993a, 2000a).

Injuries may occur from a sudden accident that causes immediate tissue damage, pain and temporary disability, or injury may arise from the slow and insidious accumulation of stress and damage that was not immediately concerning at the outset. For example, one can take a wire coat hanger and cut it with metal shears (acute) or one can bend the hanger back and forth many times creating small micro fractures in the metal until the fractures line up and the hanger breaks (chronic or overuse). Gymnasts can step on the edge of a mat and sprain an ankle, or they can jump and land repeatedly stressing an ankle slowly so that the ankle ligaments suffer small tolerable insults for some time before the pain rises to a level demanding concern. Both acute and overuse injuries can be devastating to the gymnast. Injury data regarding the distribution of these two types of injuries indicate that the majority of gymnastics injuries have a gradual onset, and roughly one-third of the injuries are acute (Caine, 2002; Caine & Maffulli, 2005; Caine et al., 1996).

Once a gymnast is injured, he/she tends to remain injured for some time. Gymnasts often continue to train and compete with an active injury (Aldridge, 1991; Aldridge, 1987; Caine

& Nassar, 2005; Chan et al., 1991; Hudash Wadley & Albright, 1993; Sands, McNeal, & Stone, 2011). A five-year study of collegiate women's gymnastics recorded the onset of injury and the duration of the injury noting that once a gymnast was injured, she tended to continue training and competition and maintained the injury for the entire season (Sands, 1993a). Further support for this was indicated by Caine and Maffulli in a review of gymnastics injury (Caine & Maffulli, 2005). These authors noted that injuries tended to be present from one to three seasons, apparently without being healed (Caine & Maffulli, 2005).

Perhaps the most clear and obvious evidence of these behaviours and consequences arises from the incidence of stress fractures (De Smet et al., 1994; DiFiori et al., 1997; Flynn, Ughwanogho, & Cameron, 2011; Hume, 2010–2014; Sands, 1993a; Vain, 2002) and stress-related soft tissue damage, such as the Achilles tendon (Leglise & Binder, 2014; Wertz, Galli, & Borchers, 2013), among gymnasts. Stress fractures appear to be common problems in the radius and wrist (Caine et al., 1992; Carek & Fumich, 1992; DiFiori et al., 1997; Roy, Caine, & Singer, 1985), spine (Brady & Vincenzino, 2002; Caine & Nassar, 2005; Ciullo & Jackson, 1985; Flynn, Ughwanogho, & Cameron, 2011; Hall, 1986; Hume, 2010–2014; Sands, 1993a; Wade et al., 2012), elbow (Chan et al., 1991; Farana et al., 2015), and clavicle (Fallon & Fricker, 2001; Fujioka et al., 2014).

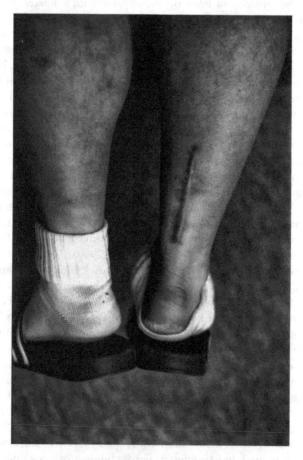

**FIGURE 35.1** Scar from Achilles tendon rupture surgery.

# 36

# INJURY LOCATION

*William A. Sands*

The anatomical location of an injury is important for sports medicine professionals and coaches. For example, Achilles tendon ruptures have stubbornly plagued both men and women in vaulting and floor exercise or tumbling events (Associated Press, 2011; Bieze Foster, 2007; Wertz, Galli, & Borchers, 2013). Achilles tendon ruptures are particularly devastating, usually requiring a year or more of rehabilitation following surgery. The injury appears to occur without warning and usually on take-offs rather than landings, although landing ruptures are not unknown. Special investigations searching for answers to this injury problem have identified potential causes, but little certainty (Arndt et al., 1998, 1999a, 1999b; Bruggemann, 1985; Self & Paine, 2001; Wertz, Galli, & Borchers, 2013). Gymnastics, like most sports, involves particular movement patterns of the limbs and torso, which often develop a corresponding pattern of acute and overuse injury. Moreover, there may be interactions among the anatomical location and the event, skills, levels and even predisposing medications (Hayem & Carbon, 1995; Kaleagasioglu & Olcay, 2012; Melhus, 2005). Treatments for infection involving fluoroquinolones (e.g. Cipro™) appear to be related to an increased likelihood of tendinopathy (Melhus, 2005), and athletes and their physicians should be warned about potential side effects.

## 36.1 Head and neck injuries

Head and neck injuries are relatively rare in gymnastics, but the potentially catastrophic nature of these injuries merits focused attention. In a study from National Electronic Injury Surveillance System of the US Consumer Product Safety Commission for 1990 through 2005, children from six years to 17 years showed a prevalence of 12.9% involving the head and neck. These data also revealed an incidence of 1.7% concussions or closed head injury (Singh et al., 2008). The relatively small incidence of head and neck injuries is supported by several other studies as reviewed by Caine and Nassar (Caine & Nassar, 2005). However, concussions suffered at some time over a career were reported by over 30% of female club gymnasts in the Seattle, Washington, USA area (O'Kane et al., 2011).

**FIGURE 36.1**    Descent of a gymnast from the vault horse to a landing on her head resulting in quadriplegia.

The cervical spine of gymnasts is particularly vulnerable because of the common inverted body positions that gymnasts perform. Unfortunately, severe spinal injuries of gymnasts can result in paralysis. A prominent catastrophic injury involving the vault event and a Chinese female gymnast have been presented as a case study (www.advancedstudyofgymnastics. com/blog/case-study-of-the-sang-lan-vault-injury-1998-goodwill-games) (Figures 36.1 and 36.2).

The incidence of spinal cord injuries may be declining (Caine & Maffulli, 2005). Recent American club level gymnastics injury studies have reported no catastrophic injuries (Caine & Maffulli, 2005). However, relatively recent historical data have demonstrated that such injuries still exist (Katoh et al., 1996; Meeusen & Borms, 1992; Schmitt & Gerner, 2001). Unfortunately, the paucity of data on these types of injuries results in few conclusions and too often a leap to judgement based on a lack of information.

## 36.2 Spine and torso injuries

Gymnastics is an unusual sport in its emphasis on extreme ranges of motion of the spine (McNeal & Sands, 2006; Sands, 2010a; Sands et al., 2015) and high internal and external loads (Bruggemann, 2010; Bruggemann, 1999; Kruse & Lemmen, 2009; Sands et al., 2015; Watts, 1985). Low back injuries tend to plague gymnasts at nearly all levels, both male and female. The performance demands placed on the spine have changed over time. In the past, it was common to observe extreme ranges of motion that showcased gymnasts', particularly female, incredible spine flexibility (Caine & Maffulli, 2005; Kruse & Lemmen,

Slightly before impact.

**FIGURE 36.2**   Descent to a head landing that resulted in a cervical fracture and paralysis.

2009; Purcell & Micheli, 2009; Sands et al., 2015) (Figure 36.3). More recently, gymnastics has emphasized power, strength, high flight and extreme difficulty over unusual contortionist body positions (Sands et al., 2015). However, many female gymnasts continue to show positions demanding extreme ranges of motion (Figure 36.4). Whether these types of positions are directly causative of injury remains unknown (Sands et al., 2015).

Do back-bends and other hyperextension exercises harm young gymnasts? Our recent investigation does not condemn skills where gymnasts stretch their spine as too dangerous (Sands et al., 2015). Training spine hyperextension usually begins in early childhood by performing a skill known as a back-bend. The extensive literature review that we undertook on spine stretching among gymnasts indicated that, within reason, spine stretching does not appear to be an unusual threat to gymnasts' health (Sands et al., 2015). However, spine stretching and loading among gymnasts, especially child gymnasts, should be undertaken via careful, thorough and long-term progression. Children should be well supervised, carefully instructed through lead-up skills, possess the strength to support themselves in the position and understand that if they feel a pain they must contact their coach immediately so that the pain can be assessed.

The change in gymnastics performance demands resulted in a shift of common injuries of the spine. Exposure to extreme hyperextension positions and posterior spine involvement has moved to the anterior spine (Caine & Maffulli, 2005) which may be the result of flexed spine positions during skills, particularly landings (Bruggemann, 1999). Compressive forces as high as 40 times body weight have been determined when gymnasts adopt a forward flexed position of the spine during a gymnastics landing (Bruggemann, 1999). The spine,

**FIGURE 36.3** A female gymnast showing a common hyperextended spine position called a "salute" following the dismount landing.

particularly lumbar postures, has shown relationships with gymnastics landing efficiency and effectiveness (Bruggemann, 2010; Cuk & Marinsek, 2013; Gittoes & Irwin, 2012; Too & Adrian, 1987; Wade et al., 2012). Certainly, repeated emphasis on these positions can result in overuse-related trauma that may lead to injuries among susceptible athletes. Of particular interest to coaches is the finding that placing an additional mat on top of a standard landing mat reduces spinal compression forces by approximately 20%. Spinal loads using modest drop jumps can reach twice the forces of running (Bruggemann, 1999). Average compression forces calculated from female gymnasts at L5/S1 were approximately 11 and seven times body weight for the forward and backward somersault, respectively (Bruggemann, 1999). Maximum compression forces at L5/S1 in the same study reached approximately 20 and 16 times body weight, respectively (Bruggemann, 1999). Another study supported these findings showing that 30% of the gymnasts landed a back somersault with the lower lumbar spine flexed and ground reaction forces of 6.8 to 13.3 times body weight (Figure 36.4). Drops from 171 cm (not as high as most gymnastics flight trajectories) showed compressive forces in the spine of over 30 times body weight (Bruggemann, 1999).

Beyond compressive forces, both sexes use techniques that rely on explosive spine hyperextension and flexion such as: the Tkatchev (Arampatzis & Bruggemann, 2001; Gervais & Tally, 1993; Sands, 1995a), tap swing techniques (Chen & Liu, 2000; Irwin & Kerwin, 2005; von Laßberg et al., 2013), stoop through (Sands, 1994c; Xin & Li, 2000), vault table support phases (Ferkolj, 2010; Hall, 1986; Penitente et al., 2010; Penitente, Sands, & McNeal, 2011), thrusting motions of the thorax (Goehler, 1977; Wiemann, 1976a, 1976b, 1979), and many others. Male gymnasts usually do not demonstrate extreme ranges of motion in the spine in aesthetic poses, but they still land from great heights and incorporate powerful hyperextension and flexion of the spine in skill techniques (Figure 36.5).

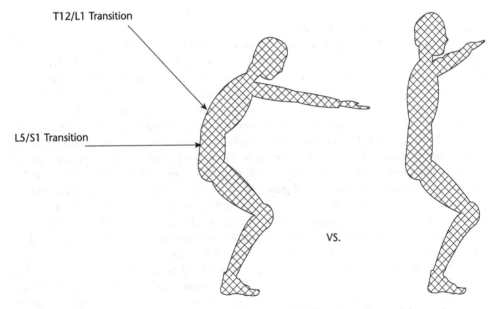

**FIGURE 36.4**    Images show the spine areas of concern (left figure), and a straighter spine (right figure).

## 36.3 Upper extremity injuries

Gymnastics is enigmatic in that the sport uses the upper extremity for weight bearing as well as reaching and grasping. The upper extremity does not behave like the lower extremity in

**FIGURE 36.5** Male gymnast using spine hyperextension during a horizontal bar dismount.

impact activities, although the underlying mechanical principles remain the same (Glasheen & McMahon, 1995; Koch, Riemann, & Davies, 2012; Sands, Shultz, & Paine, 1993; Suchomel, Sands, & McNeal, 2016). A study of human locomotion using the hands and arms (e.g. walking and running) showed that such movements demand four to five times more in metabolic cost than locomotion using the lower extremity (Glasheen & McMahon, 1995). Gymnastics is also unique in that gymnasts, like divers, can dislocate their shoulders while in mid-air (non-weight bearing) performing skills involving twisting and somersaulting (Nassar & Sands, 2008; Rubin et al., 1993). Male gymnasts on still rings and horizontal bar, and female gymnasts on uneven bars use unusual shoulder positions in the performance of essential skills such as the back lever on still rings, inlocates and dislocates, "bounces" to crosses; on horizontal bar with dorsal grips such as a German Giant, elgrips such as eagle and inverted giants; on uneven bars doing the same skills as the men on horizontal bar.

Thus, it should not be surprising that injuries to the upper extremities are common among gymnasts. From 2000 to 2004 the number of injuries to the shoulder at the U.S. club and collegiate level increased (Nassar & Sands, 2008). However, a sharp decline in these injuries occurred from 2000–2004 among U.S. national team female gymnasts and was thought to be the result of a substantial increase in conditioning as emphasized by national team training (Nassar & Sands, 2008). The upper extremity supports the entire body weight plus additional forces of impact during pushes and landings from the hands, and the upper extremity also supports the body as the only link between a hanging body and the apparatus such as observed with horizontal bar, still rings, parallel bars and uneven bars.

The wrist and elbow of the upper extremity usually rank highest of the anatomical areas for women. Men's injuries tend to focus on the shoulder. Injuries to the wrist have been the subject of numerous studies (Caine & Maffulli, 2005). These studies have shown a large incidence of wrist pain among gymnasts ranging from 46–88% (De Smet et al., 1994;

DiFiori, Caine, & Malina, 2006; DiFiori et al., 1996; Liebling et al., 1995; Mandelbaum & Teurlings, 1991). The weight bearing nature of the upper extremity in gymnastics has resulted in x-ray studies demonstrating that a gymnast's forearm, wrist and hand begin to take on the structure of a four-legged animal's forward limb – the distal upper extremity begins to look like a foot (Caine et al., 1992; Carek & Fumich, 1992; De Smet et al., 1994; DiFiori, Caine, & Malina, 2006; DiFiori et al., 1996; Liebling et al., 1995). The unusual condition noted in gymnasts' wrists has been described as a positive ulnar variance, a growth plate injury and "gymnasts' wrist". Injuries to a youngster's growth area in bone are a serious matter with bone length and thickness deficiencies a justifiably feared outcome. While gymnasts can certainly injure their wrists and upper extremities via acute falls, the upper extremity is more fragile than the lower extremity and is thereby more susceptible to overuse syndrome injuries. The incidence of distal radial physeal stress reactions ranges from 10% to 85% in studies (Caine et al., 1992; Mandelbaum, Grant, & Nichols, 1988; Mandelbaum & Teurlings, 1991). Gymnasts' complaints of wrist pain should not be ignored.

A unique injury in gymnastics is called "grip-lock". This injury involves the overlapping of the proximal and distal segments of the leather hand grips that gymnasts wear while training and competing on hanging events (Samuelson, Reider, & Weiss, 1996; Sathyendra & Payatakes, 2013). The forces present on the leather hand grips can be extraordinary reaching 2.2 times body weight (Neal et al., 1995). The injury occurs when the gymnast is swinging, and the two segments of the grip overlap and one of the segments becomes trapped under the other thereby stopping the rotation of the leather grip, the hand, the forearm and the arm of the gymnast. The resulting injury is devastating to these structures because of the large forces involved while performing large swinging movements and the "trapping" of the gymnast's hand while the body continues to rotate.

A questionnaire study of 457 female gymnasts showed that 22% of the respondents had suffered a traumatic shoulder injury (Caplan et al., 2007). The same study indicated that ligamentous laxity and shoulder instability were common although not multidirectional instability (Caplan et al., 2007). Force asymmetry was demonstrated in all gymnasts performing a forward handspring on a force platform with the asymmetry focused at the shoulder (Exell, Robinson, & Irwin, 2016). A study of hand and arm impacts showed that the impact is characterized by an initial high-frequency component followed by a low-frequency component. The high-frequency component occurs when the hands strike the supporting surface while the low-frequency component arises when the torso descends farther and decelerates (Davidson et al., 2005). Impact loads measured from hand placement on a handspring vault reached more than eight times body weight in a sample of female gymnasts (Penitente & Sands, 2015). Ground reaction forces at the hand during a back handspring were recorded at 2.37 times body weight (Koh, Grabiner, & Weiker, 1992). On the old vault horse, compressive forces at the elbow have been calculated at 2.7 times body weight (Panzer, Bates, & McGinnis, 1987). Forces on the hands during a giant swing on uneven bars can reach nearly 2000N (Sands et al., 2004b). While hanging during a giant swing, male gymnasts can create reaction forces of approximately seven times body weight at their hands (Bruggemann, 1999).

## 36.4 Lower extremity injuries

The lower extremity consistently ranks highest in anatomical locations for gymnastics injury, particularly the ankle (Kirialanis et al., 2003). Despite some noticeable gaps in reported prevalence, the ankle tends to dominate injury reports involving the lower extremity (Caine & Maffulli, 2005). Ankle injuries commonly rank highest in incidence followed closely by knee injuries (Caine & Maffulli, 2005). Interest in landing mechanics and motor strategies has led to a greater understanding of the means by which landings can be injurious (Arampatzis et al., 2004; Cuk & Marinsek, 2013; Janshen, 2000; McNitt-Gray, 1991a; McNitt-Gray et al., 2001; Self & Paine, 2001), and studies have described how landing mats can both enhance safety or contribute to injury (Alp & Bruggemann, 1992; Arampatzis, Morey-Klapsing, & Bruggemann, 2005; Gatto, Swannell, & Neal, 1992; Gros & Leikov, 1995; McNitt-Gray, Yokio, & Millward, 1993; McNitt-Gray, Yokoi, & Millward, 1994; Mills, Pain, & Yeadon, 2006; Mills, Yeadon, & Pain, 2010).

Peak ground reaction forces in gymnastics landings have been shown to range from 8.8 to 14.2 times body weight on each foot (Panzer et al., 1988) or up to 15 times body weight in laboratory controlled landing tasks (McNitt-Gray, Yokio, & Millward, 1993). Forces at the Achilles tendon have been shown to reach 15 times body weight (Bruggemann, 1999). Figure 36.6 shows the maximal dorsiflexion position of a highly trained female gymnast during a backward somersault take-off. The bone-on-bone forces at the tibiotalar joint have been measured at approximately 23 times body weight. The forces of landing at the foot talar-navicular joint can be doubled simply by incorporating a pronated or everted foot position at landing. A pronated foot may be a subtle aspect of landing positions that may increase the risk of injury (Bruggemann, 1999). Foot pronation during gymnastics tumbling has been shown by force/pressure sensors on the soles of the feet (Sands et al., 2013) (Figure 36.7) and by kinematics and high-speed video that shows a "dishing" of the foot contact area during a take-off (Figure 36.8). The combination of soft mats, dishing of the spring and floor structure, and a loading shifted toward the medial sides of the feet, especially the medial metatarsal heads. Figure 36.8 also shows an unusual foot, ankle and shank position. This athlete performed a double twisting layout backward somersault

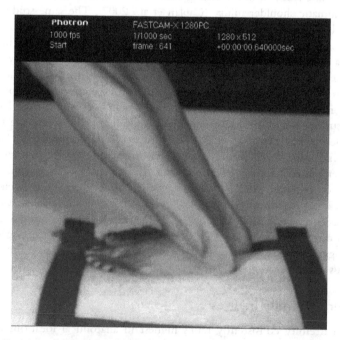

**FIGURE 36.6** Position of greatest dorsiflexion, compression of the anterior ankle and stretching of the Achilles tendon.

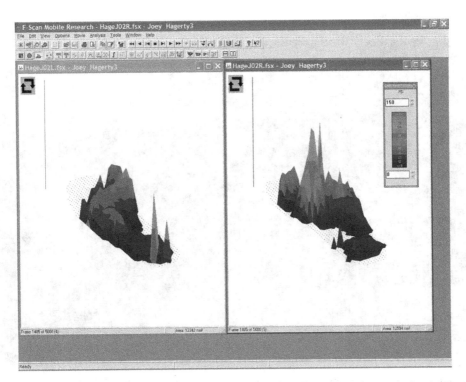

**FIGURE 36.7**  Note the high-pressure values on the medial sides of the feet with the right foot showing the highest values.

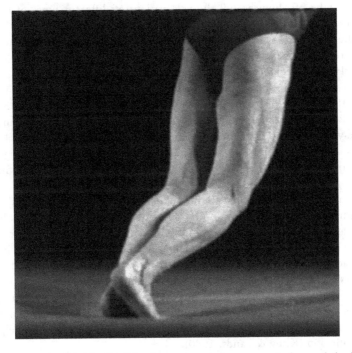

**FIGURE 36.8**  Note the "dishing" of the floor around the gymnast's feet and the lack of symmetry of the feet and ankles of the lower leg.

**FIGURE 36.9** Types and thicknesses of mats.

following this take-off. Figure 36.9 shows a variety of different types of mats used for teaching skills and landings.

During the period from 1996–2000, the number of anterior cruciate injuries in female gymnasts in the U.S. increased rapidly (Nassar & Sands, 2008). Landings are the most common skill being executed when lower extremity injuries occur and the skill strategies that gymnasts use have been the target of many investigations (Bruggemann, 2010; Cuk & Marinsek, 2013; Gittoes & Irwin, 2012; Gittoes, Irwin, & Kerwin, 2013; McNitt-Gray et al., 1997, 2001). The distance of the fall (McNitt-Gray, 1991b, 1993), landing surface (McNitt-Gray, 1991a), body segment orientations (McNitt-Gray et al., 1997, 2001), somersault rotation direction (Gittoes, Irwin, & Kerwin, 2013), and skilled joint motion abilities (Gittoes & Irwin, 2012; McNitt-Gray, 1993) interact to allow the gymnast a wide array of strategies to handle landing forces (McNitt-Gray, 1993).

A common injury involving the anterior ankle is based on "short landings" of gymnasts from tumbling and apparatus dismounts. The mechanism involves a violent contact between the anterior talus and anterior tibiofibular-talar joint. Moreover, the dome of the talus is wider anteriorly than posteriorly such that when the extreme and violent hyper-dorsiflexion action occurs, the talus jams into the mortice of the tibia and fibula thereby applying forces that tend to split the joint apart. The result of these mechanisms is an impingement injury with painful symptoms on short landings, dorsiflexion of the ankle and palpation of the region between the anterior talus and the distal tibia. A particular countermeasure was designed for this ailment called a Safe-T-Strap that helped prevent extreme dorsiflexion (Kling & Sands, 1980).

# 37

# INJURY COUNTERMEASURES

*William A. Sands*

First and foremost, gymnastics should be its own injury countermeasure. Gymnastics is injury prevention. "Medical gymnastics" was a form of both treatment and prevention of injury and health promotion through the use of calisthenics, breathing exercises and therapeutic activities (Joseph, 1949b, 1949c, 1949d; Weiker, 1985). Gymnastics Turnvereins, Sokols and other institutions involving gymnastics were brought from Europe to the U.S. These organizations were successful as community centres when manual labour was still a large part of the occupational milieu of families. A shift from participant to spectator occurred shortly after the industrial revolution along with the transition of the workforce to a more sedentary lifestyle. As such, it became harder for the general population to perform gymnastics movements because of diminished strength, particularly in the upper body, which continues to plague many people today (Girginov & Sandanski, 2004; Joseph, 1949b; Smith, 1870). This phenomenon amplifies the difficulty of gymnastics movements and the idea that modern youngsters must train purposely for several years merely to accomplish the basics of gymnastics.

Modern competitive gymnastics is considered simply too difficult and dangerous to practise without supervision, safety measures and high levels of fitness. In spite of few or no safety precautions, Parkour ("art of movement" (Wanke et al., 2013)), free running, or "street acrobatics" movements have a relatively modest injury incidence with participants reporting 1.9 injuries per sports career/year, or 5.5 injuries/1000 h training (Wanke et al., 2013). However, in spite of their laudable encouragement of physical activity, these types of acrobatics can lead to serious injury (Derakhshan & Machejefski, 2015; Grosprêtre & Lepers, 2016; Miller & Demoiny, 2008; Vivanco-Allende et al., 2013). Parkour participants, also called "traceurs", have demonstrated high levels of muscular and jumping fitness, some better than gymnasts (Grosprêtre & Lepers, 2016). Perhaps the Parkour athlete's ability to self-limit his or her activities is an important key to their ability. Moreover, a study of Parkour landing techniques (e.g. Parkour precision and Parkour roll) and traditional gymnastics landing techniques showed the superiority of the Parkour techniques in reduced ground reaction forces (Puddle & Maulder, 2013). However, in fairness, the Parkour precision technique is a typical gymnastics landing technique with added emphasis on cushioning rather than "sticking". Clearly, these activities require more research.

## 37.1 Training load

When gymnastics participation shifts from play to organized practice, many aspects of injury prevention principles come into play (Coté, Baker, & Abernethy, 2003). A monitoring study of U.S. female national team gymnasts showed that the number of skill elements performed per year was about a quarter of a million. The most decorated female gymnast at that time performed more than twice this training volume in a year (Caine et al., 1996; Sands, Henschen, & Shultz, 1989). Physical and psychological durability may be key factors in the success of a gymnast. Among male gymnasts in the U.S., all of the senior national team members have had shoulder surgery. U.S. athletes and coaches joke that surgery is a national team requirement.

Studies of monitoring and regulating training load among gymnasts are rare (Caine & Nassar, 2005; Kolt & Kirkby, 1999; Sands, 1991a,b, 2002b; Sands, Shultz, & Newman, 1993; Sartor et al., 2013; Westermann et al., 2014) with many articles coming from a single author, and only a few on men's gymnastics. Modern concepts of periodization have yet to permeate gymnastics at all competitive levels. However, there are excellent examples of these principles being espoused and incorporated into training (Arkaev & Suchilin, 2004; James, 1987; Jemni & Sands, 2011; Sands, 1999a,b; Sands, Irvin, & Major, 1995; Ubukata, 1981; Ukran, Cheburaev, & Antonov, 1970). The current gymnastics coach should be using periodization principles as opposed to "seat-of-the-pants" methods that rely on intuition, trial and error, and guessing.

## 37.2 Physical fitness

There is a common sense understanding that highly fit athletes are less likely to be injured. However, this has been stubbornly difficult to demonstrate scientifically. Of course, athletes need a threshold level of fitness to perform their skills, but beyond that threshold there is a lot of room for variability. Again, revenge effects and the fact that better gymnasts, usually those who are fitter, also spend more time exposed to training and are therefore more susceptible to injury (Caine & Maffulli, 2005; Caine et al., 1996). At least in the U.S., women's jump-specific performance does not keep pace with growth in size and mass through puberty (Sands, McNeal, & Jemni, 2002). It has been discussed in the lay media that psychological factors are responsible for setting the razor's edge of high-level performance. However, comparisons of physical abilities tests among U.S. national team women showed that the 2000 Olympic team was better than their non-Olympic team counterparts on all physical abilities test (Sands, McNeal, & Jemni, 2001b; Sands, Mikesky, & Edwards, 1991).

Gymnasts have large strength capacities, strength relative to body mass, as they are required to lift their body through unstable positions often at high speeds. Moreover, male gymnasts are expected to demonstrate their strength on floor exercise and still rings (Sands, 2006d). However, as skill levels rise, so do training exposures, heights of flights, number of repeated elements and many other factors related to winning and injury. For example, ligaments and tendons respond to loading by becoming stronger, with increased tensile strengths of approximately 10% (Bruggemann, 1999). Of course, the hypertrophy of connective tissue is beneficial for performance and injury prevention, but the loads on these

**TABLE 37.1** Comparison of U.S. women's national team members who weight train versus those who do not. Independent t-tests

| Variable | Training method | N | Mean | Std deviation | Std error | Significance |
|---|---|---|---|---|---|---|
| Body mass index | Weights | 14 | 20.33 | 1.87 | 0.50 | 0.050 |
| | Non-weights | 19 | 21.68 | 1.90 | 0.44 | |
| Body mass (kg) | Weights | 14 | 47.96 | 5.36 | 1.43 | 0.043 |
| | Non-weights | 19 | 52.12 | 5.87 | 1.35 | |
| Age (y) | Weights | 14 | 18.07 | 2.01 | 0.54 | 0.017 |
| | Non-weights | 19 | 16.53 | 1.02 | 0.23 | |
| Height (cm) | Weights | 14 | 153.49 | 4.02 | 1.08 | 0.339 |
| | Non-weights | 19 | 154.90 | 4.29 | 0.98 | |

tissues increase far more than 10%. Information on growth factors involved in tendon healing and hypertrophy is growing, but currently incomplete (Molloy, Wang, & Murrell, 2003). Common sense dictates that while connective tissue loading is perhaps the single greatest contributor to healing and tensile strength development, the demands on the tendon or ligament should not exceed the tissue's capacity for adaptation (Bruggemann, 1985; Molloy, Wang, & Murrell, 2003; Yang, Rothrauff, & Tuan, 2013; Zernicke & Loitz, 1992). Injury prediction models exist but are rarely applied outside the study that generated the models (Bale & Goodway, 1990; Beatty, McIntosh, & Frechede, 2006; Fellander-Tsai & Wredmark, 1995; Steele & White, 1986; Sward et al., 1991). Finally, while most sports use weight training to enhance strength fitness, gymnastics has been stubbornly reticent to engage fully in practice, usually for fear of "bulking up". However, at least one study of female senior national team gymnasts showed that those who practised weight training were lighter, leaner, the same height and yet older than their non-weight training counterparts (Sands & McNeal, 1997; Sands et al., 2000) (Table 37.1).

## 37.3 Injury prevention equipment

As mats become thicker, they also become heavier. Larger and thicker mats raise questions about the deceleration consistency of regional areas of the mat through repeated wear as compared to central areas. Thick mats with folds, different ages and other characteristics have been addressed (Sands et al., 1988; Sands et al., 1991b). The edges of thick mats and old worn mats are softer and more able to be completely penetrated or compressed. Areas near the seams of folding mats tend to behave differently than those areas near the centres of sections (Sands et al., 1988, 1991b). Peak accelerations of gymnasts falling on their backs from the height of the upper uneven bar rail on various areas of thick, open cell foam mats, ranged from 9.7 to 15.6 Gs with pulse durations of 0.088s to 0.137s (Cunningham, 1988; Sands et al., 1988, 1991b). Back landings are among the safest positions for landing on these types of mats. As the surface area of the body at impact is decreased, the relative force applied to that body area increases. Coaches should be aware that different areas of a soft mat behave differently and mat placement should include regard for where the gymnast is most likely to land in expected and unexpected falls (Figure 37.1).

**FIGURE 37.1**   Deepest impact frame taken from high-speed video of a gymnast landing in an open cell foam mat approximately 60cm thick. Note the conical "dishing" of the mat resulting from the impact force.

Landing "pit" areas have nearly revolutionized gymnastics training and now find themselves used in Parkour training, extreme sports, aerial skiing, mogul skiing, snowboard halfpipe, circus and even jumping motorcycles (Henderschott & Sigerseth, 1953; Normile, 1989; Sands et al., 1991a). These areas are constructed in-ground as large holes in a training room floor, built above ground using framing methods to create a large open area, or sometimes foam pieces are simply put in a large pile (Klaus, 1985; Normile, 1989; Sands et al., 1991a). Foam pits can consist of thousands of uniformly cut or randomly torn up foam pieces (Figure 37.2). The pieces can be encased in a fabric cover or used without a cover. Peak deceleration values for block foam pits have been shown to range from 4.45 Gs to 6.45 Gs, obviously less harsh than gymnastics mats. However, injuries and even deaths have occurred in foam pits.

Foam pits can also be designed by using a large solid block of foam, foam with open-air channels inside the block, and other designs. These solid foam pits are often called "resi-pits" in the U.S., after the first company to provide the large foam blocks. The blocks are nearly always encased in a breathable tightly woven net-like material or a vinyl case (Isabelle & Jones, 1990; Klaus, 1985) (Figure 37.1). Foam pits made of individual small foam blocks are usually more forgiving in landings as the gymnast can penetrate the foam blocks for several feet before coming to a stop. Solid foam pits have excellent deceleration characteristics, and they allow faster movement of the gymnast into and out of the pit.

A third type of foam pit involves either the small foam blocks or a large solid foam block which is placed on top of a trampoline bed that is anchored to the walls deep within the pit structure (http://gymnasticszone.com/gymnastics-safety-pits/). The premise behind trampoline pits is that the need for constant fluffing of loose foam is reduced, and the

**FIGURE 37.2**  Uniformly cut loose foam above ground pit beneath the uneven bars.

impact is softer. Unfortunately, there appears to be no information in the scientific literature regarding these ideas.

Unfortunately, all types of pits have seen catastrophic injury and death. Moreover, foam pits present a serious problem for emergency medical responders in extricating an injured gymnast from the pit, which is made harder by the small block piece designs. A foam pit is soft and easily compressed such that the gymnast's impact is absorbed, while at the same time simply walking on the foam is sometimes impossible. These characteristics have made the extrication of a gymnast from a foam pit an extraordinarily difficult task for emergency responders (Figure 37.3). Emergency medical responders require training to be able to reach, tend and remove an injured gymnast while not exacerbating the injury. Engaging medical help is made even more precarious when the injured gymnast has ceased breathing, has a suspected cervical spine injury or is not conscious. Training plans and methods have been produced to assist coaches in providing their assistance and emergency personnel direction on how one might go about rescuing a gymnast in a foam pit (Committee, 1995; Finkel, 2001; George, 1987; Gymnastics, 2009).

A fourth kind of landing pit is relatively new and does not involve open cell foam. This type of pit uses a large inflatable air bag that is placed in the open area. The bag is inflated by a large, powerful fan and maintains its fall protection characteristics by maintaining air pressure within the bag. The bag has vents and combined with a reverse airflow through the fan when a gymnast lands on the bag expels some air such that falls are cushioned. Air bags in automobiles, Hollywood stunt work and other activities have been around for some time, with little information beyond that provided by manufacturers. Scientific and medical literature is apparently far behind this type of equipment with only a few yet published (Olsen, 1988; Roegner, 2006). Track and field had disastrous experiences with an air bag pole vault landing area called the "Cloud-9". This author used one of these for his high school pole vaulting experience and found that landing near the edge of this type of pole vault "pit" was an almost inevitable injury. Air pressure could not be maintained high

**FIGURE 37.3**  Note the gymnast is inverted. Crawling through the foam pieces to reach the athlete will result in disturbing the foam and her body position. Attempting to reach the gymnast can exacerbate his or her injury.

enough near the edges of the bag as it was in the middle (Boden et al., 2001) (www.polevaultpower.com/forum/viewtopic.php?t=3819).

The gymnastics coach needs to understand the characteristics, limitations and uses of landing pits. Pits are not a panacea and cannot replace sound progressions. Sadly, it is tempting to allow a gymnast to simply "go for it" into a landing pit because of the reduced likelihood of injury. While coaches can often use such a tactic without injuring the gymnast, too often this method is a dismal failure.

## 37.4 Spotting

Another unique aspect of gymnastics is the role of coaches as spotters (Dunn, 1980). While many sports coaches may physically manipulate athletes to help them learn a position or motion, gymnastics is unusual in that coaches help protect gymnasts from falls and prevent injury by doing so. Spotting requires considerable skill on the part of the coach and clear communication between coach and athlete (George, 1988a,b).

There are several valuable assets to good spotting:

- Physically demonstrating proper body and limb positions
- Physically demonstrating important body position transitions and timing
- Touch-based reminders or cues for the athlete during a skill
- Physically assisting the performance of the whole skill or part of the skill
- Motivation – to reduce fear and increase confidence
- Safety – to prevent a dangerous and injurious fall (Whitlock, 1989, 1992).

Spotting is usually applied in three ways:

- One or more coaches catch a falling gymnast and thereby prevent an injury – hand spotting or safety spot.

- An external apparatus is used such as a spotting belt that is held by two coaches, or a rope system is used that can suspend the gymnast during a skill and catch him or her during a fall, thus preventing an injury – belt spotting.
- Coach and athlete participate in choreographed and unified skills where the athlete performs his or her desired skill while the coach touches the gymnast while in motion and assists the motion by skilful pushes of the gymnast with his or her hands – hand spotting with or without safety aspects.

The primary goal of spotting, no matter the type, is to prevent a gymnast from falling on his or her head and neck. There is a tacit understanding that if a skill goes horribly wrong during the motion, the spotter should be most cognizant of keeping the athlete off his or her head. The knowledge is based on the idea that an injured lower or upper extremity will usually heal with little long-term consequence. However, an injured head or neck can have consequences such as permanent injury, paralysis or even death. Spotting can be extremely effective when applied skillfully, but there are inherent limitations, particularly with hand spotting (Price, 1937; Sands, 1996, 2000a). Moreover, the spotter is also exposed to injury when trying to catch a falling person who is heavy and moving rapidly (Boone, 1979; Price, 1937; Tilley, 2013).

Spotting is not a panacea. There are serious, and often overlooked, constraints on how well any coach can function as a spotter (Sands, 1996). For example, when a gymnast falls from the height of the uneven bar rail (Figure 37.4), he or she will strike the floor in approximately 0.70s. At 0.50s of the fall the gymnast, when the gymnast is most commonly

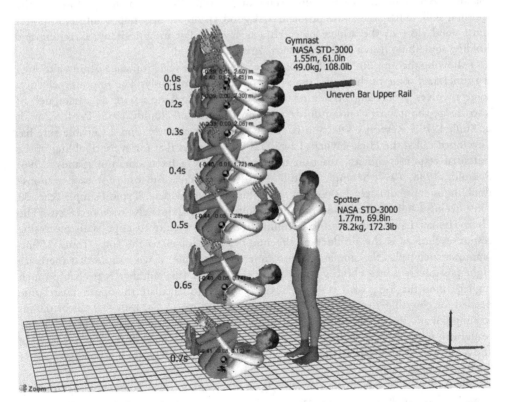

**FIGURE 37.4** Sequence images showing the time and relative positions of a body during a fall.

in a "good catch position", the athlete will have achieved a downward fall velocity of 16.2 ft·s⁻¹ or approximately 5 m·s⁻¹. Walking speeds range from 1.4 m·s⁻¹ to 2.5 m·s⁻¹. Peak vault run speeds of national team gymnasts are often between 7.0 m·s⁻¹ and 8.5 m·s⁻¹ (Penitente et al., 2008; Sands, 2000b; Sands & Cheetham, 1986). Catching and arresting the downward momentum of the gymnast in this scenario is extremely difficult, requiring precise positioning of the coach and excellent strength.

In addition to speed and strength, the spotter must react almost instantaneously, particularly during an unplanned or unexpected fall. In order to respond quickly to a situation, the spotter must adhere to motor control principles of reaction and movement time (Schmidt, 1988, 1994; Schmidt & Stull, 1970). Catching the falling gymnast requires going through a:

1. Stimulus identification stage (i.e., is the gymnast in danger and needs help?)
2. Response selection stage (i.e., of all the possible physical responses, including doing nothing, what response should I use?)
3. Response programming stage (i.e., the spotter has selected a response, the neural information regarding the motion "programme" is now communicated to the neurons influencing the important muscles).

The stimulus identification stage involves perceiving the gymnast and the situation. The duration of this stage is influenced by the clarity and magnitude of the various stimuli. Louder noises and brighter lights have been shown to influence the duration of this stage (Sands, 1996; Schmidt, 1988, 1994; Schmidt & Stull, 1970). Practice performing the spotting skill, anticipation of a problem and good vision are also important. The spotter must stand close to the athlete to catch a heavy and fast moving body. Catching and holding something heavy is difficult with arms outstretched.

Following the stimulus identification stage, the response selection stage begins, although it is unclear if some of these two stages can occur simultaneously. The response selection stage involves processing or linking the perceived information with a "catalogue" of knowledge about what to do or how to fashion a response (Schmidt, 1988, 1994; Schmidt & Stull, 1970). Response selection has been studied extensively, and a valuable rule has developed called the Hick–Hyman Law. This law indicates that for every doubling of the potential response choices, reaction time increases by a fixed duration (Sands, 1996; Schmidt, 1988, 1994; Schmidt & Stull, 1970). The increase in time for each response doubling is approximately 150ms, but this value can vary widely. Typical simple reaction time, seeing a light stimulus and pressing a button is approximately 0.18s or 180ms. The Hick–Hyman Law states that the accumulation of more time to cope with increasing response choices is the product of the number of choices to the $\log_{(2)}$ times 150ms (approximately) plus the simple reaction time. For example, if we assume that there are eight possible response choices confronting the spotter (the number is probably much higher), then the $\log_{(2)}$ of 8 = 3 (i.e., 2x2x2). Thus, to calculate the approximate time needed for choosing a response in this situation is 0.180 + (3 x 0.150) = 0.63s or 630ms to simply make the choice. The spotter has not moved yet. Looking at Figure 37.4, the reader can note that the falling gymnast is at about the knee level of the spotter, moving at over 6 m·s⁻¹, and the spotter has not yet moved. Unfortunately, the situation gets worse for the falling gymnast.

The final stage of information processing in this scenario is the response-programming stage. Unfortunately for the gymnast, it is known that the more complex the response

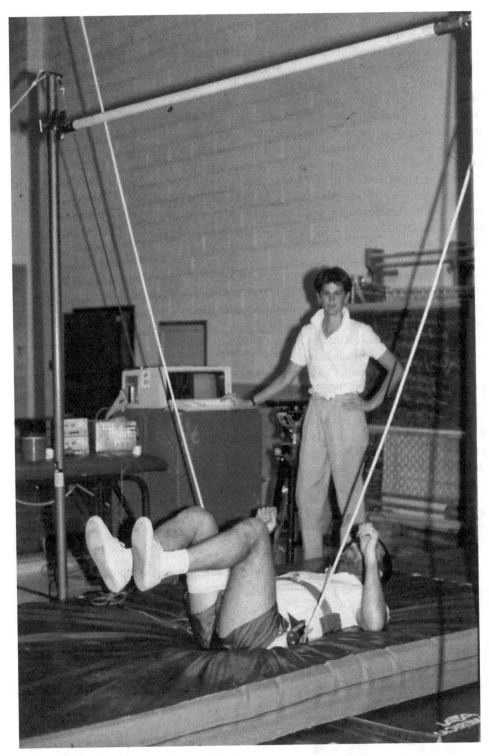

**FIGURE 37.5** Spotting belt showing ropes and belt after athlete was lowered to the mat.

programming is, the longer this stage requires. In very simple tasks like pushing a button, this stage can require as little as 0.095s or 9.5ms. However, in complex tasks of which catching a heavy falling body should qualify, the duration can be as long as 0.465s or 465ms. If we take a conservative middling value of 280ms, to add to the total duration of a response, we arrive at 910ms or 0.91s (Sands, 1996; Schmidt, 1988, 1994; Schmidt & Stull, 1970). Looking at Figure 37.4, one should note that the gymnast has already hit the floor by approximately 0.70s or 700ms. The spotter is unlikely to have even begun to move prior to the gymnast striking the floor during an unplanned and unexpected fall. As an injury countermeasure, spotting is precious, but the time, skill, strength, perception, knowledge and fault anticipation are not always up to the task.

Spotting belts, particularly overhead spotting belts, serve a valuable role in safety and injury prevention (Hiley, Apostolidis, & Yeadon, 2011; Kimball, 1990; Milem, 1990; Sands, 1990c). Unfortunately, a survey of gymnastics coaches from gymnastics schools in the U.S. showed that only a fraction of these gyms have spotting belts installed and even fewer are trained in their use and use them regularly (Sands, Crain, & Lee, 1990). Landing pits have become the primary means of catching gymnasts from falls. Overhead spotting belts are also not a panacea. Spotting belts have been known to fail, twisting-type belts can jam, spotting ropes can become entangled with the gymnast, spotting ropes can break and the attachments of the system have also been known to fail (Figure 37.5). In spite of these limitations, when properly used spotting belts provide a much larger margin for error than do either landing pits or hand spotting. When operated competently, a falling gymnast never strikes the floor or apparatus.

As a summary, gymnastics injury countermeasures are the technologies, knowledge and practices that form the "slices" of the Swiss cheese model that was presented at the beginning of this chapter. Coaches and administrators should be aware of the assets and limitations of these countermeasures (Daly, Bass, & Finch, 2001; Gittoes & Irwin, 2012; Schneier, 2006). Coaches should also understand how and why injuries happen and take preventative steps. An injured gymnast becomes a non-competitive entity, and coaches should serve as vigilant, dedicated and determined advocates of not squandering any athlete's ability because of injury.

## 37.5 CONCLUSION

Gymnastics injury is the single most important problem facing modern gymnastics. Unfortunately, knowledge of gymnastics epidemiology and countermeasures has made some progress, but never enough. As gymnastics progresses rapidly, the nature of injury also changes and this requires constant updates and retooling of nearly all aspects of gymnastics. The 2016 Rio Olympic Games televised broken leg of the French male gymnast on vaulting and the later mishandling of the injury by medical personnel who dropped his ambulance cot while loading him into the ambulance point out how horribly wrong "things" can go. Consider that all of the slices of the injury prevention cheese block lined up for this gymnast. Coaches, athletes, parents and administrators will need to do a better job of injury prevention. As the level of gymnastics becomes more "space age", perhaps our training and preparation should begin to model that of astronauts.

# 38

# CLINICAL INJURY CASES IN ARTISTIC GYMNASTICS

## The big and the small ones

*Brooke Lemmen*

## 38.1 Objective

Familiarize gymnasts, their parents, and coaches with clinical injuries affecting women's artistic gymnasts by outlining common mechanisms for those injuries, and possible preventive techniques, to allow for healthy, active participation for gymnasts of all levels.

## 38.2 Learning outcomes

- Recognize potential acute and chronic gymnastic injuries.
- Integrate preventive strategies to improve athlete performance and decrease injuries.
- Confirm integration of an Emergency Action Plan for the gymnastics facility.

## 38.3 Definition of medical terms and abbreviations

The following are some commonly used medical terms:

**Anterior** = "front" of the affected body part
**Posterior** = "back" of the affected body part
**Dorsal** = "top" of the affected body part
**Lateral** = "outside" of the affected body part
**Medial** = "inside" of the affected body part
**Extension** = to straighten out a joint, i.e. to straighten out the elbow is "Elbow extension." One area that can be confusing – "shoulder extension" in medical terms is when the hand is at the waist and the arm goes BEHIND the body, as opposed to "extending your shoulders" in gymnastics which indicates arms overhead and behind the athlete's ears. This position with "arms behind ears" is "shoulder flexion" in medical terms.
**Flexion** = to bend a joint
**Eccentric** = muscle strengthening exercise that elongates the muscle under stress
**Concentric** = muscle strengthening exercise that shortens the muscle under stress
**Inversion** = turning the ankle with the toes inside, "sickling" of the foot and ankle
**Eversion** = turning the ankle with the toes outside
**Dorsiflexion** = a "flexed" (non-pointed) foot and ankle
**Plantarflexion** = a "pointed" foot and ankle

**NSAID** = Non-Steroidal Anti-Inflammatory Drug. Medications commonly available over-the-counter for pain and inflammation. United States examples are ibuprofen or naproxen.

**OCD** = Osteochondral defect or Osteochondritis Dessicans

## 38.4 Introduction

Injuries in artistic gymnastics are common. In the United States, gymnastics is one of the most popular and injury-prone sports (Overlin, Chima, & Erickson, 2011). Injury risk varies based on the population evaluated, age, competitive level, and training hours. The challenge for any injury is definitive diagnosis and treatment. Medical care providers familiar with gymnastics, and providers well versed in "sports medicine" or "orthopaedics," aid in determining a diagnosis and treatment plan in a more expedited manner.

The athlete, parents, and coaches play an important part in recognition, management, and rehabilitation of gymnastics injuries. This is not an exhaustive summary of injuries, but a guide to injury recognition. Together with a trained medical professional (athletic trainer, physical therapist or physio, or a physician) gymnasts, parents, and coaches can be active in medical care.

Most injuries will not completely limit the athlete from any activity; a structured and informed participation plan can help keep the athlete active during recuperation. The medical professional familiar with gymnastics aids in coordinating this plan.

Although gymnastics is a high-injury sport, recent evidence shows that injury rates are declining. Kerr *et al.* (2015) showed decreased collegiate gymnastics injury rates compared to previous data. Studies by Coates *et al.* (2010) on recreational gymnasts show that the risk per hour of gymnastics participation in recreational gymnasts aged five to ten is 0.17 painful incidents per hour. As a comparison, children in day care have one incident every three hours of active play (Coates *et al.*, 2010). Thus recreational gymnastics may not be any more risky than children's play. Injury risk may be decreasing because of improvements in coaching, rule changes, and modifications to equipment and other safety initiatives (i.e. spotting and sting mats) (Kerr *et al.*, 2015).

Despite these innovations, hours of repetitive practice mean that overuse injuries are part of nearly every gymnast's career. They are also more difficult to identify. By modifying routines and focusing on different skills and events, athletes often continue to practice with an injury. Early recognition of chronic injuries, with appropriate care, can allow a more fulfilling and productive career.

Acute injuries will also occur. A well-planned and practiced Emergency Action Plan (EAP) can minimize the long-term effects both to the injured gymnast, and her teammates. EAPs should include initial recognition and stabilization of the injured athlete, activation of an emergency response system, and notification of parents/guardians. The EAP should be documented and new staff should be trained in the EAP upon hire. Regular review is necessary as emergency contacts and coaches may change. Gymnastics federations may also require basic life-saving training, such as Cardiopulmonary Resuscitation (CPR).

The loose foam block pit is a unique environment for potential injuries. While use of a foam pit is for safety, acute injuries still occur around and within the pit. The foam pit is an inherently unstable platform for emergency personnel to work when assessing and treating an injured athlete. Coaches should be familiar with the nuances of accessing a gymnast with a potential spinal cord injury, or other injury, in the foam pit to aid emergency responders. Review and practice of this scenario with local emergency medical services providers is beneficial.

Gymnasts often excel in their sport at an early age. As a result, most of their career occurs during times of growth, particularly adolescence. Unique injuries occur during periods of growth. The physeal plates, or "growth plates" of the bones in the body, can be fractured or inflamed. The growing gymnast is also at risk of muscle and tendon injury with growth. Gymnastics requires extremes of flexibility; however, as flexibility decreases during growth, the athlete is at risk of muscle and tendon injuries.

Nutritional needs required by the growing female body can affect risk. Poor nutrition slows healing, and puts the athlete at risk of further injury. Because of the aesthetic component to gymnastics, gymnasts are at risk of disordered eating with subsequent amenorrhea and decreased bone mineral density, known as the female athlete triad. Nutrition education, with emphasis on calcium, vitamin D, and iron intake, is important for lifetime health.

Following are brief summaries of common chronic and acute injuries. The list is not exhaustive; the focus is on injuries found in gymnasts, growing athletes, and diagnoses uncommon to medical providers unfamiliar with the sport.

**Do not use this summary as a substitute for consulting with your medical care provider.**

**Any injury or pain that is not resolving warrants further evaluation by a medical professional.**

## 38.5 Common medical injuries in gymnastics

### General injuries

See Table 38.1 below.

**TABLE 38.1**

| Common signs and symptoms | Possible reason for pain | Treatment options |
|---|---|---|
| Acute injury, acute pain | | |
| • Pain, immediate swelling and possibly bruising over affected bone <br> • Inability to bear weight on affected body part <br> • Surrounding muscle spasm <br> • Possible skin break from underlying injury | Fracture | • Immediate evaluation by a medical professional <br> • May necessitate activation of EAP |
| • Gross deformation of affected area <br> • Possible skin break from underlying injury <br> • Inability to move affected body part | Dislocation | • Immediate evaluation by a medical professional <br> • May necessitate activation of EAP |
| • Landing on the head, neck, or face <br> • Complaint of neck or head pain <br> • Athlete does not have to complain of symptoms into arms or other body parts to sustain this type of injury | Cervical spine injury | • Immediate evaluation by a medical professional <br> • May necessitate activation of EAP |

*(continued)*

**TABLE 38.1** *(continued)*

| Common signs and symptoms | Possible reason for pain | Treatment options |
|---|---|---|
| • Injury to the head or face<br>• May be a direct injury or force transmission (i.e. whiplash)<br>• Complaint of any concussion or brain injury symptom<br>• Athlete may have a loss of consciousness | Concussion or other brain injury | • Immediate evaluation by a medical professional<br>• May necessitate activation of EAP |
| Chronic injury, chronic pain<br>• Pain increasing in intensity and duration over time<br>• Pain initially only at the end of practice/day<br>• Occurs with change in training schedule or level<br>• Pain over the affected bone | Stress fracture | • Referral to medical professional for further evaluation<br>• Requires a high degree of suspicion for this diagnosis |

## Foot and ankle injuries

The most commonly injured area for women's artistic gymnastics is the lower extremity – primarily the ankle. Most acute injuries occur during competition, not practice (Overlin, Chima, & Erickson, 2011). Injury is most likely, either acute or overuse, on floor and vault (Overlin, Chima, & Erickson, 2011). Chronic injuries also result from compensation, or the athlete's desire to "push through" after a minor acute injury. See Table 38.2.

**TABLE 38.2**

| Common signs and symptoms | Possible reason for pain | Treatment considerations |
|---|---|---|
| HEEL – Chronic pain or injury<br>• Very common<br>• Intermittent pain on the posterior aspect of the heel<br>• Occurs around periods of growth<br>• Resolves and returns spontaneously<br>• Only occurs in growing athletes<br>• Pain with jumping, running, and even walking | Sever's Disease (apophysitis of the calcaneus) | • Stretching of the Achilles with the knee straight and bent<br>• Heel cups in and out of the gym and/or X-brace©<br>• Ice for symptom control |
| HEEL – Acute pain or injury<br>• Athlete unable to push off toes, may be unable to walk<br>• Defect in the tendon often evident<br>• May not "hurt" after initial insult<br>• Athlete often feels that they were "kicked" or "shot" in the back of the leg when injury occurs | Achilles rupture | • Referral to medical professional |

| Common signs and symptoms | Possible reason for pain | Treatment considerations |
|---|---|---|
| • Pain with walking<br>• Pain over inferior aspect of the heel<br>• May occur acutely with landing on a hard surface, or chronically with hard landings on heels (i.e. on beam) | Fat pad contusion | • Ice and NSAIDs for symptom control<br>• Heel cups in and out of the gym<br>• Limit hard landings<br>• If not improving consider stress fracture of the calcaneus |
| **FOOT – Chronic pain or injury** | | |
| • Pain, swelling over lateral foot<br>• Pain with eversion of the foot<br>• Occurs during times of growth<br>• Resolves and returns spontaneously<br>• Only occurs in athletes who are growing<br>• Pain with jumping, running | Iselin's Disease (apophysitis of the base of the fifth metatarsal) | • Stretching<br>• Ice for symptom control<br>• Limit number of painful activities<br>• If a "pop" occurred with onset of pain, evaluation for a fracture is prudent<br>• Stretches and ankle proprioception and strengthening, especially during periods of growth |
| • Pain over the bottom, "ball," of the foot<br>• Pain with plantar flexion<br>• Pain with running leaps | Turf Toe/ Sesamoid injury | • Stretching<br>• Ice<br>• Arch supports out of gym for flat feet<br>• Consider stress fracture of the sesamoid bone if not improving |
| **FOOT – Acute pain or injury** | | |
| • Mid-foot pain, often after punching with half a foot on beam or landing with foot rolled under<br>• Athlete may complain of pain into the dorsal web space between the first and second toe (Pourcho, Liu, & Milshteyn, 2013)<br>• Swelling and bruising on the dorsal mid-foot<br>• Difficulty with punching or releveé | Lis Franc injury | • Referral to medical professional<br>• Requires a high degree of suspicion for this diagnosis<br>• Weight bearing x-rays and MRI often needed for diagnosis<br>• Can be a career-ending injury for gymnasts |
| **ANKLE – Chronic pain or injury** | | |
| • Diffuse pain around ankle<br>• May have intermittent swelling<br>• "Locking" or "catching" of the ankle<br>• No specific reason for injury<br>• Pain with running, tumbling | OCD<br><br>Common site is the dome of the talus | • Referral to medical professional<br>• Requires a high degree of suspicion for this diagnosis<br>• May be visible on x-ray, MRI often needed for diagnosis |

(continued)

**TABLE 38.2** *(continued)*

| Common signs and symptoms | Possible reason for pain | Treatment considerations |
|---|---|---|
| • Occurs with repetitive short landings<br>• Pain over the anterior ankle<br>• May also occur acutely | Anterior impingement of the ankle joint | • Ice for acute symptom control<br>• Softer landings<br>• Skill review<br>• Achilles or ankle tape |
| • May occur after an ankle injury or injuries<br>• Athlete may complain of "weak" or "unstable" ankles<br>• Generalized ankle pain<br>• May have some chronic swelling | Chronic laxity | • Referral to a medical professional<br>• Strengthening and proprioceptive exercises can help<br>• Ankle support or taping<br>• May require surgery if conservative measures fail |
| Tendinopathy: Tendinopathy is a broad term encompassing painful conditions occurring in and around tendons in response to overuse (Andres & Murrell, 2008).<br><br>The following three descriptions are specific to the tendon injury listed. The treatment considerations listed here apply to all three tendinopathy injuries listed below | | • Stretches<br>• Ice and NSAIDs for acute symptoms<br>• PT modalities may be beneficial<br>• Decreased number and softer surface landings |
| • Pain in the posterior aspect of the ankle/calf with running, jumping, landings<br>• May have some swelling | Achilles tendinopathy | • Eccentric Strengthening (Andres & Murrell, 2008)<br>• Achilles tape |
| • Pain along the medial aspect of the ankle with jumping, landings<br>• Often occurs after chronic short landings on backwards skills (i.e. double backs on floor) | Posterior Tibialis tendinopathy | • Strengthening exercises for ankle and calf |
| • Pain along lateral ankle with running, jumping<br>• Occurs after an inversion ankle sprain | Peroneal tendinopathy | • Strengthening and proprioceptive exercises for the foot and ankle |

ANKLE – Acute pain or injury

| | | |
|---|---|---|
| • Occurs with an inversion mechanism<br>• Pain, swelling and bruising over the lateral ankle<br>• Athlete may have difficulty with weight bearing | Lateral ankle sprain | • Referral to a medical professional<br>• Important to differentiate from a "high ankle sprain"<br>• Ice and compression for symptom control<br>• Often improves in a matter of weeks<br>• Can result in chronic ankle laxity or pain |

| Common signs and symptoms | Possible reason for pain | Treatment considerations |
|---|---|---|
| • Occurs with an inversion and plantar flexion, or an everted and dorsiflexed injury<br>• Pain, swelling, and bruising over the lateral and anterior ankle<br>• Athlete may have difficulty with weight bearing | "High" ankle sprain | • Referral to a medical professional<br>• Important to differentiate from a "standard" lateral ankle sprain<br>• Ice and compression for symptom control<br>• Often takes six to eight weeks for recovery<br>• Optimal healing may require immobilization<br>• Can result in chronic ankle laxity or pain |

## Lower leg injuries

See Table 38.3.

## Knee injuries

Most acute trauma to the knee occurs with landings (Overlin, Chima, & Erickson, 2011). These injuries are obvious and require further evaluation by a medical professional. Chronic knee injuries result in time out of practice and competition for gymnasts. Aetiologies for chronic knee pain in the growing gymnast are listed in Table 38.4.

## Upper leg injuries

See Table 38.5.

**TABLE 38.3**

| Common signs and symptoms | Possible reason for pain | Treatment considerations |
|---|---|---|
| Chronic pain or injury<br>• Pain or numbness and/or tingling below the knee, occurring with primarily endurance activity<br>• The lower leg feels "tight"<br>• In severe cases, the skin can feel cool to the touch, or have decreased color | Exertional Compartment Syndrome | • Referral to a medical professional<br>• Evaluation includes pressure testing in the lower leg<br>• Treatment may require surgery |

**TABLE 38.4**

| Common signs and symptoms | Possible reason for pain | Treatment considerations |
|---|---|---|
| Chronic pain or injury | | |
| • Diffuse pain around the knee with jumping, running<br>• Intermittent swelling without cause<br>• "Locking" or "catching" of the knee<br>• No specific reason for injury | OCD<br><br>Common site is the femoral condyle | • Referral to medical professional<br>• Requires a high degree of suspicion for this diagnosis<br>• May be visible on an x-ray; MRI often needed for diagnosis |
| • Pain at the inferior aspect of the kneecap (patella) with straightening out the knee, tumbling, etc.<br>• May have some swelling<br>• May feel a "pop"<br>• Occurs during times of growth<br>• Occurs in athletes who are still growing | SLJ – "Sinding Larsen Johansson" Disease<br><br>Apophysitis of the inferior patellar pole | • Stretch quadriceps and hip flexor muscles, ideally in an arabesque position<br>• Ice and NSAIDs for symptom control<br>• Limit activities until pain free<br>• If a "pop" is felt, maintain high suspicion for growth plate fracture |
| • Pain at the "bump" inferior to the kneecap with running, jumping, etc.<br>• May have some swelling<br>• Occurs during times of growth<br>• Occurs in athletes who are still growing | "Osgood Schlatter's" Disease<br><br>Apophysitis of the Tibial Tubercle | • Stretch quadriceps and hip flexor muscles, ideally in an arabesque position<br>• Ice and NSAIDs for symptom control<br>• Activity as tolerated<br>• Cho-pat© type strap can be helpful |
| • No reason, or triggering event, for knee pain<br>• Pain over medial aspect of the knee<br>• Growing athlete<br>• Athlete may have a limp | Internal hip pathology | • Referral to medical professional<br>• Diagnosis requires a high degree of suspicion<br>• Testing with hip internal rotation and imaging |

**TABLE 38.5**

| Common signs and symptoms | Possible reason for pain | Treatment considerations |
|---|---|---|
| Acute pain or injury | | |
| • A "pop" with hip flexion (often with stretches for splits, or with split leaps)<br>• Pain at hamstring attachment to pelvis bone (ischial tuberosity)<br>• Pain with walking, splits, leaps<br>• Decrease motion on injured leg<br>• May be bruising and swelling<br>• Occurs in athletes who are still growing | Avulsion fracture of the Ischial Tuberosity at the hamstring origin | • Referral to a medical professional<br>• May not be evident on x-ray, may require MRI<br>• Recuperation can limit gymnast for months |

| Common signs and symptoms | Possible reason for pain | Treatment considerations |
|---|---|---|
| • Occurs when athlete falls off beam, but "slides" down the beam with their leg during the fall<br>• Most often is a contusion | Contusion or "beam bite"<br><br>Can also be a more severe Morel-Lavallee lesion | • Ice and compression for symptom control<br>• Care for overlying skin abrasion to limit risk of infection<br>• If the "beam bite" develops a "fluid-like" or "water balloon" appearance under the skin, concern for a Morel-Lavallee lesion. Seek evaluation by a medical professional |

## Injuries to the spine, low back, hip and pelvis

Low back pain is common, and often underreported, in artistic gymnastics. Prevalence ranges from 20% to 65% depending on the competitive level of the gymnast (Vanti *et al.*, 2010). Low back pain in the general adolescent population varies from 18% to 70% (Kruse & Lemmen, 2009). Injuries to the low back often occur with athlete growth, or in progression to a new competitive level. Maintenance of skill-appropriate hamstring and shoulder flexibility, in combination with strong core musculature, are key components to decrease low back pain for gymnasts of all levels. See Table 38.6.

**TABLE 38.6**

| Common signs and symptoms | Possible reason for pain | Treatment considerations |
|---|---|---|
| Chronic pain or injury | | |
| • Often occurs with changes in levels<br>• Often occurs with growth because of decreased hamstring and shoulder flexibility with compensation through the back<br>• Pain with extension, progressing to kips and landings, and even with sitting and standing<br>• Rare swelling or bruising | Spondylolysis and/or Spondylolisthesis | • Referral to a medical professional<br>• Normal x-rays will not rule out this diagnosis, requires MRI or SPECT bone scan<br>• Recuperation can limit gymnast for months |
| • Occurs in athletes who are still growing<br>• Mid or low back pain with flexion or pounding<br>• Presents during times of growth | Scheuermann's Disease | • Referral to a medical professional |
| • May occur after specific injury, or chronic exposure<br>• Often pain or numbness/tingling into the lower extremities, especially with flexion and landings | Disc pathology | • Referral to a medical professional |

*(continued)*

**TABLE 38.6** *(continued)*

| Common signs and symptoms | Possible reason for pain | Treatment considerations |
|---|---|---|
| • Result of an imbalance between flexibility and muscle strength in the athlete's core<br>• Coincides with periods of growth because of decreased flexibility<br>• Gymnast often has decreased hamstring and shoulder flexibility and core strength compared to peers<br>• Pain with flexion and extension | Generalized musculoskeletal pain<br><br>Pelvic Cross Syndrome | • Typically a diagnosis of exclusion once other concerns have been eliminated<br>• Referral to a medical professional<br>• Responds to focused physical therapy<br>• Integration of regular core strengthening beyond abdominal strengthening can be beneficial (Overlin, Chima, & Erickson, 2011) |

## Elbow and forearm injuries

Repetitive weight bearing through the upper arm results in injuries to the elbow. Though not the most common area of the upper extremity to be injured, elbow injuries often require surgery and a long period of rehabilitation to return to the previous competitive level. Recognition of early signs and symptoms indicating an elbow injury are important for both the gymnast and coach. Some authors have recommended that **any elbow pain persisting beyond two weeks should be evaluated** (Dexel *et al.*, 2014).

**TABLE 38.7**

| Common signs and symptoms | Possible reason for pain | Treatment considerations |
|---|---|---|
| Chronic pain or injury | | |
| • Diffuse pain around the elbow, especially with weight bearing<br>• "Locking" or "catching" of the elbow, possibly with decreased elbow motion<br>• Gymnast generally has no specific reason for this injury<br>• Mimics "tennis elbow" initially | OCD<br><br>Common site is the capitellum<br><br>Panner's Disease is also in this spectrum | • Referral to medical professional<br>• Requires a high degree of suspicion for this diagnosis<br>• May be visible on an x-ray, MRI often needed for diagnosis |
| • Occurs in athletes who are still growing<br>• Pain over the posterior aspect of the elbow with extension and upper body exercises | Olecranon (Triceps Insertion) apophysitis | • Stretches<br>• Limit number of painful activities |
| • Pain or numbness and/or tingling below the elbow occurring with activity, i.e. bars<br>• The lower arm can feel "tight"<br>• In severe cases, the skin can feel cool to the touch, or have decreased color | Exertional Compartment Syndrome | • Referral to a medical professional<br>• Requires a high degree of suspicion for this diagnosis<br>• Evaluation includes pressure testing of the lower arm<br>• Treatment may require surgery |

Acute elbow injuries, primarily dislocations and fractures, account for the third most common acute injury in gymnastics. In addition to activation of the EAP, assessment of general athlete well-being to limit risk of shock in these situations is important. See Table 38.7.

## Hand and wrist injuries

In women's artistic gymnastics, the wrist is the most commonly injured area of the upper extremity (Caine & Nassar, 2005). With exposure up to 16 times the athlete's body weight (Overlin, Chima, & Erickson, 2011), injury risk increases with athlete growth, advancement to higher competitive levels, and with new or increased skill training. Gymnasts aged 10–14 years have a higher reported incidence of wrist pain, likely due to growth and resulting stress on the growth plates (Chawla & Wiesler, 2015). See Table 38.8.

## 38.6 Injury prevention

Opportunities to implement conditioning and flexibility programs benefit the athlete, and may aid in decreasing their risk of injury. It is prudent to progress skill acquisition in a step-wise fashion that builds upon consistent and technically correct basic skills before progressing to more difficult maneuvres.

Gymnasts are at risk of developing both acute and chronic injuries while growing. Bones grow in length from the epiphyseal plate, or growth plate. This growth plate is cartilage and not as strong as the surrounding bone. The growth plate is susceptible to injury, including fracture. As the bone lengthens, the muscles need to stretch out to accommodate the change in bone length. Gymnasts and coaches may recognize decreased flexibility accompanying

**TABLE 38.8**

| Common signs and symptoms | Possible reason for pain | Treatment considerations |
|---|---|---|
| Chronic pain or injury | | |
| • Occurs in athletes who are still growing | "Gymnast Wrist" | • Referral to a medical professional |
| • Pain with weight bearing skills and rotation of the wrist that increases over time | Distal Radial Epiphysitis | • X-rays most helpful when compared to unaffected side<br>• Treat like a growth plate fracture |
| • Swelling may be evident over thumb side of the wrist | | |
| • Pain with weight bearing skills and rotation over pinkie side of the wrist | TFCC injury | • Referral to a medical professional<br>• May respond to conservative therapy |
| • Can result in "popping" or "cracking" with wrist rotation | (Triangular Fibrocartilage Complex) | • MRI arthrograms often needed for definitive diagnosis |
| • More common in athletes who have completed puberty | | |
| • Pain with weight bearing located at dorsal wrist | Generalized wrist pain | • Referral to a medical professional<br>• Can be a difficult diagnosis to specify |
| • Often occurs with growth and body changes | | • Limit number of painful activities<br>• Wrist bracing or taping can be beneficial |
| • Often occurs with new skill and level changes | | • Wrist strengthening exercises |

athlete growth. This places their muscles at risk of injury in the form of strains or tears, especially within the extremes of flexibility required in women's artistic gymnastics.

There are also growth plates at muscle attachments to the bone. These bone-tendon junctions grow in size with the athlete. Dynamic forces inherent in the sport of gymnastics place these apophyseal growth plates at risk of injury from the pull of the attached muscle. As listed, apophyseal "irritation" injuries occur with gymnasts (Sever's, Osgood Schlatter's, etc.). The best prevention is maintenance of flexibility during times of growth. Focus on the hamstrings, quadriceps, hip flexors (iliopsoas), and the calf muscle, including both the gastrocnemius and soleus muscles.

Vigilance with growth is important in the prevention of low back injuries. Prevention strategies to avoid or limit low back pain are beneficial, as up to 65% of female artistic gymnasts complain of low back pain sometime during their career (Vanti *et al.*, 2010; Kruse & Lemmen, 2009). Low back pain in gymnasts is multifactorial, complicating prevention programs. With puberty, body shape changes resulting in tight muscles, interrupting the muscle firing pattern necessary for proper muscle function.

As described by Jull and Janda (1987), this "Crossed Pelvic Syndrome" describes inhibited gluteal and abdominal muscles in combination with tight or over-active hip flexor and erector spinae muscles. The key to "fixing" this pelvic cross syndrome is recognition of abdominal and gluteal muscle inhibition. **Inhibition is different than weakness**. Focusing on only strengthening these muscles, without allowing for activation, results in no improvement. Activation is achieved by stretching the hip flexors and posterior lumbar fascia, continuous with the hamstring muscles. When the necessary flexibility is achieved, then strengthening can be beneficial.

This is not to say that a focus on core strengthening will not be of benefit for any athlete. Pilates-like exercises are easily incorporated into a conditioning routine. These incorporate strengthening with flexibility, and are beneficial for prevention. Focus on proper technique and muscle activation is necessary to avoid poor technique, hindering athlete performance, and deterring any preventative benefit.

Artistic gymnastics requires numerous maneuvres, utilizing a hyperextended (arched) back positioning. With limited shoulder flexibility, compensation through the low back predisposes the athlete to injury and low back pain. Correct bridging techniques, focusing on flexibility through the shoulders instead of the back. Also, stretches for shoulder flexibility not requiring bridging are encouraged.

## 38.7 CONCLUSION

Women's artistic gymnastics is a popular sport worldwide. Despite inherent risks, the incidence of injury is decreasing. Recognition of chronic injury can decrease the potential for pain and frustration on the part of the athlete, coach, and parents. Implementation of flexibility and conditioning programs, especially when athletes are growing, aid in limiting injury. Working with medical professionals familiar with the sport, and the injuries that occur, aids in establishing trust to help the athlete rehabilitate. Acute injuries are often scary; implementation of an Emergency Action Plan can greatly reduce the stress and anxiety when acute injuries occur. With recognition of both acute and chronic injuries, and implementation of strategies to reduce incidence of injury, gymnastics is a sport that can be enjoyed by athletes of all ages and competitive levels.

# 39

# MONITORING GYMNASTS FOR INJURY PREVENTION

*Elizabeth J. Bradshaw*

## 39.1 Objectives and learning outcomes

This chapter will provide a summary of athlete monitoring tools and systems that can be used in gymnastics programs. After completing this chapter, you will be able to:

- Explain the purpose of monitoring gymnasts.
- Discuss the advantages and disadvantages of monitoring gymnasts.
- Distinguish between subjective and objective monitoring methods.
- Select and use appropriate tools for monitoring specific groups of gymnasts.
- Develop an injury prevention and rehabilitation system for gymnasts.

## 39.2 Introduction

Advanced gymnastics requires the gymnast (and his/her family) to commit to a high training load at an early age, which is typically when the gymnast is of a pre-school or school age. They are faced with the challenge of growing and maturing whilst pursuing their training and competition goals, as well as maintaining their schooling. A gymnast may be fearful due to unknown outcomes of maturation and the effects it has on their gymnastics career. During fast growth they may lose some flexibility and mechanical power, directly affecting their gymnastics. The personal and financial sacrifices of the gymnast's family may also increase the stress for that child (Tofler et al., 1996). Finding the ideal training stimulus 'sweet spot' for each individual gymnast requires careful monitoring, especially during the pre-adolescent and adolescent years (Gabbett, 2016), but is necessary for avoiding overuse (high load-related) injuries through over-training and also avoiding acute injuries (low load-related or inadequate preparation through progression of skill development) and performance deficits due to under-training. Both under- and over-training impedes the gymnast from reaching their full potential, and reaching peak performance at the optimum time (e.g. Olympic Games).

Monitoring gymnasts can be subjective (surveys, questionnaires) and/or objective (physical testing). Some of the advantages of monitoring gymnasts include:

- Better communication

  o Improved communication between the gymnast, coach and athlete services (support team).

- Growth (if applicable)

  o Are they having or have they recently had a growth spurt?
  o What is their biological age (if during the adolescence years)?
  o Have they experienced their peak height velocity?

- General health and wellbeing

  o Are they generally coping with the training load on top of other demands such as school?

- Injury

  o Are load-related (inadequate or excessive training) injuries prevented?
  o Is there a pattern of acute injuries in the program that may be able to be prevented?

- Performance

  o What are the gymnast's strengths and weaknesses?
  o Are they progressing and reaching performance targets? If not, why not?
  o Can you predict their performance potential?

- Education

A well designed and facilitated monitoring system (or testing program) can provide the coaches and gymnasts with improved understanding of the demands of the sport (generally and at each level, e.g. junior), as well as the attributes required for greater success. This can also facilitate systematic planning of athlete development programs (Pyke, 2000, p. xii).

Overall, monitoring gymnasts enables the coaching team to be better informed about their gymnasts in order to maximize the gymnast's net performance potential. The coaching team can achieve this outcome by using this information to set (program) and/or periodize an appropriate training load that maximizes performance outcomes (i.e. physical preparation, skill development, competition readiness, competition performance) whilst limiting the negative consequences of training (i.e. over-training, injury, illness, fatigue) (Sands et al., 2016).

The obvious disadvantages of monitoring gymnasts are the cost and time required to develop and implement a suitable system for a group of gymnasts and/or as a part of an overall gymnastics program, and to also achieve cooperation from all of the coaches, support staff and gymnasts. Greater cooperation can be achieved by involving key stakeholders (e.g. coaches, doctors, sports scientists, program managers, parent representative) in the decision-making process when developing an athlete monitoring system, and then providing information sessions for gymnasts and their family that raise awareness of the value of a monitoring system. Regular individual or group meetings with gymnasts where feedback is provided with open discussion can also facilitate improved cooperation, as it demonstrates to the gymnasts that the information is being used to benefit their training program. This is incredibly important as monitoring systems require a time investment from the gymnast often both during (e.g. growth measures) and outside (e.g. daily wellness survey) of their training time.

Traditionally most monitoring of gymnasts (and athletes generally) has involved growth, physiological, biomechanical and medical tests; however, subjective measures through regular surveys and questionnaires are becoming more prevalent (Sands et al., 2016). Monitoring gymnasts can be achieved by secure online athlete management systems with computer-based and mobile interfaces for easy access (e.g. Smartabase Athlete Management). The advantage of centralized and more holistic athlete monitoring systems is that it creates a more gymnast-centered program that a team of people can contribute to, as illustrated in Figure 39.1, and likewise is capable of protecting sensitive information by restricting access to specific sections (e.g. medical information) to specific users. Some common methods of monitoring gymnasts will now be described.

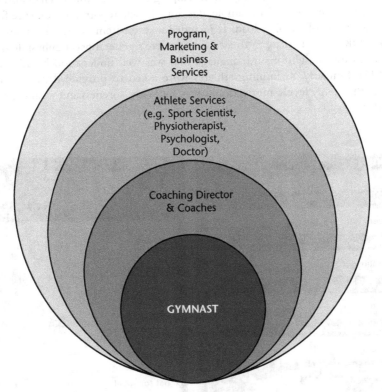

**FIGURE 39.1**   Example organizational structure for gymnast-centered programs.

(Adapted from Pyke, 2006, p. 26)

## 39.3 Gymnast self-report measures

Self-report measures are a simple and relatively effective method for athlete monitoring that is gaining increased popularity in high performance sport (Neumaier, Main and

Gastin, 2013). An advantage of athlete self-report measures is that they give the gymnast a voice, so that their experience and perspectives on their wellbeing and training can be heard (Weissensteiner, 2015). It therefore helps to ensure that the gymnast is the central focus of the training program. Other advantages are that it enables the coach to monitor the wellbeing of the gymnast as well as refining their inter-personal and intra-personal skills in effectively interpreting, understanding and responding to their gymnast, and nurture the gymnast's self-regulatory skills as they progress towards the elite senior competitive level (Weissensteiner, 2015). A recent review by Saw, Main and Gastin (2016) asserts that these subjective measures are often superior to objective measures (e.g. creatine kinase) for monitoring changes in athlete wellbeing in response to training.

In the past, surveys have typically contained physical and psychological measures of wellbeing such as the Profile of Mood States (POMS), the Recovery Stress Questionnaire for Athletes (RESTQ-S) and the Daily Analyses of Life Demands of Athletes (DALDA) (Saw, Main and Gastin, 2016). However, these published questionnaires are often too long, lack specificity to the sport and are too narrow in their focus (Saw, Main and Gastin, 2015). For that reason, many sports have begun developing their own athlete self-report measures. Foster and Lehmann (1997), for example, developed a self-report questionnaire for runners to assess the effects of training load. It included a seven-item complaint index (e.g. 0 = Perky, 5 = OK, 10 = Lethargic) to assess the athlete's general psychophysiological status, attempting to use straightforward language that was well understood by athletes (Sartor et al., 2013). Similarly, swimming athletes were asked to provide daily self-reports on general health, energy levels, motivation, stress, recovery, soreness and wellness (Crowcroft et al., 2017).

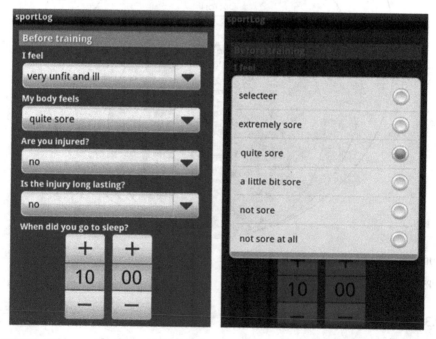

**FIGURE 39.2**  The user interface of the survey app used to monitor wellbeing and load by Simons (2014).

In gymnastics there have been few reports to date on self-report surveys. Session rating of perceived exertion (RPE) has been used with TeamGym athletes (Minganti et al., 2010) and shown to be a valid measure of internal training load. They used two instruments to assess session RPE; the CR-10 Borg's scale and the visual analog scale, with both scores then multiplied by the training duration in minutes. The Borg's scale is arguably easier to deliver and interpret in self-report survey systems. It provides a scale of 0 to 10; for example, 1 is rest, 2 is very, very easy, 5 is hard and 10 is maximal (Minganti et al., 2010). Generally, the RPE for the gymnasts in their study was higher during tumbling sessions in comparison to tramponette sessions. Simons (2014) constructed a self-report monitoring system for junior artistic gymnasts that included questions on general health, mental load, training load via a session RPE and injury, and was administered via a smartphone application (app) (Figure 39.2). When the gymnasts were grouped according to injury status (low, medium, high) during the three-month study, it revealed that the high injury group was significantly older and that the medium injury group was the happiest with the lowest levels of anger. Simons (2014) concluded that the overall monitoring system may have not been sensitive enough to detect small changes in training load and mental health that may precede injury; however, it is likely that the system needed to be assessed more longitudinally. Bradshaw and colleagues (2014a) monitored junior Australian gymnasts for one year using daily training diaries that were collected from the gymnasts each month. Self-report measures included questions on the gymnasts' wellbeing (e.g. sleep quality) and their gymnastics training. It identified that the gymnasts felt 'fit and healthy' for just over half the training period monitored (Figure 39.3). The gymnasts' 'fit and healthy' days peaked in February and March (both 20.5 days), were lowest in July (7.6 days), recovered in August (18.3 days) and September (17.5 days), then decreased again in November (12.3 days). The peaks and troughs in the gymnasts' wellbeing corresponded with their ability/inability to complete their daily training program (e.g. March – 16.6 days, July – 7.9 days) and the number of days that they were ill (e.g. March – 0.1 days, July – 3.2 days). The gymnasts' peak competition (National Championships)

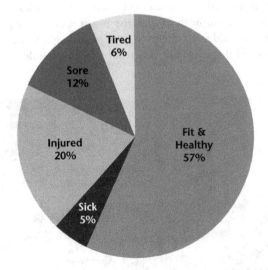

**FIGURE 39.3** Example results from one full year of monitoring Australian female junior gymnasts' wellbeing.

was in July that year, followed by a second competition (National Club Championships) in November. The Bradshaw et al. (2014a) study demonstrated that multifactorial self-report measures can identify problematic patterns in gymnasts' wellbeing, providing the opportunity for changed training strategies or interventions. The gymnasts studied were aged 9–13 years and were mainly pre-pubertal or entering puberty. In pubertal and post-pubertal female gymnasts, additional questions related to menarche and menstrual function may be warranted to detect menstrual disorders such as primary and secondary amenorrhea (Roupas and Georgopoulos, 2011).

## 39.4 Growth and maturation

Gymnasts are often grouped during training based upon chronological age and performance level. There can be considerable variation in biological maturity (also known as biological age) that indicates the gymnast's state of maturity (pre-pubertal, pubertal, post-pubertal) and, if in the pubertal state, their stage of maturation (Tanner stage 2–4; Tanner, 1962). The onset (early, average, late) and duration of maturation (slow, average, fast) can also affect physical and psychosocial development in both boys and girls (Cumming et al., in press), which can have implications for training prescription and injury risk. For these reasons, it is important to track growth and maturation.

The most common measures of growth employed in gymnastics are body mass, height and sitting height using a stadiometer, digital scales and a flat-seated stool (Figure 39.4). It is recommended that these measures are completed at least every three to six months so that periods of faster growth can be identified. In gymnastics a sitting height/standing height ratio (%) is frequently calculated as an indicator of leg length and therefore basic proportions. A review of this measure was completed by Malina et al. (2013) which provides reference data on this measure for a large cross-section of female gymnasts showing that the ratio increases in late adolescence, indicating proportionally shorter legs. They also reported that later maturing gymnasts generally have proportionally longer legs than their early maturing counterparts.

Skeletal age assessment is the most accurate maturational index and can be used across the entire growth period from birth to maturity but it is costly, requires specialist services and incurs radiation (Mirwald et al., 2002). Age of peak height velocity is a commonly used indicator for the onset of maturity. For that reason, height should also be converted into a

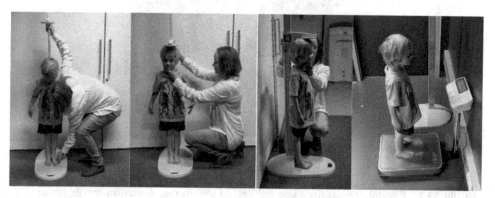

**FIGURE 39.4** Standing and body mass measures using a stadiometer and digital scales.

height velocity through dividing by year. Peak height and a maturity offset (+ or − years) with respect to the gymnasts chronological age can also be estimated from height, sitting height and body mass measures using predictive equations provided by Mirwald et al. (2002). A downloadable spread sheet with this predictive tool embedded is available from www.scienceforsport.com/peak-height-velocity/. Adult stature is identified when height growth slows to less than 1.0 cm per year or when growth slows to less than 0.5 cm every six months for a period of two years (Malina et al., 2013).

During the pubertal (adolescent) years, the Tanner puberty stages can be used to identify the onset and duration of maturation (Marshall and Tanner, 1969, 1970). The Tanner puberty stages have been converted into gender-specific line drawings and used in self-assessment questionnaires. The Tanner puberty stages consist of a series of images of female and male reproductive organs depicting development at each of the five stages of sexual maturity from a pre-pubescent child (Stage 1) through to post-pubescent adult (Stage 5). In general, whilst self-assessment instead of physician assessment of these stages is more comfortable for the gymnast, it is not as reliable (Taylor et al., 2001). Females tend to under-report pubertal development and it has been recommended that maternal assessment is completed when the gymnast is 11 years or younger for greater accuracy (Terry et al., 2016). Finally, in females the onset of menarche is a reliable measure of maturity if obtained prospectively instead of via recall (Dorn et al., 2013).

Overall, these growth and maturation measures may be useful for Bio-Banding gymnasts with respect to competition groups (instead of based purely on age), strength and conditioning approaches, and technical development (Cumming et al., 2017).

## 39.5 Blood testing

Blood testing of gymnasts is typically completed to check general health and wellbeing, as well as to examine fatigue and recovery patterns. Whilst blood testing is invasive and can be challenging for some gymnasts, many of the measures are directly related to performance and injury risk, and therefore are beneficial to the gymnast. Blood tests should be administered by a suitably qualified person (e.g. general practitioner, nurse, pathologist, phlebotomist).

Continuous monitoring of blood trace elements and vitamins at least biannually is recommended for athletes, especially before vitamin and mineral supplementation (Moran et al., 2013; Zaitseva et al., 2015). Trace elements and vitamins can affect the performance and general health of athletes (Pouramir, Haghshenas and Sorkhi, 2004). Iron is an essential element for the formation of haemoglobin, which is the protein responsible for the transport of oxygen from the respiratory organs to the peripheral tissues (Deli et al., 2013). Iron is also a vital component in the formation of myoglobin. Following iron deficiency, the concentration of myoglobin in skeletal muscle is severely reduced (40–60%) which compromises the muscle's oxidative capacity. For that reason, iron supplementation can improve physical performance in athletes whom are iron depleted but not anaemic (Deli et al., 2013). Iron deficiency has three stages as illustrated in Figure 39.5. Stage One involves clinically low serum ferritin levels, which indicates that iron stores are low. Stage Two is reduced transferrin saturation, which is when the diminished iron stores begin to affect the production of red blood cells. Stage Three is when haemoglobin production falls and the athlete is considered anaemic. In the general population, anaemia is normally associated with low iron intake, especially from food sources, and is mainly prevalent among women and children (Sureira, Amancio and Braga, 2012). In artistic gymnasts, Sureira, Amancio and Braga (2012) identified

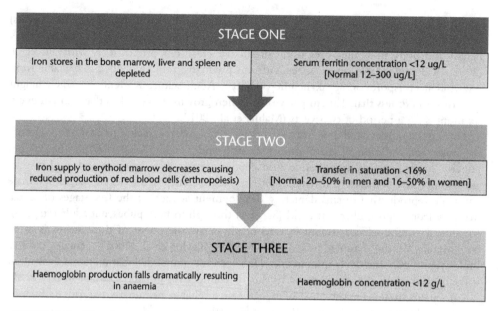

**FIGURE 39.5** The three stages of iron deficiency.

that training did not cause iron deficiency or anaemia; however, it can partially alter the gymnast's general haematological (blood trace elements and vitamins) profile.

As a result of gymnasts training indoors, they are at greater risk of vitamin D deficiency, especially during the winter months (Peeling et al., 2013). For this reason, there is a high prevalence of vitamin D deficiency in athletes and gymnasts (Bradshaw et al., 2014b; Farrokhyar et al., 2015; Jemni, 2011a; Lovell, 2008; Moran et al., 2013; Peeling et al., 2013). In gymnasts there is no relationship between vitamin D levels and bone health due to the benefits of the high (strain) loads placed on the musculoskeletal system during gymnastics training (Bradshaw et al., 2014b). However, vitamin D supports the cardiovascular, immune and musculoskeletal systems (Moran et al., 2013). Vitamin D deficiency can affect muscle strength and power, and in elite ballet dancers it has been demonstrated to have a relationship with injury including muscle spasms, strains and tears (Wyon et al., 2014). Vitamin D enhances cellular immunity, and therefore vitamin D deficiency has also been associated with increased frequency of common colds, influenza and gastroenteritis (Moran et al., 2013). In addition to vitamin D, blood trace elements associated with infection risk in athletes are summarized in Table 39.1.

Blood measures of lactate and creatine kinase (CK) are also frequently assessed in athletes as an indicator of recovery and muscle damage (Bermon, Gleeson and Walsh, 2013; Jemni et al., 2003). Blood lactate measures have been used in gymnastics to examine recovery strategies and their effects of performance. In male elite French gymnasts, Jemni et al. (2003) identified that a combined recovery of passive rest and active recovery between apparatus during competition had greater lactate clearance and led to higher performance scores (see Chapter 1, section 1.5, and Chapter 5, section 5.7). Blood lactate measures can be used to determine recovery strategies for individual gymnasts, and also as an indicator of fatigue, which is an important consideration with respect to injury risk. Further, when CK levels are extremely high (>500 U/L) it can be interpreted as an indication that the

**TABLE 39.1** Blood analysis related to immune system status in athletes, excluding iron studies and vitamin D

| Blood measure | Why? |
| --- | --- |
| Serum B12 concentration Serum foltate concentration | Water soluble B vitamins needed for the production of DNA. Important for immune cells that are dividing at rapid rates. |
| White blood cell count | The total number of white blood cells which when elevated can indicate inadequate recovery from training or competition, or when an infection is present. |
| Neutrophil count | The most abundant type of white blood cell that acts as the first line of defense against microorganisms that are capable of causing illness. |
| Lymphocyte count | Another type of white blood cell that has many functions in the immune system. During and after infection it will be elevated, however a low value can be associated with impaired immunity and increased risk of infection. |
| Monocyte count | A large white blood cell which responds to signals of inflammation in the body and contributes to wound healing. It also responds to stress and infections. |

(Bermon, Gleeson and Walsh, 2013)

muscles have not properly recovered from training, and therefore the training load and intensity should be reduced, particularly if before a competition (Bermon et al., 2013).

## 39.6 Heart rate monitoring

Heart rate (HR) monitoring is not commonly employed in gymnastics, possibly due to the inconvenience of the chest strap (which is often too large for small gymnasts) and the watch. However, it offers a non-invasive method of monitoring training intensity and recovery the following day. In gymnastics they have been used to identify the physiological demands of training and competing on each apparatus (e.g. Jemni et al., 2000). HR variability (HRV) is a further application of HR monitoring. It provides information about the responsiveness of the autonomic nervous system, which can become imbalanced in over-trained athletes (Sartor et al., 2013); therefore, HRV is a potential measure of fatigue. Sartor et al. (2013) measured HRV in elite senior male gymnasts after waking in the morning whilst lying prone and sitting. Their aim was to assess short-term HRV following the previous day's training and they found that HRV such as the mean beat-to-beat intervals was sensitive to changes in training load.

## 39.7 Criteria for selecting tests to prevent, treat and/or monitor injuries

When developing a monitoring system, it is important to consider each test selected with respect to its *relevance* to gymnastics and the specific gymnasts being assessed, the specificity to the performance characteristics of the discipline of gymnastics (e.g. artistic, rhythmic), the *validity* of the test with respect to that it should measure what it claims to measure, whether the *accuracy* of the test is within acceptable criterions, can be measured *reliably* in younger and older gymnasts (or the target age group), the *practicality* of the test which takes into account the ease of testing gymnasts (venue availability, gymnast availability, test duration, cost) as well as the time taken to deliver feedback to the coaches (if applicable) and gymnasts, and how the test results can be interpreted for guidelines for action (outcomes).

For injury prevention, it is recommended that several monitoring methods are utilized including subjective (self-report surveys) and objective (blood test) measures of the gymnast's wellbeing. Atypical responses can provide an indication of, for example, insufficient recovery from training, which could be confirmed, if necessary, by blood measures of CK or a white blood cell count. The information can then be used to adjust recovery strategies for an individual gymnast until the most effective strategy is identified. In a pre- or early-pubescent gymnast these measures should also be combined with growth and maturation measures, as well as physical attribute testing such as flexibility and musculoskeletal power (jumping). During periods of rapid growth, the gymnast's bones grow faster than ligaments and tendons. This means that there is a period where ligaments and tendons are shorter relative to bone; a period of heightened injury risk (Steinberg et al., 2011). As it is difficult to directly measure ligament and tendon growth, flexibility (see Chapter 6, section 6.3) and jump (see Chapter 7) tests provide an indication of the effects of rapid growth on the gymnast's biomechanics. Bradshaw et al. (2014c), for example, observed large biomechanical changes to the jumping patterns of female artistic gymnasts during rapid height growth. This demonstrated that continuous jumping and hopping tests are a worthwhile addition to a monitoring system for this phase of a gymnast's development for injury prevention purposes.

Similarly, during injury rehabilitation a monitoring system that includes self-report surveys provides an indication of how the gymnast is recovering both psychologically and physically. Brewer (2010) reviewed the causal links between psychological factors and injury rehabilitation, indicating multimodal treatment is effective, particularly for knee injury (more details can be found in Chapter 38). Pre-injury objective data from physical testing such as flexibility and jump testing provides an indication of how well the gymnast is biomechanically recovering once physically able to perform some basic skills and training. The use of wireless inertial measurement units (IMUs) also offers a new method of objectively measuring a gymnast's load during specific skills and/or during training sessions. It provides the opportunity to remove the guesswork when a gymnast is advised by a doctor that they are able to return to 50% of their usual training load, and for monitoring loading patterns (e.g. on high risk apparatus) in order to try and prevent injury. For example, Figure 39.6 shows IMUs being used during pommel horse training. That gymnast was found to have an average asymmetry index of 54% between wrists, indicating that he was unloading his left wrist by increasing the load on his right wrist. He subsequently revealed that he had left wrist pain which he'd not yet reported to his coaches. The use of IMUs during pommel training also revealed that the wrist loads were higher on the mushroom trainer which is equipment typically used more by younger, developing gymnasts. This is concerning, as younger male gymnasts are more susceptible to injury due to their immature musculoskeletal system and growing bodies.

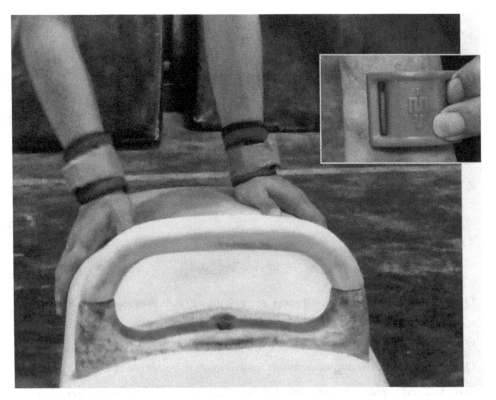

**FIGURE 39.6**    The placement of the inertial measurement units on the gymnast's wrists for pommel horse.

## 39.8 CONCLUSION

A considerable challenge in high-level gymnastics is developing healthy, capable and resilient gymnasts whilst enabling each gymnast to achieve their maximum performance potential (Bergeron et al., 2015; Mountjoy and Bergeron, 2015). A comprehensive and effective monitoring system for gymnasts (or any athlete) is usually the product of continual interaction between support staff (e.g. sport scientists) and coaches who foster a shared commitment to the testing protocols (Hahn, 2012, p. xiii). Whilst careful evaluation and changes to protocols is often necessary during the early stages of establishing a monitoring system due to the principles of sound scientific practice, decisions to change protocols thereafter should not be treated nonchalantly as testing standards (normative data) will need to be redeveloped and longitudinal trends may be more difficult to detect (Hahn, 2012, p. xiii). The creation of a monitoring system for gymnasts promotes scientific enquiry and also evidence-based practice that delivers a high performance environment which provides unwavering commitment to assisting every gymnast who has the Olympic dream (Mountjoy and Bergeron, 2015).

# PART VII REVIEW QUESTIONS

Q1.  How does an international sport governing body influence the type and extent of injuries observed in the athletes under its jurisdiction?

Q2.  List five things that can serve as injury countermeasures.

Q3.  What are the two most commonly injured joints in both male and female gymnasts?

Q4.  List and explain five guidelines for the coach that can help prevent injury.

Q5.  Explain the phenomenon of "grip lock" and how to prevent this problem.

Q6.  How does physical fitness tend to prevent injury?

Q7.  Describe a "revenge effect" and how increasing safety may result in an increase rather than decrease in injury?

Q8.  Explain why spotting is not a panacea.

Q9.  Explain how the best gymnasts may suffer the most severe injuries.

Q10.  How do foam pits and soft mats reduce the problems related to impact injuries?

Q11.  An eight-year-old female team gymnast starts to limp around the gym at the end of practice. You notice and ask her why. She responds that her heels hurt when she walks on them. You also notice that she has recently had a growth spurt and she hasn't been punching into her rebound out of her round off back handspring on floor. She admits that tumbling and running for vault hurt her heels too. She doesn't remember doing anything to hurt herself. You notice she is able to run okay during warm ups at the beginning of practice, but gets more sore as practice continues.

- What is your initial response?
- What do you think is causing her heel pain?
- What is the best thing to do with this athlete and her parents?

Q12. A 12-year-old optional gymnast is working on her switch split leaps on beam when she falls down and complains of pain in the back of her upper leg. You did not see any obvious injury, and walk over to her to see what happened.

- What is your initial response? What questions are you going to ask?
- What do you think is causing her leg pain?
- What is the next best thing to do with this athlete and her parents?

Q13. A 13-year-old recreational gymnast, who is hoping to learn some skills for her school cheerleading tryouts, is constantly rolling her wrists while waiting in line and states her wrists are hurting her more now. She has been wearing the typical wrist brace that your team girls wear for the past two to three months, but says that the past four weeks she has had increased pain. She is getting frustrated and asks you what she can do.

- What is your initial response?
- What is the next best thing to do with this athlete and her parents?

Q14. A 14-year-old team gymnast tells you that her elbow has been "catching" for the past couple of weeks and sometimes it "locks" up when she goes to straighten it or bend it. You ask if she has any pain and she tells you it has been "sore" for the past month or so, but she has just been pushing through it.

- What is your initial response?
- What do you think may be contributing to her elbow problems?
- What is the next best thing to do with this athlete and her parents?

Q15. What is the purpose of monitoring gymnasts?

Q16. What is the most common disadvantage of monitoring gymnasts?

Q17. What is the main advantage of self-report measures for gymnast monitoring?

Q18. Which of the following measures are subjective?

   a.   Session rating of perceived exertion (RPE)
   b.   Sitting height/standing height ratio (%)
   c.   Creatine kinase
   d.   Continuous hopping test on a force platform

Q19. What is the criteria for a gymnast reaching adult stature?

Q20. Why can iron supplementation improve physical performance in gymnasts who are iron depleted but not anaemic?

Q21. What blood measures are often used as an indicator of recovery and muscle damage?

   a.   Vitamin D
   b.   Lactate
   c.   Lymphocyte count
   d.   Creatine kinase
   e.   Both A and B
   f.   Both B and D

Q22. There is a period of increased injury risk following rapid growth in athletes. Why?

Q23. The wrist loads measured using an IMU (~7–16 BW) placed on the wrists were much higher than the pommel horse reaction forces that have previously been reported (<2 BW). Why is there such a large difference in results? [*Extension question.*]

Q24. What is the difference between an injury reporting system and a monitoring system to prevent injury? Discuss. [*Extension question.*]

# REFERENCES

Abernethy, B., Wann, J., & Parks, S. (1998). Training perceptual-motor skills for sport. In B. Elliott (ed.), *Training in Sport* (pp. 1–68). New York, NY: John Wiley & Sons.

Adams, J. A. (1987). Historical review and appraisal of research on learning, retention, and transfer of human motor skills. *Psychological Bulletin, 101*(1), 41–74.

Aerts, I., Crumps, E., Verhagen, E., Verschueren, J., & Meeusen, R. (2013). A systematic review of different jump landing variables in relation to injuries. *J Sports Med Phys Fitness, 53*(5), 509–519.

Agostini, T., Righi, G., Galmonte, A., & Bruno, P. (2004). The relevance of auditory information in optimizing hammer throwers performance. In P. B. Pascolo (ed.), *Biomechanics and Sports* (pp. 67–74). Vienna, Austria: Springer.

Aldridge, J. (1991). Overuse injuries affecting the developing skeleton: with reference especially to the upper limb. *FIG Scientific/Medical Symposium Proceedings, 1*(1), 39–41.

Aldridge, M. J. (1987). Overuse injuries of the distal radial growth epiphysis. In T. B. Hoshizaki, J. H. Salmela, & B. Petiot (eds), *Diagnostics, Treatment and Analysis of Gymnastic Talent* (pp. 25–30). Montreal, Canada: Sport Psyche Editions.

Alp, A., & Bruggemann, G.-P. (1992). Biomechanical analysis of landing mats in gymnastics. In G.-P. Bruggemann & J. K. Ruhl (eds), *Biomechanics in Gymnastics* (pp. 259–269). Cologne, Germany: Bundesinstitut fur sportwissenschaft.

Alter, M. J. (2004). *Science of Flexibility*. Champaign, IL: Human Kinetics.

Andres, B. M., & Murrell, G. A. C. (2008). Treatment of tendinopathy what works, what does not, and what is on the horizon. *Clinical Orthopaedics and Related Research, 466*, 1539–1554.

Anshel, M. H. (1990). *Sport Psychology: From Theory to Practice*. Scottsdale, AZ: Gorsuch Scarisbrick.

Anshel, M. H., & Payne, J. M. (2006). Application of sport psychology for optimal performance in the martial arts. In J. Dosil (ed.), *The Sport Psychologist's Handbook* (pp. 353–374). Chichester, UK: Wiley.

Arampatzis, A., & Bruggemann, G. P. (1998). A mathematical high bar-human body model for analysing and interpreting mechanical-energetic processes on the high bar. *Journal of Biomechanics, 31*, 1083–1092.

Arampatzis, A., & Bruggemann, G.-P. (1999). Energy and performance – storage and return of elastic energy by gymnastic apparatus. In M. Leglise (ed.), *Symposium Medico-Technique* (pp. 29–37). Lyss, Switzerland: International Gymnastics Federation.

Arampatzis, A., & Bruggemann, G.-P. (2001). Mechanical energetic processes during the giant swing before the Tkatchev exercise. *Journal of Biomechanics, 34*, 505–512.

Arampatzis, A., Bruggemann, G. P., & Klapsing, G. M. (2001). Leg stiffness and mechanical energetic processes during jumping on a sprung surface. *Med Sci Sports Exerc, 33*(6), 923–931.

Arampatzis, A., Morey-Klapsing, G., & Bruggemann, G.-P. (2005). Orthotic effect of a stabilising mechanism in the surface of gymnastic mats on foot motion during landing. *Journal of Electromyography and Kinesiology, 15*(5), 507–515.

Arampatzis, A., Stafilidis, S., Morey-Klapsing, G., & Bruggemann, G. P. (2004). Interaction of the human body and surfaces of different stiffness during drop jumps. *Med Sci Sports Exerc, 36*(3), 451–459.

Araujo, C. (2002). *Manual de adjudas em ginastica*. Porto, Portugal: Porto Editora, SA.

Arce, J., Haupt, H. A., Irwin, K. D., Ohle, J., Palmieri, J., & Siff, M. (1990). Training variation. *National Strength and Conditioning Association Journal, 12*(4), 14–24.

Arkaev, L. I., & Suchilin, N. G. (2004). *Gymnastics. How to Create Champions*. Oxford, UK: Meyer & Meyer.

Arkaev, L. & Suchilin, N. G. (2009). *Gymnastics: How to Create Champions; The Theory and Methodology of Training Top-Class Gymnasts*. Oxford: Meyer & Meyer.

Arndt, A., Bruggemann, G. P., Koebke, J., & Segesser, B. (1999a). Asymmetrical loading of the human triceps surae: I. Mediolateral force differences in the Achilles tendon. *Foot and Ankle International, 20*(7), 444–449.

Arndt, A., Bruggemann, G. P., Koebke, J., & Segesser, B. (1999b). Asymmetrical loading of the human triceps surae: II. Differences in calcaneal moments. *Foot and Ankle International, 20*(7), 450–455.

Arndt, A. N., Komi, P. V., Brüggemann, G. P., & Lukkariniemi, J. (1998). Individual muscle contributions to the in vivo achilles tendon force. *Clinical Biomechanics, 13*(7), 532–541. doi:http://dx.doi.org/10.1016/S0268-0033(98)00032-1

Asch, S. E., & Witkin, H. A. (1948a). Studies in space orientation: I. Perception of the upright with displaced visual field. *Journal of Experimental Psychology, 38*, 325–337.

Asch, S. E., & Witkin, H. A. (1948b). Studies in space orientation: II. Perception of the upright with displaced visual fields and with body tilted. *Journal of Experimental Psychology, 38*, 455–477.

Asmussen, E., & Bonde-Petersen, F. (1974). Storage of elastic energy in skeletal muscles in man. *Acta Physiol Scand, 91*(3), 385–392.

Asseman, F., & Gahery, Y. (2005). Effect of head position and visual condition on balance control in inverted stance. *Neuroscience Letters, 375*(2), 134–137. doi:10.1016/j.neulet.2004.10.085

Associated Press (2011). Alicia Sacramone has torn Achilles. *ESPN Olympic Sports*. Retrieved from ESPN Olympic Sports website: http://espn.go.com/olympics/gymnastics/story/_/id/7069966/alicia-sacramone-torn-achilles-surgery

Aydin, T., Yildiz, Y., Yildiz, C., Atesalp, S., & Kalyon, T. A. (2002). Proprioception of the ankle: a comparison between female teenaged gymnasts and controls. *Foot and Ankle International, 23*(2), 123–129.

Backx, F. J. G. (1996). Epidemiology of paediatric sports-related injuries. In O. Bar-Or (ed.), *The Child and Adolescent Athlete* (pp. 163–172). Oxford, England: Blackwell Science, Ltd.

Badets, A., & Blandin, Y. (2010). Feedback schedules for motor-skill learning: the similarities and differences between physical and observational practice. *J Mot Behav, 42*(4), 257–268. doi:10.1080/00222895.2010.497512

Baker, B., Kleven, S., Turnbull, S., & Dickinson, J. (1988). Transfer of training and task compatibility. *Journal of Human Movement Studies, 14*, 133–143.

Bale, P., & Goodway, J. (1987). The anthropometric and performance variables of elite and recreational female gymnasts. *N Zealand J Sports Med, 15*(3), 63–66.

Bale, P., & Goodway, J. (1990). Performance variables associated with the competitive gymnast. *Sports Medicine, 10*(3), 139–145. doi:10.2165/00007256-199010030-00001

Banister, E. W. (1991). Modeling elite athletic performance. In H. A. W. J. Duncan MacDougall, & H. J. Green (eds), *Physiological Testing of the High-Performance Athlete* (pp. 403–424). Champaign, IL: Human Kinetics.

Barantsev, S. A. (1985). Do gymnasts need to develop aerobic capacity? *Soviet Sports Review, 25*(1), 20–22.

Bardy, B. G., & Laurent, M. (1998). How is body orientation controlled during somersaulting. *Journal of Experimental Psychology, 24*(3), 963–977.

Barker-Ruchti, N. (2007). Women's artistic gymnastics: an (auto)ethnographic journey (Unpublished doctoral thesis). School of Human Movement Studies, University of Queensland, St Lucia.

Barlett, H. L., Mance, M. J., & Buskirk, E. R. (1984). Body composition and expiratory reserve volume in female gymnasts and runners. *Med Sci Sports Exerc, 16*(3), 311–315.

Bar-Or, O. (1984). The growth and development of children's physiologic and perceptional responses to exercise. In I. Ilmarinen & I. V-lim-ki (eds), *Children and Sport* (pp. 3–17). Berlin: Heidelbergh De.

Bar-Or, O. (1987). The Wingate anaerobic test: an update on methodology, reliability, and validity. *Sports Medicine, 4*, 381–394.

Batatinha, H. A., da Costa, C. E., de Franca, E., Dias, I. R., Ladeira, A. P., Rodrigues, B., . . . Caperuto, E. C. (2013). Carbohydrate use and reduction in number of balance beam falls: implications for mental and physical fatigue. *Journal of the International Society of Sports Nutrition, 10*, 32. doi:10.1186/1550-2783-10-32

Baudry, L., Leroy, D., & Chollet, D. (2005). The circle performed on a pommel horse in gymnastics: the critical role of double support phase. *Gait & Posture, 21*(Suppl 1), S34.

Baudry, L., Leroy, D., Thouvarecq, R., & Chollet, D. (2006). Auditory concurrent feedback benefits the circle performed in gymnastics. *Journal of Sports Sciences, 24*(2), 149–156.

Beatty, K. T., McIntosh, A. S., & Frechede, B. O. (2006, 14–18 July). Method for analysing the risk of overuse injury in gymnastics. Paper presented at the 24 International Symposium on Biomechanics in Sports (2006), Salzburg, Austria.

Beaudin, P. A. (1978). Prédictions de la performance en gymnastique au moyen de l'analyse d'une sélection de variable physiques, physiologiques et anthropométriques. McGill University, Montréal.

Becker, D. (1998). Gymnast's fall on routine move not sports first. *USA Today*, 10C.

Bedu, M., Fellmann, N., Spielvogel, H., Falgairette, G., van Praagh, E., & Coudert, J. (1991). Force-velocity and 30s Wingate tests in boys at high and low altitudes. *J Appl Physiol, 70*, 1031–1037.

Begon, M., Wieber, P.-B., & Yeadon, M. R. (2008). Kinematics estimation of straddled movements on high bar from a limited number of skin markers using a chain model. *Journal of Biomechanics, 41*, 581–586.

Behm, D. G. (1995). Neuromuscular implications and applications of resistance training. *Journal of Strength and Conditioning Research, 9*(4), 264–274.

Beilock, S. L., & Gonso, S. (2008). Putting in the mind versus putting on the green: expertise, performance time and the linking of imagery and action. *Journal of Experimental Psychology, 61*(6), 920–932.

Benardot, D. (1999). Nutrition for gymnasts. In N. T. Marshall (ed.), *The Athlete Wellness Book* (pp. 1–28). Indianapolis, IN: USA Gymnastics.

Benardot, D., & Czerwinski, C. (1991). Selected body composition and growth measures of junior elite gymnasts. *J Am Diet Assoc, 91*(1), 29–33.

Benardot, D., Schwarz, M., & Weitzenfeld, D. (1989). Nutrient intake in young highly competitive gymnasts. *Journal of the American Dietetic Association, 89*(3), 401–403.

Benefice, E., & Malina, R. (1996). Body size, body composition and motor performances of mild-to-moderately undernourished Senegalese children. [Research Support, Non-U.S. Gov't]. *Annals of Human Biology, 23*(4), 307–321.

Bergeron, M. F., Mountjoy, M., Armstrong, N., Chia, M., Cote, J., Malina, R. M., Pensgaard, A. M., Sanchez, A., Soligard, T., Sundgot-Borgen, J., van Mechelen, W., Weissensteiner, J. R., & Engebretsen, L. (2015). International Olympic committee consensus statement on youth athletic development. *British Journal of Sports Medicine, 49*, 843–851.

Bergh, U. (1980). Entraînement de la puissance aérobie. In P. O. Astrand & K. Rodahl (eds), *Précis de Physiologie de L'exercice Musculaire* (2 ed., pp. 303–308). Paris: Masson.

Bermon, S., Gleeson, M., & Walsh, N. P. (2013). Practical guidelines on minimizing infection risk in athletes. In M. Gleeson, N. Bishop, & N. Walsh (eds), *Exercise Immunology* (pp. 239–266). New York: Routledge.

Bernasconi, S., Tordi, N., Parratte, B., Rouillon, J.-D., & Monnier, G. (2004). Surface electromyography of nine shoulder muscles in two iron cross conditions in gymnastics. *Journal of Sports Medicine and Physical Fitness*, *44*, 240–245.

Bernasconi, S. M., Tordi, N. R., Parratte, B. M., & Rouillon, J. D. (2009). Can shoulder muscle coordination during the support scale at ring height be replicated during training exercises in gymnastics? *Journal of Strength and Conditioning Research*, *23*(8), 2381–2388. doi:10.1519/JSC.0b013e3181bac69f

Bernier, M., & Fournier, J. (2007). Mental skill evaluations of French elite athletes. In Y. Theodorakis, M. Gourdas, & A. Papaionnou (eds), *Sport and Exercise Psychology: Bridges Between Disciplines and Cultures* (pp. 89–91). Thesaloniki: University of Thesaly.

Bernstein, P. L. (ed.) (1996). *Against the Gods*. New York, NY: John Wiley & Sons.

Berthoz, A., Pavard, B., & Young, L. R. (1975). Perception of linear horizontal self-motion induced by peripheral vision (linearvection): basic characteristics and visual-vestibular interactions. *Experimental Brain Research*, *23*, 471–489.

Bessi, F. (2006). Trainingsprotokolle der Jahrgänge 1989, 1990, 1991 [Training records of gymnasts from the age groups 1989, 1990, 1991]. Herbolzheim.

Bessi, F. (2007a). Cuestionario sobre lateralidad en la Gimnasia: Documento inédito [Questionnaire about the laterality in Gymnastics. Unpublished manuscript]. Freiburg.

Bessi, F. (2007b). Fragebogen zur Lateralität für Trainer: Unveröff. Unterlage [Laterality questionnaire for coaches. Unpublished manuscript]. Freiburg.

Bessi, F. (2015). Laterality in artistic gymnastics. Poster presentation during the IV International Seminar of competitive Gymnastics. Sao Paulo.

Bessi, F. (2016a). El mundo de la Gimnasia artística: en teoría y práctica [The world of artistic gymnastics in theory and practice] (1ra Edición). Buenos Aires: Editorial Dunken.

Bessi, F. (2016b). Laterality in artistic gymnastics. *Brazilian Journal of Physical Education and Sport*, *30*(1). doi:10.1590/1807-55092016000100019

Bessi, F., Hofmann, D., Laßberg, C. v., & Heinen, T. (2016). Directional tendencies in artistic gymnastics. In T. Heinen (ed.), *Sports and Athletics Preparation, Performance, and Psychology. Gymnastics Performance and Motor Learning* (pp. 119–138). Hauppauge: Nova Science Publishers Inc.

Bessi, F., & Milbradt, J. (2015). Fragebogen zur Lateralität für Kadertrainer Gerätturnen Männer: Unveröff. Unterlage. Freiburg, Berlin.

Bessi, F., & Milbradt, J. (in preparation). Analyse der Drehpräferenzen deutscher Bundeskaderturner [Analysis of the rotational preferences of the German Juniors squad]. Freiburg, Berlin.

Bieze Foster, J. (2007). Efforts to reduce gymnastics injuries focus on spring floors. *Biomechanics*, *14*(1), 11–12.

Blochin, I. P. (1965). Energeticeskaja (gazoobmenu) upraznenij sportivnoj gimnastiki u muzsin. Teor Prakt Fiz Kult, 28, 32.

Bloom, B. S. (1985). *Developing Talent in Young People*. New York: Ballantyne.

Bobbert, M. F. (1990). Drop jumping as a training method for jumping ability. *Sports Med*, *9*(1), 7–22.

Bobbert, M. F., Huijing, P. A., & van Ingen Schenau, G. J. (1987). Drop jumping. I. The influence of jumping technique on the biomechanics of jumping. *Med Sci Sports Exerc*, *19*(4), 332–338.

Bobbert, M. F., & Van Ingen Schenau, G. J. (1988). Coordination in vertical jumping. *J Biomech*, *21*(3), 249–262.

Boden, B. P., Pasquina, P., Johnson, J., & Mueller, F. O. (2001). Catastrophic injuries in pole-vaulters. *American Journal of Sports Medicine*, *29*(1), 50–54.

Boisseau, N., Persaud, C., Jackson, A. A., & Poortmans, J. R. (2005). Training does not affect protein turnover in pre- and early pubertal female gymnasts. *Eur J Appl Physiol Occup Physiol*, *94*(3), 262–267.

Bompa, T. O. (1990). *Theory and Methodology of Training*. Dubuque, IA: Kendall/Hunt.

Bompa, T. O., & Haff, G. G. (2009). *Periodization*. Champaign, IL: Human Kinetics.

Bondarchuk, A. P. (2007). *Transfer of Training in Sports*. Michigan, USA: Ultimate Athlete Concepts.

Boone, W. T. (1979). *Better Gymnastics – How to Spot the Performer*. Mountain View, CA: World Publications.

Borgen, J. S., & Corbin, C. B. (1987). Eating disorders among female athletes. *Physician Sportsmed, 15*(2), 89–95.

Bortoleto, M. A. C. (2004). The internal logic of Men's Artistic Gymnastics: an ethnographic study in a high performance gymnasium (Unpublished doctoral thesis). University of Lleida, Lleida, Spain.

Bortoleto, M. A. C., & Peixoto, C. (2014). Qualitative video analysis as a pedagogical tool in artistic gymnastics. In L. Schiavon, T. Heinen, M. A. C. Bortoleto, M. Nunomura, E. Toledo (Org.), *High Performance Gymnastics* (pp. 99–116). Hildesheim: Arete Verlag.

Bortoleto, M. A. C., & Schiavon, L. M. (2016). Artistic gymnastics – why do coaches resist change? *Sports Coaching Review, 5*(2), 198–201.

Bosco, C. (1985). *Elasticità Muscolare e Forza Esplosiva Nelle Attività Físico-Esportive*. Roma: Società Stampa Sportiva.

Bosco, C. (1992). *La Valutazione Della Forza con el Test di Bosco*. Roma: Società Stampa Sportiva.

Bosco, C., Iacovelli, M., Tsarpela, O., Cardinale, M., Bonifazi, M., Tihanyi, J., . . . Viru, A. (2000). Hormonal responses to whole-body vibration in men. *European Journal of Applied Physiology, 81*(6), 449–454. doi:10.1007/s004210050067

Bosco, C., & Komi, P. V. (1979). Potentiation of the mechanical behavior of the human skeletal muscle through prestretching. *Acta Physiol Scand, 106*(4), 467–472.

Bosco, C., Komi, P. V., & Ito, A. (1981). Prestretch potentiation of human skeletal muscle during ballistic movement. *Acta Physiol Scand, 111*(2), 135–140.

Bosco, C., Luhtanen, P., & Komi, P. V. (1983). A simple method for measurement of mechanical power in jumping. *Eur J Appl Physiol Occup Physiol, 50*(2), 273–282.

Bosco, C., Viitasalo, J. T., Komi, P. V., & Luhtanen, P. (1982). Combined effect of elastic energy and myoelectrical potentiation during stretch-shortening cycle exercise. [Research Support, Non-U.S. Gov't]. *Acta Physiologica Scandinavica, 114*(4), 557–565.

Botkin, M. (1985). Gymnasts aid NASA in research of space-age problem. *USA Gymnastics, July/August*, 15–19.

Bortoleto, M. A. C., & Cohelho, T. F. (2016). Men's artistic gymnastics: is the use of elastic surfaces systematic in the training process? *Revista Brasileira de Educação Física e Esporte, 30*(1), São Paulo Jan./Mar. Print version ISSN 1807-5509. On-line version ISSN 1981-4690. http://dx.doi.org/10.1590/1807-55092016000100051

Bouchard, C., Dionne, F. T., Simoneau, J. A., & Boulay, M. R. (1992). Genetics of aerobic and anaerobic performances. *Exer Sport Sci Rev, 20*, 27–58.

Bourdieu, P. (1973). *Cultural Reproduction and Social Reproduction in Knowledge, Education and Cultural Change*. London: Tavistock.

Boutcher, S. H. (1993). Attention and athletic performance: an integrated approach. In T. S. Horn (ed.), *Advances in Sport Psychology* (pp. 251–265). Champaign, IL: Human Kinetics.

Bradshaw, E. (2004). Target-directed running in gymnastics: a preliminary exploration of vaulting. *Sports Biomechanics, 3*(1), 125–144.

Bradshaw, E. J., & Aisbett, B. (2006). Visual guidance during competition performance and run through training in long jumping. *Sports Biomechanics, 5*(1), 1–14.

Bradshaw, E. J., & Hume, P. A. (2012). Biomechanical approaches to identify and quantify injury mechanisms and risk factors in women's artistic gymnastics. *Sports Biomech, 11*(3), 324–341. doi: 10.1080/14763141.2011.650186

Bradshaw, E. J., & Sparrow, W. A. (2000). The speed-accuracy trade-off in human gait control when running towards targets. *Journal of Applied Biomechanics, 16*, 331–341.

Bradshaw, E. J., & Sparrow, W. A. (2001). Effects of approach velocity and foot-target characteristics on the visual regulation of step length. *Human Movement Science, 20*, 401–426.

Bradshaw, E., Thomas, K., Moresi, M., Greene, D., Braybon, W., McGillvray, K., & Andrew, K. (2014a). A longitudinal, multifactorial risk analysis of elite female gymnasts (research report). Canberra: Australian Institute of Sport.

Bradshaw, E., Thomas, K., Moresi, M., Greene, D., Braybon, W., McGillvray, K., & Andrew, K. (2014b). Is a gymnast's performance and health affected by vitamin D deficiency? In K. Sato, W. A. Sands, & S. Mizuguchi (eds), *eProceedings of the 32nd Conference of the International Society of Biomechanics in Sports, Volume 1*: Oral 19.

Bradshaw, E. J., Thomas, K., Moresi, M., Greene, D., Braybon, W., McGillivray, K., & Andrew, K. (2014c). Biomechanical field test observations of gymnasts entering puberty. In K. Sato, W. A. Sands, & S. Mizuguchi (eds), *eProceedings of the 32nd Conference of the International Society of Biomechanics in Sports, Volume 1*: Oral 19.

Brady, C., & Vincenzino, B. (2002). An investigation of the relationship between the posture of gymnastic bridge and low back pain in gymnasts. *Sports Link*, 10–12.

Bressel, E., Yonker, J. C., Kras, J., & Heath, E. M. (2007). Comparison of static and dynamic balance in female collegiate soccer, basketball, and gymnastics athletes. *Journal of Athletic Training, 42*(1), 42–46.

Brewer, B. (2010). The role of psychological factors in sport injury rehabilitation outcomes. *International Review of Sport and Exercise Psychology, 3*(1), 40–61.

Bridgeman, G., Hendry, D., & Start, L. (1975). Failure to detect displacement of visual world during saccadic eye movements. *Vision Research, 15*, 719–722.

Brooks, T. (2003). Women's collegiate gymnastics: a multifactorial approach to training and conditioning. *Strength and Cond J, 25*(2), 23–37.

Brotherhood, J. R. (1984). Nutrition and sports performance. *Sports Med, 1*, 350–389.

Brown, E. W., Witten, W. A., Weise, M. J., Espinoza, D., Wisner, D. M., Learman, J., & Wilson, D. J. (1996). Attenuation of ground reaction forces in salto dismounts from the balance beam. J. M. C. S. Abrantes Proceedings XIV International Symposium on Biomechanics in Sports (pp. 336–338). Lisbon, Portugal: Edicoes FMH Universidade Tecnica de Lisbon.

Bruggemann, G. P. (1985). Mechanical load on the achilles tendon during rapid dynamic sport movements. In S. M. Perren & E. Schneider (eds), *Biomechanics: Current Interdisciplinary Research* (pp. 669–674). Dordrecht, Netherlands: Martinus Nijhoff.

Bruggemann, G. P. (1987). Biomechanics in gymnastics. *Medicine and Sport Science, 25*, 142–176.

Bruggemann, G.-P. (1994). Biomechanics of gymnastic techniques. *Sport Science Review, 3*(2), 79–120.

Bruggemann, G. P. (1999). Mechanical load in artistic gymnastics and its relation to apparatus and performance. In M. Leglise (ed.), *Symposium Medico-Technique* (pp. 17–27). Lyss, Switzerland: International Gymnastics Federation.

Bruggemann, G.-P. (2010). Neuromechanical load of the biological tissue and injury in gymnastics. Paper presented at the Proceedings of the 28th Conference of the International Society of Biomechanics in Sports, Northern Michigan University, Marquette, Michigan, USA.

Bunc, V., & Petrizilkova, Z. (1994). Energy cost of selected exercise in elite femnale gymnasts. *Acat Univ Carol Kinanthrop, 30*, 11–18.

Burton, D. (1988). Do anxious swimmers swim slower? Reexamining the elusive anxiety–performance relationship. *Journal of Sport and Exercise, 10*, 45–61.

Burton, D. (1993). Goal setting in sport. In R. N. Singer, M. Murphey, & L. K. Tennant (eds), *Handbook of Research on Sport Psychology* (pp. 467–491). New York, NY: Macmillan.

Caine, C. G., Caine, D. J., & Lindner, K. J. (1996). The epidemiologic approach to sports injuries. In D. J. Caine, C. G. Caine, & K. J. Lindner (eds), *Epidemiology of Sports Injuries* (pp. 1–13). Champaign, IL: Human Kinetics.

Caine, D. J. (2002). Injury epidemiology. In W. A. Sands, D. J. Caine, & J. Borms (eds), *Scientific Aspects of Women's Gymnastics* (pp. 72–109). Basel, Switzerland: Karger.

Caine, D., Cochrane, B., Caine, C., & Zemper, E. (1989). An epidemiologic investigation of injuries affecting young competitive female gymnasts. *American Journal of Sports Medicine, 17*(6), 811–820.

Caine, D., Knutzen, K., Howe, W., Keeler, L., Sheppard, L., Henrichs, D., & Fast, J. (2003). A three-year epidmiological study of injuries affecting young female gymnasts. *Physical Therapy in Sport, 4*, 10–23.

Caine, D. J., Lindner, K. J., Mandelbaum, B. R., & Sands, W. A. (1996). Gymnastics. In D. J. Caine, C. G. Caine, & K. J. Lindner (eds), *Epidemiology of Sports Injuries* (pp. 213–246). Champaign, IL: Human Kinetics.

Caine, D. J., & Maffulli, N. (2005). Gymnastics injuries. In D. J. Caine & N. Maffulli (eds), *Epidemiology of Pediatric Sports Injuries* (Vol. 48, pp. 18–58). Basel, Switzerland: Karger.

Caine, D. J., & Nassar, L. (2005). Gymnastics injuries. *Medicine and Sport Science*, *48*, 18–58. doi:10.1159/000084282

Caine, D., Roy, S., Singer, K. M., & Broekhoff, J. (1992). Stress changes of the distal radial growth plate. *American Journal of Sports Medicine*, *20*(3), 290–298.

Calmels, C., & Fournier, J. F. (2001). Duration of physical and mental execution of gymnastic routines. *The Sport Psychologist*, *15*(2), 142–150.

Calmels, C., Pichon, S., & Grezes, J. (2014). Can we simulate an action that we temporarily cannot perform? *Neurophysiologie Clinique*, *44*(5), 433–445. doi:10.1016/j.neucli.2014.08.004

Calvert, G., Spence, C., & Stein, B. E. (2004). *Handbook of Multisensory Processing*. Cambridge, MA: MIT Press.

Calvo, M., Rodas, G., Vallejo, M., Estruch, A., Arcas, A., Javierre, C., . . . Ventura, J. L. (2002). Heritability of explosive power and anaerobic capacity in humans. *European Journal of Applied Physiology*, *86*(3), 218–225.

Caplan, J., Julien, T. P., Michelson, J., & Neviaser, R. J. (2007). Multidirectional instability of the shoulder in elite female gymnasts. *American Journal of Orthopedics*, *36*(12), 660–665.

Carek, P. J., & Fumich, R. M. (1992). Stress fracture of the distal radius: not just a risk for elite gymnasts. *The Physician and Sportsmedicine*, *20*(5), 115–118.

Carl, J., & Gellmann, R. (1987). Human smooth pursuit: stimulus-dependent responses. *Journal of Neurophysiology*, *57*, 1446–1463.

Carlock, J. M., Smith, S. L., Hartman, M. J., Morris, R. T., Ciroslan, D. A., Pierce, K. C., et al. (2004). The relationship between vertical jump power estimates and weightlifting ability: a field-test approach. *J Strength Cond Res*, *18*(3), 534–539.

Carnahan, H., & Lee, T. D. (1989). Training for transfer of a movement timing skill. *J Mot Behav*, *21*(1), 48–59.

Carpenter, R. H S. (1988). *Movements of the Eyes*. London: Plion.

Cassell, C., Benedict, M., & Specker, B. (1996). Bone mineral density in elite 7- to 9-yr-old female gymnasts and swimmers. *Medicine & Science in Sports & Exercise*, *28*(10), 1243–1246.

Chamari, K., Hachana, Y., Ahmed, Y. B., Galy, O., Sghaier, F., Chatard, J. C., . . . Wisloff, U. (2004). Field and laboratory testing in young elite soccer players. *Br J Sports Med*, *38*(2), 191–196.

Chan, D., Aldridge, M. J., Maffulli, N., & Davies, A. M. (1991). Chronic stress injuries of the elbow in young gymnasts. *The British Journal of Radiology*, *64*, 1113–1118.

Chase, M. A., Magyar, M. T., & Drake, B. M. (2005). Fear of injury in gymnastics: self-efficacy and psychological strategies to keep on tumbling. *J. Sport Sci*, *23*(5), 465–475.

Chawla, A., & Wiesler, E. R. (2015) Nonspecific wrist pain in gymnasts and cheerleaders. *Clinics in Sports Medicine*, *34*, 143–149.

Chen, C.-K., & Liu, Y. (2000). The kinematic analysis of giant swing and dismount of double salto backward stretched with 720§ turns on horizontal bar. In Y. Hong & D. P. Johns (eds), *Proceedings of XVIII International Symposium on Biomechanics in Sports* (I ed., pp. 225–228). Hong Kong, China: The Chinese University of Hong Kong, International Society for Biomechanics in Sports.

Chen, J. D., Wang, J. F., Li, K. J., Zhao, Y. W., Wang, S. W., Jiao, Y., et al. (1989). Nutritional problems and measures in elite and amateur athletes. *Am J Clin Nutr*, *49*(5 Suppl), 1084–1089.

Chilvers, M., Donahue, M., Nassar, L., & Manoli, A. (2007). Foot and ankle injuries in elite female gymnasts. *Foot & Ankle International*, *28*(2), 214–218.

Chiviacowsky, S., Wulf, G., & Lewthwaite, R. (2012). Self-controlled learning: the importance of protecting perceptions of competence. *Frontiers in Psychology*, *3*, 458. doi:10.3389/fpsyg.2012.00458

Christina, R. W., & Davis, G. (1990). Diving skill progressions, Part 1 Principles of teaching skill progressions. In J. L. Gabriel & G. S. George (eds), *U.S. Diving Safety Manual* (pp. 89–103). Indianapolis, IN: U.S. Diving Publications.

Christou, M., Smilios, I., Sotiropoulos, K., Volaklis, K., Pilianidis, T., & Tokmakidis, S. P. (2006). Effects of resistance training on the physical capacities of adolescent soccer players. *J Strength Cond Res*, *20*(4), 783–791.

Ciullo, J. V., & Jackson, D. W. (1985). Pars interarticularis stress reaction, spondylolysis, and spondylolisthesis in gymnasts. *Clinics in Sports Medicine*, *4*(1), 95–110.

Claessens, A. L., Lefevre, J., Beunen, G., & Malina, R. M. (1999). The contribution of anthropometric characteristics to performance scores in elite female gymnasts. *Journal of Sports Medicine and Physical Fitness*, *39*(4), 355–360.

Claessens, A. L., Malina, R. M., Lefevre, J., Beunen, G., Stijnen, V., Maes, H., et al. (1992). Growth and menarcheal status of elite female gymnasts. / Croissance et menstruation des gymnastes d'elite feminines. *Medicine & Science in Sports & Exercise*, *24*(7), 755–763.

Clarke, K. S. (1998). On issues and strategies. In H. Appenzeller (ed.), *Risk Management in Sport* (pp. 11–22). Durham, NC: Carolina Academic Press.

Coates, C., McMurtry, C. M., Lingley-Pottie, P., & McGrath, P. J. (2010). The prevalence of painful incidents among young recreational gymnasts. *Pain Research & Management*, *15*(3), 179–184.

Cogan, K. D. (2006). Sport psychology in gymnastics. In J. Dosil (ed.), *The Sport Psychologist's Handbook* (pp. 641–661). Chichester, UK: Wiley.

Cogan, K. D., & Vidmar, P. (2000). *Sport Psychology Library: Gymnastics*. Morgantown, WV: Fitness Information Technology.

Cometti, G. (1988). *La Pliometrie*. Dijon: UFR Staps de Dijon.

Committee, U. S. O. (1995). Gymnastics pit removal for cervical injuries – Video Tape. *U.S. Olympic Committee Publications, Colorado Springs, CO, 7:17, DUB 10/19/00*, 1.

Corbin, C. (1972). Mental practice. In W. Morgan (ed.), *Ergogenic Aids and Muscular Performance* (pp. 688–784). New York: Academic Press.

Coren, S. (1993). The lateral preference inventory for measurement of handedness, footedness, eyedness, and earedness: norms for your adults. *Bulletin of the Psychonomic Society*, *31*(1), 1–3.

Cormie, P., Sands, W. A., & Smith, S. L. (2004). A comparative case study of Roche vaults performed by elite male gymnasts. *Technique*, *24*(8), 6–9.

Côté, J. (1999). The influence of the family in the development of talent in sports. *The Sport Psychologist*, *13*, 395–417.

Côté, J., Baker, J., & Abernethy, B. (2003). From play to practice: a developmental framework for the acquisition of expertise in team sports. In J. S. Starkes & K. A. Ericsson (eds), *Expert Performance in Sports: Advances in Research on Sport Expertise* (pp. 89–113). Champaign, IL: Human Kinetics.

Côté, J., & Hay, J. (2002). Children's involvement in sport: a developmental perspective. In J. M. Silva & D. Stevens (eds), *Psychological Foundations in Sport* (2 ed., pp. 484–502.). Boston: Merrill.

Courteix, D., Lespessailles, E., Obert, P., & Benhamou, C. L. (1999). Skull bone mass deficit in prepubertal highly-trained gymnast girls. *Int J Sports Med*, *20*(5), 328–333.

Courteix, D., Rieth, N., Thomas, T., Van Praagh, E., Benhamou, C., Collomp, K., et al. (2007). Preserved bone health in adolescent elite rhythmic gymnasts despite hypoleptinemia. *Hormone Research*, *68*, 20–27.

Cousineau, P. (2003). *The Olympic Odyssey*. Wheaton, IL: Quest Books.

Coventry, E., Sands, W. A., & Smith, S. L. (2006). Hitting the vault board: implications for vaulting take-off – a preliminary investigation. *Sports Biomechanics*, *5*(1), 63–75.

Crielaard, J. M., & Pirnay, F. (1981). Anaerobic and aerobic power of top athletes. *Eur J Appl Physiol*, *47*, 295–300.

Crowcroft, S., McCleave, E., Slattery, K., & Coutts, A. J. (2017). Assessing the measurement sensitivity and diagnostic characteristics of athlete monitoring tools in national swimmers. *International Journal of Sports Physiology and Performance*, *12*, S2-95–S2-100.

Crumley, K. (1998). Optimal skill continuity. *Technique*, *18*(6).

Csikszentmihali, M. (1975). *Beyond Boredom and Anxiety*. San Francisco, CA: Jossey-Bass.

Csikszentmihalyi, M., Rathunde, K., & Whalen, S. (1993). *Talented Teenagers: The Roots of Success and Failure*. Cambridge: Cambridge University Press.

Cuk, I., & Marinsek, M. (2013). Landing quality in artistic gymnastics is related to landing symmetry. *Biology of Sport, 30*(1), 29–33. doi:10.5604/20831862.1029818

Cumming, S. P., Lloyd, R. S., Oliver, J. L., Eisenmann, J. C., & Malina, R. M. (2017). Bio-banding in sport: applications to competition, talent identification, and strength and conditioning of youth athletes. *Strength and Conditioning Journal, 39*(2), 34–47.

Cunningham, S. J. (1988). *A Model of Gymnastic Landing Mats and Pits and Human Subject Response.* University of Utah.

Cureton, T. K. (1941). Flexibility as an aspect of physical fitness. *The Research Quarterly, 12*, 381–390.

Dallas, G., & Kirialanis, P. (2010). Judges' evaluation of routines in men's artistic gymnastics. *Sci Gymnastics J, 2*(2), 49–58. doi:UDC 796.412.2.012.2

Daly, R., Bass, S., Finch, C., & Corral, A.-M. (1998). Balancing gymnastics and injury risk. Retrieved from Victoria, Canada: Deakin University School of Movement.

Daly, R. M., Bass, S. L., & Finch, C. F. (2001). Balancing the risk of injury to gymnasts: how effective are the counter measures? *British Journal of Sports Medicine, 35*, 8–20.

Dargent-Paré, C., Agostini, M. de, Mesbah, M., & Dellatolas, G. (1992). Foot and eye preferences in adults: relationship with handedness, sex and age. *Cortex, 28*(3), 343–351.

Davidson, P. L., Mahar, B., Chalmers, D. J., & Wilson, B. D. (2005). Impact modeling of gymnastic back-handsprings and dive-rolls in children. *Journal of Applied Biomechanics, 21*, 115–128.

Davis, L. M. (1974). *The History of Gymnastics on and with Apparatus since World War II.* Los Angeles, CA: UCLA.

Davlin, C. D., Sands, W. A., & Shultz, B. B. (2001a). Peripheal vision and back tuck somersaults. *Perceptual and Motor Skills, 93*, 465–471.

Davlin, C. D., Sands, W. A., & Shultz, B. B. (2001b). The role of vision in control of orientation in a back tuck somersault. *Motor Control, 3*, 337–346.

Davlin, C. D., Sands, W. A., & Shultz, B. B. (2002). Influence of vision on kinesthetic awareness while somersaulting. *International Sports Journal, 6*(2), 172–177.

Davlin, C. D., Sands, W. A., & Shultz, B. B. (2004). Do gymnasts "spot" during a back tuck somersault. *International Sports Journal, 8*(2), 72–79.

De Smet, L., Claessens, A., Lefevre, J., & Beunen, G. (1994). Gymnast wrist: an epidemiologic survey of ulnar variance and stress changes of the radial physis in elite female gymnasts. *American Journal of Sports Medicine, 22*(6), 846–850.

Debu, B., & Woollacott, M. (1988). Effects of gymnastics training on postural responses to stance perturbations. *J Mot Behav, 20*(3), 273–300.

Deli, C. K., Fatouros, I. G., Koutedakis, Y., & Jamurtas, A. Z. (2013). Iron supplementation and physical performance. In M. Hamlin, S. Draper, & Y. Kathiravel (eds), *Current Issues in Exercise and Sports Medicine.* Rijeka: Intech.

Denadai, B. S., Figuera, T. R., Favaro, O. R., & M., Gonçalves (2004). Effect of the aerobic capacity on the validity of the anaerobic threshold for determination of the maximal lactate steady state in cycling. *Braz J Med Biol Res, 37*(10), 1551–1556.

Derakhshan, N., & Machejefski, T. (2015). Distinction between parkour and freerunning. *Chinese Journal of Traumatology. Zhonghua Chuang Shang Za Zhi, 18*(2), 124.

Dexel, J., Marschner, K., Beck, H., Platzek, I., Wasnik, S., Schuler, M., Nasreddin, A., & Kasten, P. (2014). Comparative study of elbow disorders in young high-performance gymnasts. *International Journal of Sports Medicine, 35*, 960–965.

DiFiori, J. P., Caine, D. J., & Malina, R. M. (2006). Wrist pain, distal radial physial injury and ulnar variance in the young gymnast. *American Journal of Sports Medicine, 10*(10), 1–10.

DiFiori, J. P., Puffer, J. C., Mandelbaum, B. R., & Dorey, F. (1997). Distal radial growth plate injury and positive ulnar variance in nonelite gymnasts. *American Journal of Sports Medicine, 25*(6), 763–768.

DiFiori, J. P., Puffer, J. C., Mandelbaum, B. R., & Mar, S. (1996). Factors associated with wrist pain in the young gymnast. *American Journal of Sports Medicine, 24*(1), 9–14.

Dion, D. E. (1985). Prediction de la performance de l'athlete d'elite en plongeon par l'entremise de parametres anthropometriques et biomecaniques. University of Laval.

Dodd, S., Powers, S., Callender, T., & Brooks, E. (1984). Blood lactate disappearance at various intensities of recovery exercise. *J. Appl. Physiol. Resp. Environ. Exercise Physiol, 57*(5), 1462–1465.

Dorn, L. D., Sontag-Padilla, L. M., Pabst, S., Tissot, A., & Susman, E. J. (2013). Longitudinal reliability of self-reported age at menarche in adolescent girls: variability across time and setting. *Developmental Psychology, 49*(6), 1187–1193.

Douda, H., & Toktnakidis, S. P. (1997). Muscle strength and flexibility of the lower limbs between rhythmic sports and artistic female gymnasts. Paper presented at the Second Annual Congress of the European College of Sport Science.

Douda, H., Avloniti, A., Kasabalis, A., Smilitis, H., & Toktnakidis, S. P. (2006). Application of ratings of perceived exertion and physiological responses to maximal effort in rhythmic gymnasts. *Int J Appl Sports Sciences, 18*(2), 78–88.

Dowthwaite, J. N., & Scerpella, T. A. (2009). Skeletal geometry and indices of bone strength in artistic gymnasts. *J Musceuronal Interact, 9*(4), 198–214.

Drabik, J. (1996). *Children & Sports Training.* Island Pond, VT: Stadion Publishing Co.

Drinkwater, B., Nison, K., Chesnut, C. H., Bremmer, W. J., Shainholtz, S., & Southworth, M. B. (1984). Bone mineral content of amenorrheic and eumenorrheic athletes. *N Engl J Med, 311*(5), 277–281.

Driss, T., Vandewalle, H., & Monod, H. (1998). Maximal power and force-velocity relationships during cycling and cranking exercises in volleyball players. Correlation with the vertical jump test. *J Sports Med Phys Fitness, 38*(4), 286–293.

Ducher, G., Hill, B. L., Angeli, T., Bass, S. L., & Eser, P. (2009). Comparison of pQCT parameters between ulna and radius in retired elite gymnasts: the skeletal benefits associated with long-term gymnastics are bone- and site-specific. *J Musculoskelet Neuronal Interact, 9*(4), 247–255.

Dukalsky, V. V., & Dukalsky, A. V. (1977). Use of a phonogoniometer and tilting device to teach diving. *Yessis Review of Soviet Physical Education and Sports, 12*(4), 94–97.

Dunlavy, J. K., Sands, W. A., McNeal, J. R., Stone, M. H., Smith, S. L., Jemni, M., & Haff, G. G. (2007). Strength performance assessment in a simulated men's gymnastics still rings cross. *Journal of Sports Science and Medicine, 6*, 93–97.

Dunn, K. (1980). Spotters, thicker mats may prevent injury. *The Physician and Sportsmedicine, 8*(9), 20.

Durand-Bush, N. (1995). Validity and reliability of the Ottawa Mental Skills Assessment Tool (OMSAT-3) (Unpublished masters thesis). School of Human Kinetics, University of Ottawa.

Durand-Bush, N., & Salmela, J. H. (2002). The development and maintenance of expert athletic performance: perceptions of world and Olympic champions. *Journal of Applied Sport Psychology, 14*, 154–171.

Durand-Bush, N., Salmela, J. H., & Green-Demers, I. (2001). The Ottawa Mental Skills Assessment Tool (OMSAT-3★). *The Sport Psychologist, 15*, 1–19.

Easterbrook, J. A. (1959). The effect of emotion on cue utilization and the organization of behavior. *Psychological Review, 66*, 183–201.

Effenberg, A. O., Fehse, U., Schmitz, G., Krueger, B., & Mechling, H. (2016). Movement sonification: effects on motor learning beyond rhythmic adjustments. *Frontiers in Neuroscience, 10*, 219.

Effenberg, A. O., & Mechling, H. (2005). Movement sonification: a new approach in motor control and learning. *Journal of Sports and Exercise Psychology, 27*, 58–68.

Elbæk, L., & Froberg, K. (1992). Specific physical training parameters in relation to Danish team gymnastics. Biomechanics in gymnastics. Conference Proceedings. First International Conference, 8–10 September, Cologne, 431–441.

Eloranta, V. (1997). Programming leg muscle activity in vertical jumps. *Coaching and Sport Science Journal, 2*(3), 17–28.

Ericksen, H. M., Gribble, P. A., Pfile, K. R., & Pietrosimone, B. G. (2013). Different modes of feedback and peak vertical ground reaction force during jump landing: a systematic review. *J Athl Train, 48*(5), 685–695.

Ericsson, K. A. (1996). The road to excellence: the acquisition of expert performance in the arts and sciences, sports, and games. In K. A. Ericsson (ed.), *The Acquisition of Expert Performance: An Introduction to Some of the Issues* (pp. 1–50). Mahwah, NJ: Erlbaum.

Ericsson, K. A. (2003). Development of elite performance and deliberate practive: an update from the perspective of the expert performance approach. In J. L. Starkes & K. A. Ericsson (eds), *Expert Performance in Sport* (pp. 49–84). Champaign, IL: Human Kinetics.

Ericsson, K. A. (2007). Deliberate practice and the modifiability of body and mind: toward a science of the structure and acquisition of expert and elite performance. *International Journal of Sport Psychology, 38*(1), 4–34.

Ericsson, K. A. (2008). Deliberate practice and acquisition of expert performance: a general overview. *Academic Emergency Medicine, 15*(11), 988–994. doi:10.1111/j.1553-2712.2008.00227.x

Ericsson, K. A. (2013). Training history, deliberate practice and elite sports performance: an analysis in response to Tucker and Collins review—what makes champions? *British Journal of Sports Medicine, 47*(9), 533–535. doi:10.1136/bjsports-2012-091767

Ericsson, K. A. (2016). Summing up hours of any type of practice versus identifying optimal practice activities: commentary on Macnamara, Moreau, & Hambrick. *Perspectives on Psychological Science, 11*(3), 351–354. doi:10.1177/1745691616635600

Ericsson, K. A., Krampe, R. T., & Tesch-Römer, C. (1993). The role of deliberate practice in the acquisition of expert performance. *Psychological Review, 100*, 363–406.

Ericsson, K. A., & Poole, R. (2016). *Peak: Secrets from the New Science of Expertise.* New York, NY: Houghton Mifflin Harcourt Publishing.

Estape, E., Lopez, M., & Grande, I. (1999). *Las Habilidades Gimnasticas Y Acrobatcas En El Ambito Educativo.* Barcelona, Spain: Inde Publicaciones.

Estape Tous, E. (2002). *La Acrobacia En Gimnasia Artistica.* Barcelona, Spain: Inde Publicaciones.

Exell, T. A., Robinson, G., & Irwin, G. (2016). Asymmetry analysis of the arm segments during forward handspring on floor. *European Journal of Sport Science, 16*(5), 545–552. doi:10.1080/1746 1391.2015.1115558

Falgairette, G., Bedu, M., Fellmann, N., Van praagh, E., & Coudert, J. (1991). Bio-energetic profile in 144 boys aged from 6 to 15 years with special reference to sexual maturation. *Eur J Appl Physiol, 62*, 151–156.

Fallon, K. E., & Fricker, P. A. (2001). Stress fracture of the clavicle in a young female gymnast. *British Journal of Sports Medicine, 35*(6), 448–449.

Farana, R., Janezckova, P., Uchytil, J., & Irwin, G. (2015). Effect of different hand positions on elbow loading during the round off in male gymnastics: a case study. *Gymnastics Science Journal, 7*(2), 5–13.

Faria, I. E., & Faria, E. W. (1989). Relationship of the anthropometric and physical characteristics of male junior gymnasts to performance. *J Sports Med Phys Fitness, 29*(4), 369–378.

Faria, I. E., & Pillips, A. (1970). A study of telemetered cardiac response of young boys and girls during gymnastic participation. *J Sports Med, 10*, 145–160.

Farrokhyar, F., Tabasinejad, R., Dao, D., Peterson, D., Ayeni, O.R., Hadioonzadeh, R., & Bhandari, M. (2015). Prevalence of vitamin D inadequacy in athletes: a systematic review and meta-analysis. *Sports Medicine, 45*, 365–378.

Federation, C. G. (1986). Coaching certification manual, Introductory gymnastics, 1. Vanier City, Ontario, Canada: Canadian Gymnastics Federation.

Federation, I. G. (2013). *2013 Code of Points Women's Artistic Gymnastics.* Lausanne, Switzerland: Federation International de Gymnastique.

Federation_Internationale_de_Gymnastique. (2009). *FIG Apparatus Norms.* Lausanne, Switzerland: International Gymnastics Federation.

Feigley, D. A. (1987). Coping with fear in high level gymnastics. In J. H. Salmela, B. Petiot, & T. B. Hoshizaki (eds), *Psychological Nurturing and Guidance of Gymnastic Talent* (pp. 13–27). Montreal: Sport Psyche.

Fellander-Tsai, L., & Wredmark, T. (1995). Injury incidence and cause in elite gymnasts. *Archives of Orthopedic Trauma Surgery, 114*, 344–346.

Ferkolj, M. (2010). A kinematic analysis of the handspring double salto forward tucked on a new style of vaulting table. *Science of Gymnastics*, *2*(1), 35–48.

Feynman, R. (1965). *The Character of Physical Law*. Cambridge, MA: MIT Press.

F.I.G. (2009). Code of Points For men's artistic gymnastics competitions (3rd). Lausanne: Fédération Internationale de Gymnastique (FIG).

F.I.G. (2013). Code of Points For men's artistic gymnastics competitions (4th). Lausanne: Fédération Internationale de Gymnastique (FIG).

FIG (International Gymnastics Federation) (2015). Apparatus Norms. Lausanne: FIG Press. Retrieved from: www.fig-gymnastics.com/site/rules/app-norms

FIG (International Gymnastics Federation) (2016). Bulletin n. 237. Lausanne: FIG Press, Sept.

Filaire, E., & Lac, C. (2002). Nutritional status and body composition of juvenile elite female gymnasts. *J Sports Med Phys Fitness*, *42*(1), 65–70.

Finch, C. F. (1997). An overview of some definitional issues for sports injury surveillance. *Sports Medicine*, *24*(3), 157–163.

Fink, H. (1985). Some considerations for gymnastics conditioning. Budapeste: World Gymnastics, FIG and AIPS Press, n° 25, p. 48.

Finkel, C. B. (2001). Removal of a gymnast with suspected cervical injuries from a soft foam pit. *Technique*, *21*(9), 5–7, 11–15, 39.

Fischer, B. (1987). The preparation of visually guided saccades. *Reviews of Physiology, Biochemistry, and Pharmacology*, *106*, 2–35.

Fitzpatrick, R. C., Marsden, J., Lord, S. R., & Day, B. L. (2002). Galvanic vestibular stimulation evokes sensations of body rotation. *NeuroReport*, (13), 2379–2383.

Flynn, J. M., Ughwanogho, E., & Cameron, D. B. (2011). The growing spine and sports. In B. A. Akbarnia, M. Yazici, & G. H. Thompson (eds), *The Growing Spine* (pp. 151–162). Heidelberg, Germany: Springer-Verlag.

Fogarty, G. J. (1995). Some comments on the use of psychological tests in sport settings. *International Journal of Sport Psychology*, *26*, 161–170.

Fogelholm, G. M., Kukkonen-Harjula, T. K., Taipale, S. A., Sievänen, H. T., Oja, P., & Vuori, I. (1995). Resting metabolic rate and energy intake in female gymnasts, figure-skaters and soccer players. *Int J Sports Med*, *16*(8), 551–556.

Foster, C., & Lehmann, M. (1997). Overtraining syndrome. In G. Guten (ed.), *Running Injuries* (pp. 173–188). Orlando: W.B. Saunders Co.

Foster, G. M. (1973). *Traditional Cultures and the Impact of Technological Change*. New York: Harper & Row.

Fournier, J., Calmels, C., Durand-Bush, N., & Salmela, J. H. (2005). Effects of a season-long PST program on gymnastic performance and on psychological skill development. *ISJEP*, *1*, 7–25.

French, D. N., Gómez, A. L., Volek, J. S., Rubin, M. R., Ratamess, N. A., Sharman, M. J., et al. (2004). Longitudinal tracking of muscular power changes of NCAA division I collegiate women gymnasts. *J Strength Cond Res*, *18*(1), 101–107.

Fry, A. C., Ciroslan, D., Fry, M. D., LeRoux, C. D., Schilling, B. K., & Chiu, L. Z. (2006). Anthropometric and performance variables discriminating elite American junior men weightlifters. *J Strength Cond Res*, *20*(4), 861–866.

Fujihara, T., & Gervais, P. (2012). Circles with a suspended aid: reducing pommel reaction forces. *Sports Biomech*, *11*(1), 34–47. doi:10.1080/14763141.2011.637124

Fujioka, H., Nishikawa, T., Koyama, S., Yamashita, M., Takagi, Y., Oi, T., . . . Yoshiya, S. (2014). Stress fractures of bilateral clavicles in an adolescent gymnast. *Journal of Shoulder and Elbow Surgery*, *23*(4), e88–90. doi:10.1016/j.jse.2014.01.004

Gabbett, T. J. (2006). Performance changes following a field conditioning program in junior and senior rugby league players. *J Strength Cond Res*, *20*(1), 215–221.

Gabbett, T. J. (2016). The training-injury prevention paradox: should athletes be training smarter and harder? *British Journal of Sports Medicine*, *50*, 273–280.

Garcia, C., Barela, J. A., Viana, A. R., & Barela, A. M. (2011). Influence of gymnastics training on the development of postural control. *Neuroscience Letters*, *492*(1), 29–32. doi:10.1016/j.neulet.2011.01.047

Gardiner, E. N. (ed.) (1930). *Athletics of the Ancient World*. Chicago, IL: Ares Publishers, Inc.

Garnier, P., Mercier, B., Mercier, J., Anselme, F., & Préfaut, C. (1995). Aerobic and anaerobic contribution to Wingate test performance in sprint and middle-distance runners. *Eur J Appl Physiol*, *70*, 58–65.

Gatto, F., Swannell, P., & Neal, R. (1992). A force-indentation relationship for gymnastic mats. *Journal of Biomechanical Engineering*, *114*, 338–345.

Gautier, G., Thouvarecq, R., & Chollet, D. (2007). Visual and postural control of an arbitrary posture: the handstand. *Journal of Sports Sciences*, *25*(11), 1271–1278. doi:10.1080/02640410601049144

Gautier, G., Thouvarecq, R., & Larue, J. (2008). Influence of experience on postural control: effect of expertise in gymnastics. *J Mot Behav*, *40*(5), 400–408. doi:10.3200/jmbr.40.5.400-408

George, G. S. (1987). Remediation procedures applicable to victims of suspected catastrophic injury in landing pits. *Gymnastics Safety Update*, *2*(2), 1–2.

George, G. S. (1988a). Reflections on spotting. *Gymnastics Safety Update*, *3(3)*, 2.

George, G. S. (1988b). Spotting – a sacred trust. *Gymnastics Safety Update*, *3*(1), 1–2.

Georgopoulos, N. A., Theodoropoulou, A., Leglise, M., Vagenakis, A. G., & Markou, K. B. (2004). Growth and skeletal maturation in male and female artistic gymnasts. *J Clin Endocrinol Metab.*, *89*(9), 4377–4382.

Gerstein, M. (2008). *Flirting with Disaster*. New York, NY: Union Square Press.

Gervais, P., & Tally, F. (1993). The beat swing and mechanical descriptors of three horizontal bar release-regrasp skills. *Journal of Applied Biomechanics*, *9*(1), 66–83.

Gillingham, K. K., & Wolfe, J. W. (1985). Spatial orientation in flight. In R. L. Deltart (ed.), *Fundamentals of Aerospace Medicine* (pp. 299–381). Philadelphia, PA: Lea & Febiger.

Girginov, V., & Sandanski, L. (2004). From participants to competitors: the transformation of British gymnastics and the role of the Eastern European model of sport. *International Journal of the History of Sport*, *21*(5), 815–832.

Gittoes, M., Jr., & Irwin, G. (2012). Biomechanical approaches to understanding the potentially injurious demands of gymnastic-style impact landings. *Sports Medicine, Arthroscopy, Rehabilitation, Therapy & Technology*, *4*(1), 4. doi:10.1186/1758-2555-4-4

Gittoes, M. J., Irwin, G., & Kerwin, D. G. (2013). Kinematic landing strategy transference in backward rotating gymnastic dismounts. *Journal of Applied Biomechanics*, *29*(3), 253–260.

Glasheen, J. W., & McMahon, T. A. (1995). Arms are different from legs: mechanics and energetics of human hand running. *Journal of Applied Physiology*, *78*(4), 1280–1297.

Goehler, J. (1977). The mechanical effect of the forward "leg snap". *International Gymnast*, *19*(10), 56–59.

Goswami, A., & Gupta, S. (1998). Cardiovascular stress and lactate formation during gymnastic routines. *J Sports Med Physical Fitness*, *38*, 317–322.

Gould (1998). Goal-setting for peak performance. In J. Williams (ed.), *Personal Growth to Peak Performance* (2 ed., pp. 182–196) Mountain View, CA: Mayfield.

Gould, D., Guinan, D., Greenleaf, C., Medbery, R., Strickland , M., Lauer, L., et al. (1998). Positive and negative factors influencing U.S. Olympic athletes and coaches: Atlanta Games assessment. Colorado Springs, CO: U.S. Olympic Committee.

Gould, D., & Krane, V. (1993). The arousal-athletic relationship: current status and future directions. In T. S. Horn (ed.), *Advances in Sport Psychology* (pp. 119–141). Champaign, IL: Human Kinetics.

Grabiner, M. D., & McKelvain, R. (1987). Implementation of a profiling/prediction test battery in the screening of elite men gymnasts. In B. Petiot, J. Salmela, & T. Hoshizak (eds), *World Identification Systems for Gymnastics Talent* (pp. 121–125). Montreal: Sport Psyche Editions. ISSN: 1577-0354 71.

Gros, H. J., & Leikov, H. (1995). Safety considerations for gymnastics landing mats. In A. Barabas & G. Fabian (eds), *Biomechanics in Sports XII* (pp. 194–197). Budapest, Hungary: International Society of Biomechanics in Sports, ITC Plantin.

Grosprêtre, S., & Lepers, R. (2016). Performance characteristics of Parkour practitioners: who are the traceurs? *European Journal of Sport Science, 16*(5), 526–535. doi:10.1080/17461391.2015.1060263

Grossfeld, A. (2014). Changes during the 110 years of the world artistic gymnastics championships. *Sci Gymnastics J, 6*(2), 5–27.

Groussard, C. Y., & Delamarche, P. (2000). Physiological profile of young male gymnasts of national and international level. In B. G. Bardy, T. Pozzo, P. Nouillot, N. Tordi, P. Delemarche, C. Ferrand et al. (eds), *2nd International Study Days of AFRAGA* (pp. 48–51). Rennes: France.

Guillot, A., & Collet, C. (2005). Duration of mentally simulated movement: a review. *J Mot Behav, 37*(1), 10–20. doi:10.3200/jmbr.37.1.10-20

Gymnastics, U. (2009). *Gymnastics Risk Management.* Indianapolis, IN: USA Gymnastics.

Hacker, P., Malmberg, E., & Nance, J. (1996). *Gymnastics Fun & Games.* Champaign, IL: Human Kinetics.

Haguenauer, M., Legreneur, P., & Monteil, K. M. (2005). Vertical jumping reorganization with aging: a kinematic comparison between young and elderly men. *J Appl Biomech, 21*(3), 236–246.

Hahn, A. (2012). Introduction. In R. Tanner (ed.), *Physiological Tests for Elite Athletes* (2 ed.). Champaign, IL: Human Kinetics.

Hakkinen, K., Mero, A., & Kauhanen, H. (1989). Specificity of endurance, sprint and strength training on physical performance capacity in young athletes. *J Sports Med Phys Fitness, 29*(1), 27–35.

Hall, C., Varley, I., Kay, R., & Crundall, D. (2014). Keeping your eye on the rail: gaze behavior of horse riders approaching a jump. *PLoS ONE, 9*(5), e97345.

Hall, S. J. (1986). Mechanical contribution to lumbar stress injuries in female gymnasts. *Medicine & Science in Sports & Exercise, 18*(6), 599–602.

Halsband, U., Ito, N., Tanji, J., & Freund, H.-J. (1993). The role of premotor cortex and the supplementary motor area in the temporal control of movement in man. *Brain, 116*(1), 243–266. doi:10.1093/brain/116.1.243

Halsband, U., & Lange, R. K. (2006). Motor learning in man: a review of functional and clinical studies. *Journal of Physiology, Paris, 99*(4–6), 414–424. doi:10.1016/j.jphysparis.2006.03.007

Hanin, Y., & Hanina, M. (2009). Optimization of performance in top-level athletes: an action-focused coping approach. *International Journal of Sports Science & Coaching, 4*(1), 47–55.

Hardy, L. (1990). A catastrophe model of performance in sport. In J. G. Jones & L. Hardy (eds), *Stress and Performance in Sport* (pp. 81–106). Chichester, UK: John Wiley & Sons.

Harre, D. (1982). *Principles of Sports Training.* Berlin, German Democratic Republic: Sportverlag.

Harris, D. V., & Williams, J. M. (1993). Relaxation and energizing techniques for regulation of arousal. In J. M. Williams (ed.), *Applied Sport Psychology: Personal Growth to Peak Performance* (2 ed., pp. 185–199). Mountain View, CA: Mayfield.

Hars, M., Holvoet, P., Barbier, F., Gillet, C., & Lepoutre, F. X. (2008). Study of impulses during a walkover backward on the balance beam in women gymnasts. Paper presented at the 1st Scientific Symposium of the Asian Gymnastics Union.

Hatze, H. (1998). Validity and reliability of methods for testing vertical jumping performance. *Journal of Applied Biomechanics, 14,* 127–140.

Hay, J. G. (1973). *The Biomechanics of Sports Techniques.* Englewood Cliffs, NJ: Prentice Hall.

Hay, J. G. (1988). Approach strategies in the long jump. *International Journal of Sport Biomechanics, 4,*114–129.

Hayem, G., & Carbon, C. (1995). A reappraisal of quinolone tolerability. The experience of their musculoskeletal adverse effects. *Drug Safety, 13*(6), 338–342.

Heinen, T., Vinken, P., & Fink, H. (2011). The effects of directing the learner's gaze on skill acquisition in gymnastics. *Athletic Insight Journal, 3*(2), 165–181.

Heinen, T. (2011). Evidence for the spotting hypothesis in gymnasts. *Motor Control, 15*(2), 267–284.

Heinen, T., Bermeitinger, C., & Laßberg, C. v. (2016). Laterality in individualized sports. In F. Loffing, N. Hagemann, & B. Strauss (eds), *Laterality in Sports. Theories and Applications.* Academic Press.

Heinen, T., Jeraj, D., Thoeren, M., & Vinken, P. M. (2011). Target-directed running in gymnastics: the role of the springboard position as an additional source to regulate handsprings on vault. *Biology of Sport, 28*, 215–221.

Heinen, T., Jeraj, D., Vinken, P. M., & Velentzas, K. (2012a). Rotational preference in gymnastics. *Journal of Human Kinetics, 33*, 33–43. doi:10.2478/v10078-012-0042-4

Heinen, T., Jeraj, D., Vinken, P. M., & Velentzas, K. (2012b). Land where you look? – Functional relationships between gaze and movement behaviour in a backward salto. *Biology of Sport, 29*, 177–183.

Heinen, T., Koschnick, J., Schmidt-Maass, D., & Vinken, P. M. (2014). Gymnasts utilize visual and auditory information for behavioural synchronization in trampolining. *Biology of Sport, 31*(3), 223–226. doi:10.5604/20831862.1111850

Heinen, T., Vinken, P. M., & Velentzas, K. (2010). Does laterality predict twist direction in gymnastics? *Science of Gymnastics Journal, 2*(1), 5–14.

Hellebrandt, F. A., Parrish, A. M., & Houtz, S. J. (1947). Cross education. *Archives of Physical Medicine and Rehabilitation, 28*, 76–85.

Heller, J., Tuma, Z., Dlouha, R., Bunc, V., & Novakova, H. (1998). Anaerobic capacity in elite male and female gymnasts. *Acta Universitatis Carolinae. Kinanthropologica, 34*, 75–81.

Henderschott, R., & Sigerseth, P. O. (1953). Landing force in a portable collapsible jumping pit compared with that in conventional jumping pits. *The Research Quarterly, 24*(4), 410–413.

Henry, F. M., & Rogers, D. E. (1960). Increased response latency for complicated movements and a "Memory Drum" theory of neuromotor reaction. *The Research Quarterly, 31*(3), 448–458.

Hernández, T. T., Balón, G. N., & Galarraga, A. L. (2009). Relationship between lactate, heart rate and duration in selections modes of men's artistic gymnastics. *Rev Cubana Med Deporte Cultura Fis, 4*(3), 36–47.

Hickson, R. C., Dvorack, B. A., Gorostiaga, E. M., Kurowski, T. T., & Foster, C. (1988). Potential for strength and endurance training to amplify endurance performance. *J Appl Physiol, 65*, 2285–2290.

Hiley, M. J., Apostolidis, A., & Yeadon, M. R. (2011). Loads on a gymnastics safety support system during maximal use. *Journal of Sports Engineering and Technology, 225*(1), 1–7. doi:10.1177/17543371JSET84

Hiley, M. J., & Yeadon, M. R. (2008). Optimisation of high bar circling technique for consistent performance of a triple piked somersault dismount. *Journal of Biomechanics, 41*, 1730–1735.

Hodgkins, J. (1963). Reaction time and speed of movement in males and females of various ages. *The Research Quarterly, 34*(3), 335–343.

Hoeger, W. W. K., & Fisher, G. A. (1981). Energy costs for men's gymnastic routines. *International Gymnast, 23*(1), TS1–TS3.

Hofmann, D. (2015). Interview by F. Bessi. Freiburg.

Holt, J., Holt, L. E., & Pelham, T. W. (1995). Flexibility redefined. Paper presented at the XIII International Symposium on Biomechanics in Sports. International Society of Biomechanics in Sports.

Holvoet, P., Lacouture, P., & Duboy, J. (2002). Practical use of airborne simulation in a release-regrasp skill on the high bar. *J Appl Biomech, 18*, 332–344.

Holvoet, P., Lacouture, P., Duboy, J., Junqua A., & Bessonnet G. (2002). Joint forces and moments involved in giant swings on the high bar. *Science & Sports, 17*, 26–30.

Hondzinski, J. M., & Darling, W. G. (2001). Aerial somersault performance under three visual conditions. *Motor Control, 5*(3), 281–300.

Hooke, R. (1678). Lectures de potentia restitutive. Or of Spring: Explaining the power of springing bodies. Reprinted in Early Science in Oxford, R. T. Gunther, Ed., Vol. 8, pp. 331–356. Oxford: Oxford University Press 1931.

Horak, J. (1969). The performance of top sportsmen. *Teor. Praxe. Teel. Vych., 16*, 18–20.

Horswill, C. A., Miller, J. E., Scott, J. R., Smith, C. M., Welk, G., & Van Handel, P. (1992). Anaerobic and aerobic power in arms and legs of elite senior wrestlers. *Int J Sports Med, 13*, 558–561.

Hrysomallis, C. (2007). Relationship between balance ability, training and sports injury risk. *Sports Medicine, 37*(6), 547–556.

Huang, R., Lu, M., Song, Z., & Wang, J. (2015). Long-term intensive training induced brain structural changes in world class gymnasts. *Brain Struct Funct, 220*(2), 625–644. doi:10.1007/s00429-013-0677-5

Hubel, D. H. (ed.) (1988). *Eye, Brain, and Vision*. New York, NY: Scientific American Library.

Hudash Wadley, G., & Albright, J. P. (1993). Women's intercollegiate gymnastics injury patterns and "permanent" medical disability. *American Journal of Sports Medicine, 21*(2), 314–320.

Hume, P. (2010–2014). *Minimizing Injuries in Gymnastics Activities*. CoachesInfo.com (Internet) 2005. Cited 12 August 2011. Retrieved from: www.coachesinfo.com/index.php?option=com_content&view=article&id=185:gymnastics-isbs-minimising&catid=62:gymnastics-isbs&Itemid=108

Humphy, C. (2010). B vitamins and exercise. Retrieved March 2010 from: www.sagf.co.za/ipage.php?id=65&type_id=9>

Inbar, O., & Bar-Or, O. (1977). Relationships of anaerobic and aerobic arm and leg capacities to swimming performance of 8–12 years old children. In R. J. Shephard & H. Lavallée (eds), *Frontiers of Physical Activities and Child Health* (pp. 238–292). Québec: Du Pélican.

International Gymnastics Federation, F. I. G. (ed.) (2000). Norms for men's and women's vaulting table. Moutier, Switzerland: International Gymnastics Federation.

Irurtia, A. A., Marina, M. A., Galilea, P. A. B., & Busquets, A. F. (2007). Heart rate rating during training in young gymnasts. *Apunts: Educ Fisy deporte, 89*(3), 64–74.

Irwin, D. (1993). *Behind the Bench: Coaches Talk About Life in the NHL*. Toronto: McClelland & Stewart.

Irwin, D., & Brockmole, J. R. (2004). Suppressing where but not what. The effects of saccades on dorsal and ventral stream visual processing. *Psychological Science, 15*(7), 467–473.

Irwin, G., & Kerwin, D. G. (2005). Biomechanical similarities of progressions for the longswing on high bar. *Sports Biomechanics, 4*(2), 163–178.

Irwin, G., & Kerwin, D. G. (2009). The influence of the vaulting table on the handspring front somersault. *Sports Biomechanics, 8*(2), 114–128.

Irwin, G., Williams, G. K. R., & Kerwin, D. G. (2014). Gymnastics coaching and science: biomechanics perspectives. In L. M. Schiavon, T. Heinen, M. A. C. Bortoleto, M. Nunomura, & E. Toledo (eds), *High Performance Gymnastics* (pp. 163–176). Hildesheim, Germany: Arete-Verlag.

Isabelle, E., & Jones, T. (1990). Solid-foam training pits. In G. S. George (ed.), *USGF Gymnastics Safety Manual* (pp. 52–56). Indianapolis, IN: U.S. Gymnastics Federation.

Iteya, M., & Gabbard, C. (1996). Laterality patterns and visual-motor coordination of children. *Perceptual and Motor Skills, 83*(1), 31–34.

Jackson, A. S., Beard, E. F., Wier, L. T., Ross, R. M., Stuteville, J. E., & Blair, S. N. (1995). Changes in aerobic power of men, ages 25–70 yr. *Med Sci Sports Exerc, 27*(1), 113–120.

Jackson, A. S., Wier, L. T., Ayers, G. W., Beard, E. F., Stuteville, J. E., & Blair, S. N. (1996). Changes in aerobic power of women, ages 20–64 yr. *Med Sci Sports Exerc, 28*(7), 884–891.

Jackson, S. A., & Cziksentmihalyi, M. (1999). *Flow in Sports: The Keys to Optimal Experiences and Performances*. Champaign, IL: Human Kinetics.

Jacobson, E. (1938). *Progressive Relaxation*. Chicago, IL: University of Chicago Press.

James, S. (1987). Periodization of weight training for women's gymnastics. *National Strength and Conditioning Association Journal, 9*(1), 28–31.

Jankarik, A., & Salmela, J. H., (1987). Longitudinal changes in physical, organic and perceptual factors in Canadian male gymnasts. In B. Petiot, J. H. Salmela, & T. B. Hoshizaki (eds), *Rev. int. med. cienc. act. fis. deporte* (vol. 13 – número 49) World Identification Systems for Gymnastics Talent. Montreal, Canada: Sport Psyche Editions.

Jankauskiené, R., & Kardelis, K. (2005). Body image and weight reduction attempts among adolescent girls involved in physical activity. *Medicina (Kaunas), 41*(9), 796–801.

Janshen, L. (2000). Neuromuscular control during gymnastic landings. In Y. Hong & D. P. Johns (eds), *Proceedings of XVIII International Symposium on Biomechanics in Sports* (1 ed., pp. 155–157). Hong Kong: International Society of Biomechanics in Sports.

Janssen, J. M. (2007). Netherlands Patent No. Bulletin 2007/02: E. P. Office.

Jansson, E., Sjodin, B., & Tesch, P. (1978). Changes in muscle fiber type distribution in man after physical training. A sign of fiber type transformation? *Acta Physiol Scand, 104,* 235–237.

Jemni, M. (2001). Etude du profil bioénergétique et de la récupération chez des gymnastes. Université Rennes 2 Haute Bretagne, Rennes – France.

Jemni, M. (2010). Recovery modalities: effects on hormones' balance and performance. 5th International Scientific Congress "Sport, Stress, Adaptation". National Sport Academy Vasil Levski, Sofia, Bulgaria. 23–25 April 2010

Jemni, M. (2011a). Physiology for gymnastics. In M. Jemni (ed.), *The Science of Gymnastics* (p. 202). London, New York, Delhi: Routledge, Taylor & Francis.

Jemni, M. (ed.) (2011b). *The Science of Gymnastics.* London: Routledge.

Jemni, M., Friemel, F., Le Chevalier, J. M., & Origas, M. (1998). Bioénergétique de la gymnastique de haut niveau. *Education Physique et Sportive, 39,* 29–34.

Jemni, M., Friemel, F., Le Chevalier, J. M., & Origas, M. (2000). Heart rate and blood lactate concentration analysis during a high level men's gymnastics competition. *J Strength Cond Res, 14*(4), 389–394.

Jemni, M., Friemel, F., & Sands, W. (2002). Etude de la récupération entre les agrès lors de quatre séances d'entraînement de gymnastique masculine. *Education Physique et Sportive, 57,* 57–61.

Jemni, M., Friemel, F., Sands, W. A., & Mikesky, A. (2001). Evolution of the physiological profile of gymnasts over the past 40 years. (Review). *Can J Appl Physiol, 26*(5), 442–456.

Jemni, M., Keiller, D., & Sands, W. A. (2008). Are there any health risks associated to high training loads in highly trained gymnasts? Paper presented at the 13th Annual Congress of the European College of Sport Science (ECSS).

Jemni, M., & Robin, J. F. (2005). Proceeding of the 5th International Conference of the AFRAGA (Association Française de Recherche en Activités Gymniques et Acrobatiques), Hammamet, Tunisia, 11–13 April.

Jemni, M., & Sands, W. (2000). La planification de l'entraînement en gymnastique. Exemple: la dernière semaine avant la compétition. *Gym technic, 31,* 17–20.

Jemni, M., & Sands, W. A. (2011). Training principles in gymnastics. In M. Jemni (ed.), *The Science of Gymnastics* (pp. 26–31). London, UK: Routledge.

Jemni, M., Sands, W. A., Friemel, F., Cooke, C., & Stone, M. (2006). Effect of gymnastics training on aerobic and anaerobic components in elite and sub elite men gymnasts. *J Strength Cond Res, 20*(4), 899–907.

Jemni, M., Sands, W., Friemel, F., & Delamarche, P. (2003). Effect of active and passive recovery on blood lactate and performance during simulated competition in high level gymnasts. *Can. J. Appl. Physiol., 28*(2), 240–256.

Jerome, W., Weese, R., Plyley, M., Klavora, P., & Howley, T. (1987). The Seneca gymnastics experience. In J. H. Salmela, B. Petiot, & T. B. Hoshizaki (eds), *Psychological Nurturing and Guidance of Gymnastic Talent* (pp. 90–119). Montreal: Sport Psyche.

Joch, W. (1990). Dimensions of motor speed. *Modern Athlete and Coach, 28*(2), 25–29.

Jones, G. P. L. (1988). *Expert Systems: Knowledge, Uncertainty and Decision.* Chapman and Hall.

Joseph, L. H. (1949a). Gymnastics during the renaissance as a part of the human educational program. *CIBA Symposia, 10*(5), 1034–1040.

Joseph, L. H. (1949b). Gymnastics in the pre-revolutionary eighteenth century. *CIBA Symposia, 10*(5), 1054–1060.

Joseph, L. H. (1949c). Medical gymnastics in the sixteenth and seventeenth centuries. *CIBA Symposia, 10*(5), 1041–1053.

Joseph, L. H. (1949d). Physical education in the early middle ages. *CIBA Symposia, 10*(5), 1030–1033.

Jull, G., & Janda, V. (1987). Muscles and motor control in low back pain. In L. T. Twomey & J. R. Taylor (eds), *Physical Therapy for the Low Back: Clinics in Physical Therapy* (pp. 253–278). New York: Churchill Livingstone.

Jurimae, J., & Abernethy, P. J. (1997). The use of isoinertial, isometric and isokinetic dynamometry to discriminate between resistance and endurance athletes. *Biology of Sport, 14*(2), 163–171.

Kaleagasioglu, F., & Olcay, E. (2012). Fluoroquinolone-induced tendinopathy: etiology and preventive measures. *Tohoku Journal of Experimental Medicine, 226*(4), 251–258.

Karacsony, I., & Cuk, I. (2005). *Floor Exercises: Methods, Ideas, Curiosities, History*. Ljubljana: University of Ljubljana Press.

Katoh, S., Shingu, H., Ikata, T., & Iwatsubo, E. (1996). Sports-related spinal cord injury in Japan (from the nationwide spinal cord injury registry between 1990 and 1992). *Spinal Cord, 34*(7), 416–421.

Kawamori, N., & Haff, G. G. (2004). The optimal training load for the development of muscular power. *J Strength Cond Res, 18*(3), 675–684.

Kennel, C., Hohmann, T., & Raab, M. (2014). Action perception via auditory information: agent identification and discrimination with complex movement sounds. *Journal of Cognitive Psychology, 26*(2), 157–165.

Kennel, C., Streese, L., Pizzera, A., Justen, C., Hohmann, T., & Raab, M. (2015). Auditory reafferences: the influence of real-time feedback on movement control. *Frontiers in Psychology, 6*(69).

Kerr, G. (1990). Preventing gymnastic injuries. *Canadian Journal of Sport Sciences, 15*(4), 227.

Kerr, G., & Minden, H. (1988). Psychological factors related to the occurrence of athletic injuries. *Journal of Sport & Exercise Psychology, 10*, 167–173.

Kerr, J. H. (1997). *Motivation and Emotion in Sport: Reversal Theory*. East Sussex, UK: Psychology Press.

Kerr, Z. Y., Hayden, R., Barr, M., Klossner, D. A., & Dompier, T. P. (2015). Epidemiology of National Collegiate Athletic Association Women's Gymnastics Injuries, 2009–2010 through 2013–2014. *Journal of Athletic Training, 50*(8), 870–878.

Kerwin, D. G., & Trewartha, G. (2001). Strategies for maintaining a handstand in the anterior-posterior direction. *Medicine & Science in Sports & Exercise, 33*(7), 1182–1188.

Kerwin, D. G., Yeadon, M. R., & Harwood, M. J. (1993). High bar release in triple somersault dismounts. *Journal of Applied Biomechanics, 9*(4), 279–286.

Kimball, D. (1990). Spotting safety. In J. L. Gabriel (ed.), *U.S. Diving Safety Manual* (pp. 141–148). Indianapolis, IN: U.S. Diving, Inc.

King, M. A., & Yeadon, M. R. (2004). Maximizing somersault rotation in tumbling. *Journal of Biomechanics, 37*, 471–477.

King, M. A., & Yeadon, M. R. (2005). Factors influencing performance in the hecht vault and implications for modelling. *Journal of Biomechanics, 38*, 145–151.

King Hogue, M. (1990). Body awareness and spatial orientation. In J. L. Gabriel (ed.), *U.S. Diving Safety Manual* (pp. 97–103). Indianapolis, IN: U.S. Diving Publications.

Kirby, R. L., Simms, F. C., Symington, V. J., & Garner, G. B. (1981). Flexibility and musculo-skeletal symptomatology in female gymnasts and age-matched controls. *Am J Sports Med, 9*, 160–164.

Kirialanis, P., Malliou, P., Beneka, A., & Giannakopoulos, K. (2003). Occurrence of acute lower limb injuries in artistic gymnasts in relation to event and exercise phase. *British Journal of Sports Medicine, 37*(2), 137–139.

Kirkendall, D. T. (1985). Physiologic aspects of gymnastics. *Clinics in Sports Medicine, 4*(1), 17–22.

Klaus, B. (1985). In-ground training pits. In G. George (ed.), *USGF Gymnastics Safety Manual* (pp. 56–58). Indianapolis, IN: U.S. Gymnastics Federation.

Kling, S. C., & Sands, W. A. (1980). Safety wrapper and strap: Google Patents. Retrieved from: www.google.com/patents/US4227321

Klostermann, A., & Küng, P. (2016). Gaze strategies in skateboard trick jumps: spatiotemporal constraints in complex locomotion. *Research Quarterly for Exercise and Sport*. Published online on October 12, 2016. doi:10.1080/02701367.2016.1229864

Know, Y. H., Fortney, V. L., & Shin, I. S. (1990). Analysis of Yurchenko vaults performed by female gymnasts during the 1988 Seoul Olympic Games. *Int J Sport Biomech, 2*(6), 157–177.

Knuttgen, H. G., & Komi, P. V. (1992). Basic definitions for exercise. In P. V. Komi (ed.), *Strength and Power in Sport* (pp. 3–6). Oxford, UK: Blackwell Scientific Publications.

Koch, J., Riemann, B. L., & Davies, G. J. (2012). Ground reaction force patterns in plyometric push-ups. *Journal of Strength and Conditioning Research, 26*(8), 2220–2227. doi:10.1519/JSC.0b013e318239f867

Koh, M., & Jennings, L. (2007). Strategies in preflight for an optimal Yurchenko Layout Vault. *Journal of Biomechanics, 40*, 471–477.

Koh, T. J., Grabiner, M. D., & Weiker, G. G. (1992). Technique and ground reaction forces in the back handspring. *American Journal of Sports Medicine, 20*(1), 61–66.

Kolt, G. S., & Kirkby, R. J. (1999). Epidemiology of injury in elite and subelite female gymnasts: a comparison of retrospective and prospective findings. *British Journal of Sports Medicine, 33*(4), 312–318.

Konstantin, F. F., Subic, A., & Mehta, R. (2008). The impact of technology on sport — new frontiers. *Sports Technology, 1*(1), 1–2.

Koscielny, B. (2009). Analyse der Mehrkampffinalistinnen und Mehrkampffinalisten bei der FIG Turn WM 2007 hinsichtlich eines Drehschemas (Zulassungsarbeit). Albert-Ludwigs-Universität Freiburg, Freiburg.

Komi, P. V. (1986). Training of muscle strength and power: interaction of neuromotoric, hypertrophic, and mechanical factors. *Int J Sports Med, 7*(1), 10–15.

Komi, P. V., & Bosco, C. (1978). Utilization of stored elastic energy in leg extensor muscles by men and women. *Med Sci Sports, 10*(4), 261–265.

Koyama, K., Nakazato, K., Min, S., Gushiken, K., Hatakeda, Y., Seo, K., & Hiranuma, K. (2012). COL11A1 gene is associated with limbus vertebra in gymnasts. *International Journal of Sports Medicine, 33*(7), 586–590. doi:10.1055/s-0031-1299752

Krejcova, H., Jer bek, J., Bojar, M., Tutzk , E., Cerny, R., & Polechov, P. (1987). Influence of sports load on the vestibular apparatus. In M. D. Graham & J. L. Kemink (eds), *The Vestibular System: Neurophysiologic and Clinical Research* (pp. 133–139). New York, NY: Raven Press.

Krestovnikov, A. N. (1951). Ocerki po fisiologii fiziceskich upraznenij. Moskova: FIS.

Krug, J., Reiss, S., & Knoll, K. (2000). Training effects of rapid rotations in a "somersault simulator". In Y. Hong & D. P. Johns (eds), *Proceedings of XVIII International Symposium on Biomechanics in Sports* (pp. 667–671). Hong Kong, China: The Chinese University of Hong Kong, International Society of Biomechanics in Sports.

Kruse, D., & Lemmen, B. (2009). Spine injuries in the sport of gymnastics. *Current Sports Medicine Reports, 8*(1), 20–28. doi:10.1249/JSR.0b013e3181967ca6

Laßberg, C. v. (2008). Aspekte der mentalen Raumrepräsentation im Kunstturnen [Mental representation of spatial orientation in artistic gymnastics]. Unveröff: Unterlage.

Lange, B., Halkin, A. S., & Bury, T. (2005). Exigences physiologiques necessaires a la pratique de la gymnastique de haut niveau. *RMLG. Revue médicale de Liège, 60*(12), 939–945.

Latash, M. L. (2012). *Fundamentals of Motor Control.* Boston, MA: Elsevier.

Lazarus, R. S., & Folkman, S. (1984). *Stress, Appraisal, and Coping.* New York: Springer.

Le Breton, D. (1995). *La Sociologie du Risque.* Paris: PUF.

Lechevalier, J. M., Origas, M., Stein, J. F., Fraisse, F., Barbierie, L., Mermet, P., et al. (1999). Comparaison de 3 séances d'entraînement-type chez des gymnastes espoirs: Confrontation avec les valeurs du métabolisme enregistrées en laboratoire. *Gym Technic, 27*, 24–31.

Lee, D. N. (1980). Visuomotor coordination in space-time. In G. E. Stelmach & J. Requin (eds), *Tutorials in Motor Behavior* (pp. 281–295). Amsterdam: North-Holland.

Lee, D. N., Young, D. S., Reddish, P. E., Lough, S., & Clayton, T. M. H. (1983). Visual timing in hitting an acceleration ball. *Quarterly Journal of Experimental Psychology, 35A*, 333–346.

Lee, D. N., Young, D. S., & Rewt, D. (1992). How do somersaulters land on their feet? *Journal of Experimental Psychology: Human Perception and Performance, 18*(4), 1195–1202.

Lee, T. D., Swanson, L. R., & Hall, A. L. (1991). What is repeated in a repetition? Effects of practice conditions on motor skill acquisition. *Physical Therapy, 71*(2), 150–156.

Leglise, M. (1985). Some medical observations on the development of high-level gymnastics. Budapes: World Gymnastics, FIG and AIPS Press, n° 23, p. 27.

Leglise, M., & Binder, M. (2014). *Gymnastics Injuries.* Retrieved from FIG publications, Lausanne, Switzerland.

León-Prados, J. A. (2006) Estudio del uso de tests físicos, psicológicos y fisiológicos para estimar el estado de rendimiento de la selección nacional de Gimnasia Artística Masculina (Doctoral thesis). Dpto Deporte e Informática (Universidad Pablo de Olavide), Sevilla.

Lescura, N. S., & Bagesteiro, L. B. (2011). Study and project of 2D system design for simplified kinematics gait [Estudo e Projeto de Sistema 2D para Avaliação Cinemàtica Simplificada da Marcha]. XX Encontro de Iniciação Científica, ENCERRADO, Sao Paolo, Brazil, p. 122.

Lewthwaite, R., & Wulf, G. (2012). Motor learning through a motivational lens. In N. J. Hodges & A. M. Williams (eds), *Skill Acquisition in Sport* (pp. 173–191). London, UK: Routledge.

Lidor, R., Hershko, Y., Bilkevitz, A., Arnon, M., & Falk, B. (2007). Measurement of talent in volleyball: 15-month follow-up of elite adolescent players. *J Sports Med Phys Fitness, 47*(2), 159–168.

Liebling, M. S., Berdon, W. E., Ruzal-Shapiro, C., Levin, T. L., Roye, D., & Wilkinson, R. (1995). Gymnast's wrist (pseudorickets growth plate abnormality) in adolescent athletes: findings on plain films and MR imaging. *American Journal of Radiology, 164*, 157–159.

Lindholm, C., Hagenfeldt, K., & Hagman, U. (1995). A nutrition study in juvenile elite gymnasts. *Acta Paediatr, 84*(3), 273–277.

Lindner, K. J., & Caine, D. J. (1990). Injury patterns of female competitive club gymnasts. *Canadian Journal of Sport Sciences, 15*(4), 254–261.

Lindner, K. J., Caine, D. J., & Johns, D. P. (1991). Withdrawal predictors among physical and performance characteristics of female competitive gymnasts. *J Sports Sci., 9*(3), 259–272.

Linge, S., Halllingstad, O., & Solberg, F. (2006). Modeling the parallel bars in men's artistic gymnastics. *Human Movement Science, 25*, 221–237.

Loehr, J. E. (1983). The ideal performance state. *Science Periodical on Research and Technology in Sport,* 1–8.

Lohse, K. R., Wulf, G., & Lewthwaite, R. (2012). Attentional focus affects movement efficiency. In N. J. Hodges & A. M. Willams (eds), *Skill Acquisition in Sport* (2 ed.). London, UK: Routledge.

Loko, J., Aule, R., Sikkut, T., Ereline, J., & Viru, A. (2000). Motor performance status in 10 to 17-year-old Estonian girls. *Scand J Med Sci Sports, 10*(2), 109–113.

Loucks, A. B., & Redman, L. M. (2004). The effect of stress on menstrual function. *Trends Endocrinol Metab, 15*, 466–471.

Lovell, G. (2008). Vitamin D status of females in an elite gymnastics program. *Clinical Journal of Sports Medicine, 18*(2), 159–161.

Luhtanen, P., & Komi, R. V. (1978). Segmental contribution to forces in vertical jump. *Eur J Appl Physiol Occup Physiol., 38*(3), 181–188.

Luis, M., & Tremblay, L. (2008). Visual feedback use during a back tuck somersault: evidence for optimal visual feedback utilization. *Motor Control, 12*, 210–218.

McArdle, W., Katch, F., & Katch, V. (2005). *Essentials of Exercise Physiology* (3 ed.). Philadelphia, PA: Lippincott Williams & Wilkins.

McClements, J. D., & Sanderson, L. K. (1998). What do athletes learn when they learn a motor skill? *New Studies in Athletics, 13*(1), 31–40.

Macnamara, B. N., Hambrick, D. Z., & Moreau, D. (2016). How important is deliberate practice? Reply to Ericsson. *Perspectives on Psychological Science, 11*(3), 355–358. doi:10.1177/1745691616635614

Macnamara, B. N., Moreau, D., & Hambrick, D. Z. (2016). The relationship between deliberate practice and performance in sports: a meta-analysis. *Perspectives on Psychological Science, 11*(3), 333–350. doi:10.1177/1745691616635591

McNeal, J. R., & Sands, W. A. (2006). Stretching for performance enhancement. *Current Sports Medicine Reports, 5*, 141–146.

McNitt-Gray, J. L. (1991a). The influence of joint flexion, impact velocity, rotation, and surface characteristics on the forces and torques experienced during gymnastics landings. *FIG Scientific/Medical Symposium Proceedings, 1*(1), 17–18.

McNitt-Gray, J. L. (1991b). Kinematics and impulse characteristics of drop landings from three heights. *International Journal of Sport Biomechanics, 7*(2), 201–224.

McNitt-Gray, J. L. (1993). Kinetics of the lower extremities during drop landings from three heights. *Journal of Biomechanics, 26*(9), 1037–1046.

McNitt-Gray, J. L., Hester, D. M. E., Mathiyakom, W., & Munkasy, B. A. (2001). Mechanical demand and multijoint control during landing depend on orientation of the body segments relative to the reaction force. *Journal of Biomechanics, 34*, 1471–1482.

McNitt-Gray, J. L., Irvine, D. M. E., Munkasy, A., Eagly, J., Smith, T., & Chen, Y. T. (1997). Mechanics and motor control of gymnastics take-offs and landings. Report to USOC Sport Science Division.

McNitt-Gray, J. L., Requejo, P. S., Flashner, H., & Held, L. (2004). Modeling the musculoskeletal behavior of gymnasts during landings on gymnastics mats. In M. Hubbard, R. D. Metha, & J. M. Pallis (eds), *The Engineering of Sport 5* (pp. 402–408). Sheffield, UK: International Sports Engineering Association.

McNitt-Gray, J. L., Yokio, T., & Millward, C. (1993). Landing strategy adjustments made by female gymnasts in response to drop height and mat composition. *Journal of Applied Biomechanics, 9*(3), 173–190.

McNitt-Gray, J. L., Yokoi, T., & Millward, C. (1994). Landing strategies used by gymnasts on different surfaces. *Journal of Applied Biomechanics, 10*(3), 237–252.

MacPherson, A. C., Collins, D., & Obhi, S. S. (2009). The importance of temporal structure and rhythm for the optimum performance of motor skills: a new focus for practitioners of sport psychology. *Journal of Applied Sport Psychology, 21*, 48–61.

Magakian, A. (1978). *La Gymnastique*. Paris, France: Chiron-Sports.

Mahoney, J. L., Vandell, D. L., Simpkins, S., & Zarrett, N. (2009). Adolescent out-of-school activities. Contextual influences on adolescent development. In R. M. Lerner & L. Steinberg (eds), *Handbook of Adolescent Psychology*. Hoboken, NJ: John Wiley & Sons, Inc.

Mahoney, M. J. (1989). Psychological predictors of elite and non-elite performance in Olympic weightlifting. *Int J Sport Psy, 20*, 1–12.

Mahoney, M. J., & Avener, M. (1977). Psychology of the elite athlete: an exploratory study. *Cognitive Therapy and Research, 3*, 361–366.

Mahoney, M. J., Gabriel, T. J., & Perkins, T. S. (1987). Psychological skills and exceptional athletic performance. *The Sport Psychologist, 1*, 189–199.

Malina, R. M. (1994). Physical activity and training: effects on stature and the adolescent growth spurt. *Medicine & Science in Sports & Exercise, 26*(6), 759–766.

Malina, R. M. (1996). Growth and maturation of female gymnasts. *Spotlight on Youth Sports, 19*(3), 1–3.

Malina, R. M., Baxter-Jones, A. D. G., Armstrong, N., Beunen, G. P., Caine, D., Daly, R. M., Lewis, R. D., Rogol, A. D., & Russell, K. (2013). Role of intensive training in the growth and maturation of artistic gymnasts. *Sports Medicine, 43*, 783–802.

Malmberg, E. (1978). Science, innovation, and gymnastics in the USSR. *International Gymnast, 20*(2), 63.

Mandelbaum, B. R., Grant, T. T., & Nichols, A. W. (1988). Wrist pain in a gymnast. *The Physician and Sportsmedicine, 16*(1), 80–84.

Mandelbaum, B. R., & Teurlings, L. (1991). The gymnast's wrist pain syndrome. *FIG Scientific/Medical Symposium Proceedings, 1*(1), 34–36.

Marcinik, E. J., Potts, J., Scholabach, G., Will, S., Dawson, P., & Hurley, B. F. (1991). Effects of strength training on lactate threshold and endurance performance. *Med Sci Sports Exerc, 23*(6), 739–743.

Marina, M., & Jemni, M. (2014). Plyometric training performance in elite-oriented prepubertal female gymnasts. *Journal of strength and conditioning research/National Strength & Conditioning Association, 28*(4), 1015–1025. doi:10.1519/JSC.0000000000000247

Marina, M., Jemni, M., & Rodríguez, F. A. (2012). Plyometric jumping performances of elite male and female gymnasts. *J Strength Cond Res., 26*(7), 1879–1886. doi:10.1519/JSC.0b013e31823b4bb8

Marina, M., Jemni, M., & Rodriguez, F. (2013). Jumping performance profile of male and female gymnasts. *The Journal of Sports Medicine and Physical Fitness, 53*(4), 378–386.

Marina, M., Jemni, M., & Rodriguez, F. A. (2014). A two-season longitudinal comparative study of jumps with added weights and counter movement jumps in well-trained pre-pubertal female gymnasts. *The Journal of Sport Medicine and Physical Fitness, 54*(6), 730–741.

Marina, M., Jemni, M., Rodriguez, F. A., & Jimenez, A. (2012). Plyometric jumping performances of male and female gymnasts from different heights. *Journal of strength and conditioning research / National Strength & Conditioning Association, 26*(7), 1879–1886. doi:10.1519/JSC.0b013e31823b4bb8

Marina, M., & Rodríguez, F. A. (2013). Usefulness and metabolic implications of a 60-second repeated jumps test as a predictor of acrobatic jumping performance in gymnasts. *Biology of Sport, 30*(1), 9–15. doi:10.5604/20831862.1029815

Marina, M., & Rodríguez, F. A. (2014). Physiological demands of young women's competitive gymnastic routines. *Biol Sport, 31*(3), 217–222.

Marina, M., & Torrado, P. (2013). Does gymnastics practice improve vertical jump reliability from the age of 8 to 10 years? *Journal of Sports Sciences, 31*(11), 1177–1186. doi:10.1080/02640414.2013.771816

Markou, K. B., Mylonas, P., Theodoropoulou, A., Kontogiannis, A., Leglise, M., Vagenakis, A. G., et al. (2004). The influence of intensive physical exercise on bone acquisition in adolescent elite female and male artistic gymnasts. *J Clin Endocrinol Metab, 89*(9), 4383–4387.

Markovic, G., Dizdar, D., Jukic, I., & Cardinale, M. (2004). Reliability and factorial validity of squat and countermovement jump tests. *J Strength Cond Res, 18*(3), 551–555.

Markovic, G., & Jaric, S. (2005). Scaling of muscle power to body size: the effect of stretch-shortening cycle. *Eur J Appl Physiol, 95*(1), 11–19.

Marshall, S. W., Covassin, T., Dick, R., Nassar, L. G., & Agel, J. (2007). Descriptive epidemiology of collegiate women's gymnastics injuries: National Collegiate Athletic Association Injury Surveillance System, 1988–1989 through 2003–2004. *Journal of Athletic Training, 42*(2), 234–240.

Marshall, W. A., & Tanner, J. M. (1969). Variations in pattern of pubertal changes in girls. *Archives of Disease in Childhood, 44*(235), 291–303.

Marshall, W. A., & Tanner, J. M. (1970). Variations in the pattern of pubertal changes in boys. *Archives of Disease in Childhood, 45*(239), 13–23.

Martens, R. (1977). *Sport Competition Anxiety Test.* Champaign, IL: Human Kinetics Publishers.

Martens, R., Christina, R. W., Harvey, J. S., & Sharkey, B. J. (1981). *Coaching Young Athletes.* Champaign, IL: Human Kinetics.

Martin, K. A. (2002). Development and validation of the coaching staff cohesion scale. *Measurement in Physical Education and Exercise Science, 6*(1), 23–42.

Massidda, M., Toselli, S., & Calo, C. M. (2015). Genetics and artistic gymnastics: 2014 update. *Austin Biomarkers Diagnostic, 2*(1), 1–8.

Matejek, N., Weimann, E., Witzel, C., Mölenkamp, S., Schwidergall, S., & Böhles, H. (1999). Hypoleptinemia in patients with anorexia nervosa and in elite gymnasts with anorexia athletica. *Int J Sports Med, 20,* 451–456.

Matveyev, L. (1977). *Fundamentals of Sports Training.* Moscow, USSR: Progress Publishers.

Mayhew, J. L., & Salm, P. C. (1990). Gender differences in anaerobic power tests. *European Journal of Applied Physiology and Occupational Physiology, 60,* 133–138.

Meeusen, R., & Borms, J. (1992). Gymnastic injuries. *Sports Medicine, 13*(5), 337–356.

Meeuwisse, W. H., & Love, E. J. (1997). Athletic injury reporting. *Sports Medicine, 24*(3), 184–204.

Melhus, A. (2005). Fluoroquinolones and tendon disorders. *Expert Opinion on Drug Safety, 4*(2), 299–309.

Melrose, D. R., Spaniol, F. J., Bohling, M. E., & Bonnette, R. A. (2007). Physiological and performance characteristics of adolescent club volleyball players. *J Strength Cond Res, 21*(2), 481–486.

Mero, A. (1998). Power and speed training during childhood. In V. P. E. (ed.), *Pediatric Anaerobic Performance* (pp. 241–267). Champaign, IL: Human Kinetics.

Meyers, D. (2016). *The End of the Perfect 10.* New York, NY: Touchstone.

Mikulas, S. (1994). Evolution du niveau de l'état fonctionnel de l'analyseur vestibulaire en gymnastique sportive (garçons). In M. Ganzin (ed.), *Gymnastique Artistique et GRS. Communications Scientifiques et Techniques d'Experts Étrangers.* Paris, France: INSEP.

Milem, D. J. (1990). Spotting belts. In G. S. George (ed.), *USGF Gymnastics Safety Manual* (pp. 47–49). Indianapolis, IN: U.S. Gymnastics Federation.

Milev, N. (1994). Analyse cinématique comparative du double salto arrière tendu avec et sans vrille (360°) à la barre fixe. In M. Ganzin (ed.), *Gymnastique Artistique et GRS. Communications Scientifiques et Techniques d'Experts Étrangers* (pp. 115–124). Paris, France: INSEP.

Miller, J. R., & Demoiny, S. G. (2008). Parkour: a new extreme sport and a case study. *Journal of Foot and Ankle Surgery, 47*(1), 63–65.

Mills, C., Yeadon, M. R., & Pain, M. T. (2010). Modifying landing mat material properties may decrease peak contact forces but increase forefoot forces in gymnastics landings. *Sports Biomech, 9*(3), 153–164. doi:10.1080/14763141.2010.524244

Mills, M. T. G., Pain, R., & Yeadon, F. (2006). Modelling a viscoelastic gymnastics landing mat during impact. *Journal Applied Biomechanics, 22*, 103–111.

Mills, M. T. G., Pain, R., & Yeadon, F. (2009). Reducing ground reaction forces in gymnastics' landing may increase internal loading. *J Biomech, 42*(6), 671–678.

Milosis, D. C., & Siatras, T. A. (2012). Sex differences in young gymnasts' postural steadiness. *Perceptual and Motor Skills, 114*(1), 319–328.

Minganti, C., Capranica, L., Meeusen, R., Amici, S., & Piacentini, M. F. (2010). The validity of session rating of perceived exertion method for quantifying training load in teamgym. *The Journal of Strength and Conditioning Research, 24*, 3063–3068.

Mirwald, R. L., Baxter-Jones, A. D. G., Bailey, D. A., & Beunen, G. P. (2002). An assessment of maturity from anthropometric measurements. *Medicine and Science in Sport and Exercise, 34*(4), 689–694.

Mkaouer, B., Jemni, M., Amara, S. M., Abahnini, K., Agrebi, B., Tabka, Z., et al. (2005). Analyse des paramètres déterminants de la performance lors du grand jeté lancer-rattraper en GR. Paper presented at the 5th International Conference of the Association Française pour la Recherche en Activités Gymniques et Acrobatiques (AFRAGA).

Mkaouer, B., Jemni, M., Amara, S., Chaabène, H., & Tabka, Z. (2013). Kinematic and kinetic analysis of two gymnastics acrobatic series to performing the backward stretched somersault. *Journal of Human Kinetics, 37*(1), 17–26.

Mkaouer, B., Jemni, M., Amara, S., & Tabka, Z. (2008). Kinematics study of jump in backward rotation. Paper presented at the 1st Scientific Symposium of the Asian Gymnastics Union.

Mkaouer, B., Jemni, M., Chaabene, H., Amara, S., Njah, A., & Chtara, M. (2017). Effect of two Olympic rotation orders on cardiovascular and metabolic variables in men's artistic gymnastics. *Journal of Human Kinetics*, in press.

Moffroid, M., & Whipple, R. H. (1970). Specificity of speed of exercise. *Physical Therapy, 50*, 1693–1699.

Mohr, C., Brugger, P., Bracha, H. S., Landis, T., & Viaud-Delmon, I. (2004). Human side preferences in three different whole-body movement tasks. *Behavioral Brain Research, 151*, 321–326.

Molloy, T., Wang, Y., & Murrell, G. (2003). The roles of growth factors in tendon and ligament healing. *Sports Medicine, 33*(5), 381–394.

Montgomery, D. L., & Beaudin, P. A. (1982). Blood lactate and heart rate response of young females during gymnastic routines. *J Sports Medicine, 22*, 358–364.

Montgomery, H. E., Marshall, R., Hemingway, H., Myerson, S., Clarkson, P., Dollery, C., . . . Humphries, S. E. (1998). Human gene for physical performance. *Nature, 393*, 221.

Montpetit, R. (1976). Physiology of gymnastics. In J. Salmela (ed.), *The Advanced Study of Gymnastics*. Springfield, IL: C. Thomas Publisher.

Montpetit, R., & Matte, G. (1969). Réponses cardiaques durant l'exercice de gymnastique. *Kinanthropologie, 1*, 211–222.

Moraes, L. C., Salmela, J. H., Rabelo, A. S., & Vianna, Jr., N. S. (2004). Le dévloppement des jeunes joueurs braziliens au football et au tennis: Le role des parents. *STAPS, 64*, 108–126.

Moran, D. S., McClung, J. P., Kohen, T., & Lieberman, H. R. (2013). Vitamin D and physical performance. *Sports Medicine, 43*, 601–611.

Moritani, T., & DeVries, H. A. (1979). Neural factors versus hypertrophy in the time course of muscle strength gain. *American Journal of Physical Medicine & Rehabilitation, 58*(3), 115–130.

Moschos, S., Chen, J. L., & Mantzoros, C. S. (2002). Leptin and reproduction; a review. *Fertil Steril, 77*, 433–444.

Mountjoy, M., & Bergeron, M. F. (2015). Youth athletic development: aiming high while keeping it healthy, balanced and fun. *British Journal of Sports Medicine, 49*(13), 841–842.

Müller, E., Raschner, C., & Schwameder, H. (1999). The demand profile of modern high-performance training. In F. L. E. Müller & G. Zallinger (eds), *Science in Elite Sport* (pp. 1–31). London, UK: E & FN Spon.

Mulloy Forkin, D., Koczur, C., Battle, R., & Newton, R. A. (1996). Evaluation of kinesthetic deficits indicative of balance control in gymnasts with unilateral chronic ankle sprains. *The Journal of Orthopaedic and Sports Physical Therapy, 23*(4), 245–250.

Muñoz, M. T., de la Piedra, C., Barrios, V., Garrido, G., & Argente, J. (2004). Changes in bone density and bone markers in rhythmic gymnasts and ballet dancers: implications for puberty and leptin levels. *Eur J Endocrinol, 151*(4), 491–496.

Munzert, J., Zentgraf, K., Stark, R., & Vaitl, D. (2008). Neural activation in cognitive motor processes: comparing motor imagery and observation of gymnastic movements. *Experimental Brain Research, 188*(3), 437–444. doi:10.1007/s00221-008-1376-y

Murphy, S. M., & Jowdy, D. P. (1993). Imagery and mental practice. In T. S. Horn (ed.), *Advances in Sport Psychology* (pp. 221–250). Champaign, IL: Human Kinetics.

Murphy, S. M., Woolfolk, R. L., & Budney, A. J. (1988). The effects of emotive imagery on strength performance. *Journal of Sport and Exercise Psychology, 10*, 334–345.

Murray, J. (1989). An investigation of competitive anxiety versus positive affect (Unpublished master's thesis). University of Virginia, Charlottesville, VA.

Myers, D. G. (1998). *Psychology* (5 ed.). New York, NY: Worth Publishers.

Nakasone, M. (2015). *Research Regarding the Occurrence of New Elements for Vaulting Table in Women's Artistic Gymnastics from Japanese Literatures*. Berlin: Lambert Academic Publishing.

Nassar, L., & Sands, W. A. (2008). The artistic gymnast's shoulder. In K. E. Wilk, J. R. Reinold, & M. D. Andews (eds), *The Athlete's Shoulder* (2 ed., pp. 491–506). Burlington, MA: Elsevier.

Naundorf, F., Brehmer, S., Knoll, K., Bronst, A., & Wagner, R. (2008). Development of the velocity for vault runs in artistic gymnastics for the last decade. Paper presented at the XXVI International Conference on Biomechanics in Sports, Seoul, Korea (14–18 July 2008), 481–484.

Neal, R. J., Kippers, V., Plooy, D., & Forwood, M. R. (1995). The influence of hand guards on forces and muscle activity during giant swings on the high bar. *Medicine and Science in Sports and Exercise, 27*(11), 1550–1556.

Neumaier, A., Main, L., & Gastin, P. (2013). Factors influencing the implementation of self-report measures for athlete monitoring. *Journal of Science and Medicine in Sport, 165*, e65.

Nideffer, R. M. (1987). Issues in the use of psychological tests in applied settings. *The Sport Psychologist, 1*, 18–28.

Niemi, M. B. (2009). Cure in the mind. *Scientific American Mind*, (February/March), 42–49.

Noble, L. (1975). Heart rate and predicted VO2 during women's competitive gymnastic routines. *Journal of Sports Medicine & Physical Fitness, 15*(2), 151–157.

Normile, D. (1989). Inside the USSR. *International Gymnast, 31*(1), 16–25.

Noyes, F. R., Lindenfeld, T. N., & Marshall, M. T. (1988). What determines an athletic injury (definition)? Who determines an injury (occurrence)? *American Journal of Sports Medicine, 16*(1), S65–68.

Obert, P., Stecken, F., Courteix, D., Germain, P., Lecoq, A. M., & Guenon, P. (1997). Adaptations myocardiques chez l'enfant prépubère soumis à un entraînement intensif. Etude comparative entre une population de gymnastes et de nageurs. *Science et Sports, 12*, 223–231.

Oda, S., & Moritani, T. (1994). Maximal isometric force and neural activity during bilateral and unilateral elbow flexion in humans. *European Journal of Applied Physiology and Occupational Physiology, 69*, 240–243.

Ogawa, S., Asakawa, Y., Akutsu, K., & Watanabe, T. (1956). On the energy metabolism in gymnastics events. *Japanese J. Physical Fitness, 5*, 243.

Ogilvie, B. C., & Tutko, T. A. (1966). *Problem Athletes and How to Handle Them.* London: Pelham Books.

O'Kane, J. W., Levy, M. R., Pietila, K. E., Caine, D. J., & Schiff, M. A. (2011). Survey of injuries in Seattle area levels 4 to 10 female club gymnasts. *Clinical Journal of Sport Medicine, 21*(6), 486–492. doi:10.1097/JSM.0b013e31822e89a8

Olbrecht, J. (2000). *The Science of Winning.* Luton, UK: Swimshop.

Oliveira, M. S. (2014). The training gym microculture of women's artistic gymnastics at a high level sport (Unpublished doctoral thesis). São Paulo University, São Paulo, Brazil.

Oliveira, M. S., & Bortoleto, M. A. C. (2011). Notes on historical, material and morphological evolution of men's artistic gymnastics apparatus. *UEM: Journal of Physical Education, 22*, 10–20.

Olsen, P. A. (1988). Injuries in children associated with trampolinelike air cushions. *Journal of Pediatric Orthopaedics, 8*, 458–460.

Orlick, T. (2008). *In Pursuit of Excellence* (4 ed.). Champaign, IL: Human Kinetics.

Orlick, T., & Partington, J. (1986). *Psyched: Inner Views of Winning.* Ottawa, ON, Canada: The Coaching Association of Canada.

Orlick, T., & Partington, J. (1988). Mental links to excellence. *The Sport Psychologist, 2*, 105–130.

Örsel, A., Vieru, N., Weber, J., Schweizer, L., Titov, V., Ashmore, J., . . . Bänfer, J. (2011). *Apparatus Norms.* Lausanne, Switzerland: Federation Internationale de Gymnastique.

Oudejans, R. R. D. (2008). Reality-based practice improves handgun shooting performance of police officers. *Ergonomics, 81*(3), 261–273.

Overlin, A. J. F., Chima, B., & Erickson, S. (2011) Update on artistic gymnastics. *Current Sports Medicine Reports, 10*(5), 304–309.

Pain, M. T., Mills, C. L., & Yeadon, M. R. (2005). Video analysis of the deformation and effective mass (density) of gymnastics landing mats. *Med Sci Sports Exerc, 37*(10), 1754–1760.

Paine, D. D. (1998). *Spring Floor Resilience and Compliance Modeling* (PhD), University of Utah, Salt Lake City, UT.

Panzer, V. P., Bates, B. T., & McGinnis, P. M. (1987). A biomechanical analysis of elbow joint forces and technique differences in the Tsukahara vault. In T. B. Hoshizaki, J. H. Salmela, & B. Petiot (eds), *Diagnostics, Treatment and Analysis of Gymnastic Talent* (pp. 37–46). Montreal, Canada: Sport Psyche Editions.

Panzer, V. P., Wood, G. A., Bates, B. T., & Mason, B. R. (1988). Lower extremity loads in landings of elite gymnasts. In G. de Groot, A. P. Hollander, P. A. Huijing, & G. J. van Ingen Schenau (eds), *Biomechanics XI-B* (pp. 727–735). Amsterdam, Netherlands: Free University Press.

Papadopoulos, G., Kaimakamis, V., Kaimakamis, D., & Proios, M. (2014). Main characteristics of rules and competition systems in gymnastics from 1896 to 1912. *Sci Gymnastics J, 6*(2), 29–40.

Patla, A. E., & Vickers, J. N. (2003). How far ahead do we look when required to step on specific locations in the travel path during locomotion? *Experimental Brain Research, 148*, 133–138.

Peeling, P., Fulton, S. K., Binnie, M., & Goodman, C. (2013). Training environment and vitamin D status in athletes. *International Journal of Sports Medicine, 34*, 248–252.

Penitente, G., & Sands, W. A. (2015). Exploratory investigation of impact loads during the forward handspring vault. *J Hum Kinet, 46*, 59–68. doi:10.1515/hukin-2015-0034

Penitente, G., Sands, W. A., McNeal, J., Smith, S. L., & Kimmel, W. (2010). Investigation of hand contact forces of female gymnasts performing a handspring vault. *International Journal of Sports Science and Engineering, 4*(1), 015–024.

Penitente, G., Sands, W. A., & McNeal, J. R. (2011). Vertical impact force and loading rate on the gymnastics table vault. *Portuguese Journal of Sport Sciences, 11*(Suppl. 2), 668–670.

Penitente, G., Sands, W. A., Smith, S. L., Kimmel, W. L., & Wurtz, B. R. (2008). Vault and sprint run-ups: male junior national team gymnasts. *Technique, 28*(1), 6–8, 42.

Perel, E., & Killinger, D. W. (1979). The interconversion and aromatization of androgens by human adipose tissue. *J Steroid Biochem, 10*, 623–627.

Pérez, P., Llana, S., & Alcántara, E. (2008). Standard tests ability to measure impact forces reduction on mats. *International Journal of Sports Science and Engineering*, 2(3), 162–168.

Pérez-Soriano, P., Llana-Belloch, S., Morey-Klapsing, G., Perez-Turpin, J. A., Cortell-Tormo, J. M., & van den Tillaar, R. (2010). Effects of mat characteristics on plantar pressure patterns and perceived mat properties during landing in gymnastics. *Sports Biomech*, 9(4), 245–257.

Peterson, M. D., Alvar, B. A., & Rhea, M. R. (2006). The contribution of maximal force production to explosive movement among young collegiate athletes. *J Strength Cond Res*, 20(4), 867–873.

Petrov, V. (1994a). Modèle expérimental d'exécution du double salto avant groupe avec reprise de barre à la barre fixe. In M. Ganzin (ed.), *Gymnastique Artistique et GRS. Communications Scientifiques et Techniques d'Experts Étrangers* (pp. 135–142). Paris, France: INSEP.

Petrov, V. (1994b). Technique et méthode d'exécution d'un salto avant jambes écartées à partir d'un grand tour jusqu'à la reprise de la barre en suspension arrière. In M. Ganzin (ed.), *Gymnastique Artistique et GRS. Communications Scientifiques et Techniques d'Experts Étrangers* (pp. 169–175). Paris, France: INSEP.

Piard, C. (1982). *Fondements de la Gymnastique: Technologie et Pedagogie*. Paris: Vigot.

Pool, J., Binkhorst, R. A., & Vos, J. A. (1969). Some anthropometric and physiological data in relation to performance of top female gymnasts. *Internationale Zeitschrift Feur Angewante Physiologie*, 27, 329–338.

Porac, C. (2016). *Laterality: Exploring the Enigma of Left-Handedness*. London: Academic Press.

Potiron-Josse, M., & Bourdon, A. (1989). Le gros cœur du sportif. *Science & Sports*, 4(4), 305–316.

Pouramir, M., Haghshenas, & O., Sorkhi, H. (2004). Effects of gymnastic exercise on the body iron status and hematologic profile. *Iranian Journal of Medicine and Science*, 29(3), 140–141.

Pourcho, A. M., Liu, Y. H., & Milshteyn, M. A. (2013). Electrodiagnostically confirmed posttraumatic neuropathy and associated clinical exam findings with lisfranc injury. *Foot & Ankle International*, 34(8), 1068–1073.

Pozzo, T., & Studeny, C. (1987). *Théorie et Pratique des Sports Acrobatiques*. Paris: Vigot.

Prados, J. A. L. (2005). Analysis of gymnasts lactate concentration: guidelines for action in reference to inter-exercises rest and post-workout intake. *Apunts: Educ Fis Deporte*, 79(1), 86–93.

Prassas, S. G. (1988). Biomechanical model of the press handstand in gymnastics. *International Journal of Sport Biomechanics*, 4(4), 326–341.

Previc, F. H. (1991). A general theory concerning the prenatal origins of cerebral lateralization in humans. *Physiological Reviews*, (98), 299–334.

Price, H. D. O. (1937). The art of guarding or "spotting". *The Journal of Health and Physical Education*, 8(3), 151–199. doi:10.1080/23267240.1937.10619730

Puddle, D. L., & Maulder, P. S. (2013). Ground reaction forces and loading rates associated with parkour and traditional drop landing techniques. *Journal of Sports Science & Medicine*, 12(1), 122–129.

Purcell, L., & Micheli, L. (2009). Low back pain in young athletes. *Sports Health*, 1(3), 212–222.

Puthucheary, Z., Skipworth, J. R., Rawal, J., Loosemore, M., Van Someren, K., & Montgomery, H. E. (2011). Genetic influences in sport and physical performance. *Sports Medicine*, 41(10), 845–859. doi:10.2165/11593200-000000000-00000

Pyke, F. (2000). Strength assessment by isokinetic dynamometry. In C. J. Gore (ed.), *Physiological Tests for Elite Athletes*. Lower Mitcham: Australian Sports Commission.

Pyke, F. (2006). *Champions in Sport and Life*. Artarmon: ETN Communications.

Rabelo, A. S. (2001).The role of families in the development of aspiring expert soccer players (Unpublished masters thesis) Federal University of Minas Gerais, Brazil.

Rabinovitch, P., McLean, E. B., Beck, G. R., & Brown, A. C. (1978). Recurrent pre-retinal hemorrhages following a negative "g" maneuver on school playground equipment. *The Journal of Pediatrics*, 92, 846–853.

Radcliffe, J. C., & Osternig, L. R. (1995). Effects on performance of variable eccentric loads during depth jumps. *J Sports Rehabil.*, 4(1), 31–41.

Ravizza, K. H. (2002). A philosophical construct: a framework for performance enhancement. *International Journal of Sport Psychology*, 33, 4–18.

Ravizza, K., & Rotella, R. (1982). Cognitive somatic behavioral interventions in gymnastics. In L. Zaichkowsky & W. E. Sime (eds), *Stress Management for Sport* (pp. 25–35). Reston, VA: AAHPERD.

Reason, J. (2000). Human error: models and management. *BMJ: British Medical Journal, 320*(7237), 768–770.

Régnier, G., & Salmela, J. H. (1987). Predictors of success in Canadian male gymnasts. In J. H. S. B. Petiot & H. B. Hoshizaki (eds), *World Identification Systems for Gymnastic Talent* (pp. 141–150). Montreal: Sport Psyche.

Reinhard, P. (1998). Local coach hounded for tape of Chinese gymnast's injury ★ Jack Carter happened to film Sang Lan's fateful routine at Goodwill Games. Suddenly, everyone is looking for him. Retrieved from http://articles.mcall.com/1998-08-26/news/3211111_1_vault-carter-goodwill-games-officials

Rézette, D., & Ablard, B. (1985). Orientation versus motion visual cues to control sensorimotor skills in some acrobatic leaps. *Human Movement Science, 4*, 297–306.

Richards, J. E., Ackland, T. R., & Elliott, B. C. (1999). The effect of training volume and growth on gymnastic performance in young women. *Pediatric Exercise Science, 11*(4), 349–363.

Rodríguez, F. A., Marina, M., & Boucharin, E. (1999). Physiological demands of women's competitive gymnastic routines. Paper presented at the 4th Annual Congress of the European College of Sport Science. 430. Rome.

Roegner, R. (2006). Inflatable amusement rides. *Consumer Product Safety Review, 10*(3), 6.

Rosalie, S. M., & Müller, S. (2012). A model for the transfer of perceptual-motor skill learning in human behaviors. *Research Quarterly for Exercise and Sport, 83*(3), 413–421. doi:10.1080/0270136 7.2012.10599876

Rosen, L. W., & Hough, D. O. (1988). Pathogenic weight-control behaviors of female college gymnasts. *Physician & Sportsmedicine, 16*(9), 140–143, 146.

Roskamm, H. (1980). Le système de transport de l'oxygène. In P. O. Astrand & K. Rodahl (eds), *Précis de Physiologie de l'Exercice Musculaire* (2 ed., pp. 316–317). Paris: Masson.

Rotella, R. J., & Lerner, J. D. (1993). Responding to competitive pressure. In R. N. Singer, M. Murphey, & L. K. Tennant (eds), *Handbook of Research on Sport Psychology* (pp. 528–541). New York, NY: Macmillan.

Roupas, N. D., & Georgopoulos, N. A. (2011). Menstrual function in sports. *Hormones, 10*(2), 104–116.

Roy, S., Caine, D., & Singer, K. M. (1985). Stress changes of the distal radial epiphysis in young gymnasts. *American Journal of Sports Medicine, 13*(5), 301–308.

Rubin, B. D., Anderson, S. J., Chandler, J., & Kibler, W. B. (1993). A physiological and shoulder injury profile of elite divers. In R. Malina & J. L. Gabriel (eds), *U.S. Diving Sport Science Seminar 1993 Proceedings* (pp. 158–164). Indianapolis, IN: U.S. Diving Publications.

Sadowski, J., Mastalerz, A., & Niznikowski, T. (2013). Benefits of bandwidth feedback in learning a complex gymnastic skill. *J Hum Kinet, 37*, 183–193.

Sale, D. G. (1986). Neural adaptation in strength and power training. In N. M. N. L. Jones & A. J. McComas (eds), *Human Muscle Power* (pp. 289–308). Champaign, IL: Human Kinetics.

Sale, D. G. (1989). Strength training in children. In C. V. Gisolfi & D. R. Lamb (eds), *Perspectives in Exercise Science and Sports Medicine* (Vol. 2, pp. 165–216). Traverse City, MI: Cooper Publishing Group.

Sale, D. G. (1992). Neural adaptation to strength training. In P. V. Komi (ed.), *Strength and Power in Sport* (pp. 249–265). Oxford, UK: Blackwell Scientific Publications.

Sale, D., & MacDougall, D. (1981). Specificity in strength training: a review for the coach and athlete. *Canadian Journal of Applied Sport Sciences, 6*(2), 87–92.

Sale, D. G., & Norman, R. W. (1982). Testing strength and power. In J. D. MacDougall, H. A. Wenger, & G. H.J. (eds), *Physiological Testing of the Elite Athlete* (pp. 7–38). Ithaca, NY: Mouvement Publications.

Salmela, J. H. (1976). Psychomotor task demands of gymnastics. In J. H. Salmela (ed.), *The Advanced Study of Gymnastics* (pp. 5–19). Springfield, IL: C. C. Thomas.

Salmela, J. H. (1989). Long term intervention with the Canadian Men's gymnastics team. *The Sport Psychologist*, *3*, 340–349,

Salmela, J. H. (1996). *Great Job Coach!* Ottawa: Potentium.

Salmela, J. H., Marques, M. P., & Machado, R. (2004). The informal structure of football in Brazil. *Insight*, *7*(1), 17–19

Salmela, J. H., Monfared, S. F., Mosayebi, S. S., & Durand-Bush, N. (2009). Mental skill profiles and expertise levels of elite Iranian athletes. *Int J Sport Psy*, *40*(2), 229–248.

Salmela, J. H., & Moraes, L. C. (2003). Development of expertise: the role of coaching, families, and cultural contexts. In J. L. Starkes & K. A. Ericsson (eds), *Expert Performance in Sports* (pp. 275–294). Champaign, IL: Human Kinetics.

Salmela, J. H., Mosayebi, F., & Monfared, S. S. (2007). Perceptions of Iranian athletes and coaches of the effectiveness of mental training interventions at the Asian Games. In Y. Theodorakis, M. Goudas, & A. Papaionnou (eds), *Sport and Exercise Psychology: Bridges Between Disciplines and Cultures* (pp. 92–96). Thesaloniki: University of Thesaly.

Salmela, J. H., Petiot, B., Hallé, M., & Régnier, G. (1980). *Competitive Behaviors of Olympic Gymnasts*. Springfield, IL. C. C. Thomas.

Saltin, B., & Astrand, P. O. (1967). Maximal oxygen uptake in athletes. *J. Appl. Physiol.*, *23*, 353–358.

Samuelson, M., Reider, B., & Weiss, D. (1996). Grip lock injuries to the forearm in male gymnasts. *American Journal of Sports Medicine*, *24*(1), 15–18.

Sands, W. A. (1981a). *Beginning Gymnastics*. Chicago, IL: Contemporary Books.

Sands, W. A. (1981b). Competition injury study: a preliminary report. *USGF Technical Journal*, *1*(3), 7–9.

Sands, W. A. (1984). *Coaching Women's Gymnastics*. Champaign, IL: Human Kinetics.

Sands, W. A. (1985). Conditioning for gymnastics: a dilemma. *Technique*, *5*(3), 4–7.

Sands, W. A. (1987). The edge of the envelope. *Gymnastics Safety Update*, *2*(3), 2–3.

Sands, W. A. (1990a). Determining skill readiness. *Technique*, *10*(3), 24–27.

Sands, W. A. (1990b). National women's tracking program. *Technique*, *10*, 23–27.

Sands, W. A. (1990c). Spotting belts. In G. S. George (ed.), *USGF Gymnastics Safety Manual* (2 ed., pp. 47–50). Indianapolis, IN: U.S. Gymnastics Federation.

Sands, W. A. (1991a). Monitoring elite gymnastics athletes via rule based computer systems. In *Masters of Innovation III* (p. 92). Northbrook, IL: Zenith Data Systems.

Sands, W. A. (1991b). Monitoring the elite female gymnast. *National Strength and Conditioning Association Journal*, *13*(4), 66–71.

Sands, W. A. (1991c). Science puts the spin on somersaulting. *RIP*, *2*(2), 18–20.

Sands, W. A. (1991d). Spatial orientation while somersaulting. *Technique*, *11*(1), 16–19.

Sands, W. A. (1993a). The role of science in sport. *Technique*, *13*(10), 17–18.

Sands, W. A. (1993b). *Talent Opportunity Program*. Indianapolis, USA: United States Gymnastics Federation.

Sands, W. A. (1994a). Physical abilities profiles – 1993 National TOPs testing. *Technique*, *14*(8), 15–20.

Sands, W. A. (1994b). The role of difficulty in the development of the young gymnast. *Technique*, *14*(3), 12–14.

Sands, W. A. (1994c). The German giant. *Technique*, *14*(8), 22–23.

Sands, W. A. (1995a). Tkatchev Drill. *Technique*, *16*(1), 9.

Sands, W. A. (1995b). How can coaches use sport science? *Track Coach*, *134*(winter), 4280–4283.

Sands, W. A. (1996). How effective is rescue spotting? *Technique*, *16*(9), 14–17.

Sands, W. A. (1998). A look at training models. *Technique*, *19*, 6–8.

Sands, W. A. (1999a). Communicating with coaches: envisioning data. In S. Prassas (ed.), *International Smposium on Biomechanics in Sports* (pp. 11–20). Perth, Australia: Edith Cowan University, School of Biomedical and Sports Sciences.

Sands, W. A. (1999b). A look at training models. *Technique*, *19*(9), 6–8.

Sands, W. A. (2000a). Injury prevention in women's gymnastics. *Sports Medicine*, *30*(5), 359–373.

Sands, W. A. (2000b). Olympic preparation camps 2000 physical abilities testing. *Technique, 20*, 6–19.

Sands, W. A. (2000c). Physiological aspects of gymnastics. Paper presented at the 2emes Journees Internationales d'Etude de L'Association Française de Recherche en Activités Gymniques et Acrobatiques (A.F.R.A.G.A.).

Sands, W. A. (2000d). Vault run speeds. *Technique, 20*(4), 5–8.

Sands, W. A. (2002a). *Gymnastics Risk Management: Safety Handbook 2002 Edition*. Indianapolis, IN: USA Gymnastics.

Sands, W. A. (2002b). Monitoring gymnastics training. Paper presented at the 3èmes Journées Internationales d'Etude de l'AFRAGA, 7–9 November 2002, Lille, France.

Sands, W. A. (2003). Physiology. In W. A. Sands, D. J. Caine, & J. Borms (eds), *Scientific Aspects of Women's Gymnastics* (pp. 128–161). Basel, Switzerland: Karger.

Sands, W. A. (2007). Skill learning and performance – physiological aspects. In R. M. Malina & J. L. Gabriel (eds), *USA Diving Coach Development Reference Manual* (pp. 105–114). Indianapolis, IN: USA Diving.

Sands, W. A. (2010a). Flexibility. In M. Cardinale, R. Newton, & K. Nosaka (eds), *Strength and Conditioning Biological Principles and Practical Applications* (pp. 391–400). Hoboken, NJ: John Wiley & Sons, Ltd.

Sands, W. A. (2010b). Puzzles and paradoxes – gymnastics. Paper presented at the Anais do II Seminario Internacional de Ginastica Artistica e Rithmica de Competicao, 29–30 June 2010, Campinas, Brazil.

Sands, W. A. (2011a). Linear kinetics applied to gymnastics. In M. Jemni (ed.), *The Science of Gymnastics*. London: Routledge.

Sands, W. A. (2011b). Talent identification in women's artistic gymnastics, the talent opportunity program. In J. Baker, S. Cobley, & J. Schorer (eds), *Talent Identification and Development in Sport* (pp. 83–94). New York, NY: Routledge.

Sands, W. A., Abramowitz, R., Hauge Barber, L., Irvin, R., & Major, J. A. (1992). A comparison of routine error distributions. *Technique, 12*(6), 7–10.

Sands, W. A., Caine, D. J., & Borms, J. (2003). Scientific aspects of women's gymnastics. *Med Sport Sci. Basel, Karger, 45*, 128–161.

Sands, W. A., & Cheetham, P. J. (1986). Velocity of the vault run: junior elite female gymnasts. *Technique, 6*(3), 10–14.

Sands, W. A., Crain, R. S., & Lee, K. M. (1990). Gymnastics coaching survey – 1989. *Technique, 10*(1), 22–27.

Sands, W., Cunningham, S. J., Johnson, S. C., Meek, S. G., & George, G. S. (1988). Levels of protection gymnastics safety equipment: a summary for coaches. *Technique, 8*(3–4), 22–25.

Sands, W. A., Cunningham, S. J., Johnson, S. C., Meek, S. G., & George, G. S. (1991a). Deceleration characteristics of foam pit landing areas in gymnastics. *FIG Scientific/Medical Symposium Proceedings, 1*(1), 19–23.

Sands, W. A., Cunningham, S. J., Johnson, S. C., Meek, S. G., & George, G. S. (1991b). Deceleration characteristics of gymnastics landing mats. *FIG Scientific/Medical Symposium Proceedings, 1*(1), 24–27.

Sands, W. A., Dunlavy, J. K., Smith.S. L., Stone, M. H., & McNeal, J. R. (2006a). Understanding and training the Maltese. *Technique, 26*(5), 6–9.

Sands, W. A., Eisenman, P., Johnson, S., Paulos, L., Abbot, P., Zerkel, S., et al. (1987). Getting ready for '88. *Technique, 7*, 12–18.

Sands, W. A., & George, G. S. (1988). Somersault trajectory differences: foam block versus coil spring floor. *Technique, 8*(1), 8–9.

Sands, W. A., Henschen, K. P., & Shultz, B. B. (1989). National women's tracking program. *Technique, 9*(4), 14–19.

Sands, W. A., Hofman, M. G., & Nattiv, A. (2002). Menstruation, disordered eating behavior, and stature: a comparison of female gymnasts and their mothers. *International Sports Journal, 6*(1), 1–13.

Sands, W. A., Irvin, R. C., & Major, J. A. (1995). Women's gymnastics: the time course of fitness acquisition. A 1-year study. *Journal of Strength and Conditioning Research, 9*(2), 110–115.

Sands, W. A., Jemni, M., Stone, M., McNeal, J., Smith, S. L., & Piacentini, T. (2005). Kinematics of vault board behaviours – a preliminay comparison. Paper presented at the 5th International Conference of the Association Française pour la Recherche en Activités Gymniques et Acrobatiques (AFRAGA).

Sands W. A., Kavanaugh, A., Murray, S., McNeal, Jr., & Jemni, M. (2016). Modern techniques and technologies applied to training and performance monitoring. *Inter J of Sports Physiol and Perf,* (Dec 5), 1–29. Epub ahead of print. doi:10.1123/ijspp.2016-0405

Sands, W. A., Kimmel, L. W., Mcneal, R. J., Smith, S. L., Penitente, G., Murray, S. R., Sato, K., Mizuguchi, S., & Stone, M. H. (2013). Kinematic and kinetic tumbling take-off comparisons of a spring-floor and an air floor: a pilot study. *Science of Gymnastics Journal, 5*(3), 31–46.

Sands, W. A., & McNeal, J. R. (1995a). *Drills for Skills (V1.0).* Carmichael, CA: U.S. Elite Coaches Association for Women's Gymnastics.

Sands, W. A., & McNeal, J. R. (1995b). The relationship of vault run speeds and flight duration to score. *Technique, 15*(5), 8–10.

Sands, W. A., & McNeal, J. R. (1997). A minimalist approach to conditioning for women's gymnastics. In S. Whitlock (ed.), *1997 USA Gymnastics Congress Proceedings Book* (pp. 78–80). Indianapolis, IN: USA Gymnastics.

Sands, W. A., & McNeal, J. R. (1999a). Consequences of the round-off twist direction. *Technique, 19*(2), 26–28.

Sands, W. A., & McNeal, J. R. (1999b). Judging gymnastics with biomechanics. *SportScience, 3*(1). Retrieved from: sportsci.org/jour/9901/was.html

Sands, W. A., & McNeal, J. R. (1999c). Body size and sprinting characteristics of 1998 National TOP's athletes. *Technique, 19*(5), 34–35.

Sands, W. A., & McNeal, J. R. (2002). Some guidelines on the transition from the old horse to the new table. *Technique, 22*(1), 22–23.

Sands, W. A., McNeal, J., & Jemni, M. (2001a). Anaerobic power profile: talent-selected female gymnasts age 9–12 years. *Technique, 21,* 5–9.

Sands, W. A., McNeal, J., & Jemni, M. (2001b). Fitness profile comparisons: USA women's junior, senior, and Olympic gymnastics teams. *Journal of Strength and Conditioning Research, 15*(3), 398.

Sands, W. A., McNeal, J. R., & Jemni, M. (2002). Does average jumping power keep pace with increasing age and size in U.S. National Team female gymnasts. *Medicine and Science in Sports and Exercise, 34*(5), S143.

Sands, W. A., McNeal, J. R., Jemni, M. (2005). A look at the sprint test. Talent opportunity program. *Technique,* January, 6–7.

Sands, W. A., McNeal, J. R., Jemni, M., & Delong, T. H. (2000). Should female gymnasts lift weights? *SportScience, 4*(3), Retrieved from: www.sportsci.org/jour/0003/was.html

Sands, W. A., McNeal, J. R., Jemni, M., & Penitente, G. (2011). Thinking sensibly about injury prevention and safety. *Science of Gymnastics Journal, 3*(3), 43–58.

Sands, W. A., McNeal, J. R., Ochi, M. T., Urbanek, T. L., Jemni, M., & Stone, M. H. (2004a). Comparison of the Wingate and Bosco anaerobic tests. *J Strength Cond Res, 18*(4), 810–815.

Sands, W. A., McNeal, J. R., Penitente, G., Murray, S. R., Nassar, L., Jemni, M., . . . Stone, M. H. (2015). Stretching the spines of gymnasts: a review. *Sports Medicine, 46*(3), 315–327. doi:10.1007/s40279-015-0424-6

Sands, W. A., McNeal, J. R., & Stone, M. H. (2011). Thermal imaging and gymnastics injuries: a means of screening and injury identification. *Science of Gymnastics Journal, 3*(2), 5–12.

Sands, W. A., McNeal, J., Stone, M., Russell, E., & Jemni, M. (2006b). Flexibility enhancement with vibration: acute and long-term. *Med Sc Sports Exer, 38*(4), 720–725.

Sands, W. A., McNeal, J. R., Stone, M. H., Smith, S. L., Dunlavy, J. K., Jemni, M., et al. (2006c). Exploratory relationship of drop jump performance with gymnastics vaulting and floor exercise scores. Paper presented at the 11th Annual Congress of the ECSS.

Sands, W. A., McNeal, J. R., & Urbanek, T. (2003). On the role of "Functional Training" in gymnastics and sports. *Technique, 23*(4), 12–13.

Sands, W. A., Mikesky, A. E., & Edwards, J. E. (1991). Physical abilities field tests U.S. Gymnastics Federation Women's National Teams. *USGF Sport Science Congress Proceedings, 1*(1), 39–47.

Sands, W. A., Shultz, B. B., & Newman, A. P. (1993). Women's gymnastics injuries. A 5-year study. *American Journal of Sports Medicine, 21*(2), 271–276.

Sands, W. A., Shultz, B. B., & Paine, D. D. (1993). Gymnastics performance characterization by piezoelectric sensors and neural networks. *Technique, 13*(2), 33–38.

Sands, W. A., Smith, S. L., Westenburg, T. M., McNeal, J. R., & Salo, H. (2004b). Kinematic and kinetic case comparison of a dangerous and superior flyaway dismount – women's uneven bars. In M. Hubbard, R. D. Metha, & J. M. Pallis (eds), *The Engineering of Sport 5* (pp. 414–420). Sheffield, UK: International Sports Engineering Association.

Sands, W. A., & Stone, M. H. (2006). Monitoring the elite athlete. *Olympic Coach, 17*(3), 4–12.

Sands, W. A., Stone, M. H., McNeal, J. R., Smith, S. L., Jemni, M., Dunlavy, J. K., . . . Haff, G. G. (2006d). A pilot study to measure force development during a simulated maltese cross for gymnastics still rings. Paper presented at the XXIV International Symposium on Biomechanics in Sports, 14–18 July 2006, Salzburg, Austria.

Sano, S., Ikegami, Y., Nunome, H., Apriantono, T., & Sakurai, S. (2007). The continuous measurement of the springboard reaction force in gymnastic vaulting. *Journal of Sports Sciences, 25*(4), 381–391.

Sartor, F., Vailati, E., Valsecchi, V., Vailati, F., & La Torre, A. (2013). Heart rate variability reflects training load and psychophysiological status in young elite gymnasts. *The Journal of Strength and Conditioning Research, 27*(10), 2782–2790.

Sathyendra, V., & Payatakes, A. (2013). Grip lock injury resulting in extensor tendon pseudorupture: case report. *Journal of Hand Surgery, 38*(12), 2335–2338. doi:10.1016/j.jhsa.2013.09.010

Savchin, S., & Biskup, L. (2003) Aerobic and anaerobic capacities of young gymnasts as a factor of training loads. Kharkov State Academy of Design and Arts (KSADA) (HHPI). Ukraine, 6 - S. 14–20.

Savelsbergh, G., van der Kamp, J., Williams, A. M., & Ward, P. (2005). *Ergonomics, 48,* 1686–1697.

Savelsbergh, G., Williams, A. M., van der Kamp, J., & Ward, P. (2002). Visual search, anticipation and expertise in soccer goalkeepers. *Journal of Sports Sciences, 20,* 279–287.

Saw, A. E., Main, L. C., & Gastin, P. B. (2015). Monitoring athletes through self-report: factors influencing implementation. *Journal of Sports Medicine, 14,* 137–146.

Saw, A. E., Main, L. C., & Gastin, P. B. (2016). Monitoring the athlete training response: subjective self-reported measures trump commonly used objective measures: a systematic review. *British Journal of Sports Medicine, 50,* 281–291.

Schack, T., Essig, K., Frank, C., & Koester, D. (2014). Mental representation and motor imagery training. *Frontiers in Human Neuroscience, 8,* 328. doi:10.3389/fnhum.2014.00328

Schaffert, N., Mattes, K., & Effenberg, A. O. (2011). An investigation of online acoustic information for elite rowers in on-water training conditions. *Journal of Human Sport and Exercise, 6,* 392–405.

Schembri, G. (1983). *Introductory Gymnastics.* Sydney, Australia: Australian Gymnastics Federation, Inc.

Schiavon, L. M. (2009). Women's artistic gymnastics and oral history: the sport formation of Brazilian gymnasts taking part in Olympic games (1980–2004) (Unpublished doctoral thesis). University of Campinas, Campinas, Brazil.

Schindler, A. E., Ebert, A., & Friedrich, E. (1972). Conversion of androstenedione to estrone by human fat tissue. *J Clin Endocrinol Metab, 35,* 627.

Schindler, S. (2016). Analyse der Mehrkampffinalisten bei der FIG Turn WM 2014 in Nanning hinsichtlich eines Drehschemas (Zulassungsarbeit). Albert-Ludwigs-Universität Freiburg, Freiburg.

Schmidt, R. A. (1988). *Motor Control and Learning.* Champaign, IL: Human Kinetics.

Schmidt, R. A. (1994). Movement time, movement distance, and movement accuracy: a reply to Newell, Carlton, and Kim. *Human Performance, 7*(1), 23–28.

Schmidt, R. A., & Lee, T. D. (2008). *Motor Control and Learning.* Champaign, IL: Human Kinetics.

Schmidt, R. A., & Stull, G. A. (1970). Premotor and motor reaction time as a function of preliminary muscular tension. *J Mot Behav*, *11*(2), 96–110.

Schmidt, R. A., & Wulf, G. (1997). Continuous concurrent feedback degrades skill learning: implications for training and simulation. *Human Factors*, *39*(4), 509–525.

Schmidt, R. A., & Young, D. E. (1991). Methodology for motor learning: a paradigm for kinematic feedback. *Journal of Motor Behavior*, *23*(1), 13–24.

Schmidtbleicher, D. (1992). Training for power events. In P. V. Komi (ed.), *Strength and Power for Sports* (pp. 381–395). Oxford: Blackwell Scientific.

Schmitt, H., & Gerner, H. J. (2001). Paralysis from sport and diving accidents. *Clinical Journal of Sport Medicine*, *11*(1), 17–22.

Schneier, B. (2006). *Beyond Fear*. New York, NY: Springer Science+Business.

Schone, H. (ed.) (1984). *Spatial Orientation*. Princeton, NJ: Princeton University Press.

Schone, H., & Lechner-Steinleitner, S. (1978). The effect of preceding tilt on the perceived vertical. *Acta Otolaryngologica*, *85*, 68–73.

Schweizer, L. (2008). Biomechanische Grundlagen von Schraubenbewegungen beim Bodenturnen: Vortrag während der Freiburger Gerätturntage, Freiburg.

Scott, R. (2012). Shawn Johnson says sport can be 'brutal,' but it's good. *USA Today, Sports Section, 3C*. Retrieved from: http://content.usatoday.com/communities/gameon/post/2012/06/shawn-johnson-says-sport-can-be-brutal-but-its-good/1

Seck, D., Vandewalle, H., Decrops, N., & Monod, H. (1995). Maximal power and torque-velocity relationship on a cycle ergometer during the acceleration phase of a single all-out exercise. *Eur J Appl Physiol.*, *70*, 161–168.

Seeley, M. K., & Bressel, E. A. (2005). Comparison of upper-extremity reaction forces between the Yurchenko vault and floor exercise. *Journal of Sports Science & Medicine*, *4*(2), 85–94.

Self, B. P., & Paine, D. (2001). Ankle biomechanics during four landing techniques. *Medicine & Science in Sports & Exercise*, *33*(8), 1338–1344.

Seliger, V., Budka, I., Buchberger, J., Dosoudil, F., Krupova, J., Libra, M., et al. (1970). Métabolisme énergétique au cours des exercices de gymnastique. *Kinanthropologie*, *2*, 159–169.

Selye, H. (1974). *Stress Without Distress*. New York, NY: New American Library.

Shaghlil, N. (1978). La gymnastique et son action sur l'appareil circulatoire et respiratoire. Paper presented at the 1er Colloque médical international de gymnastique.

Sheets, A. L. (2008). Evaluation of a subject-specific female gymnast model and simulation of an uneven parallel bar swing. *Journal of Biomechanics*, *4*, 326–341.

Sherald, M. (1989). Neural nets versus expert systems. *PC AI*, *3*(4), 10–15.

Shiffrin R. M. (1976). Capacity limitations in information in information processing, attention and memory. In W. K. Estes (ed.) *Handbook of Learning and Cognitive Processes* (Vol. 4, Attention and Memory, pp. 177–236). Hillsdale, NJ: Erlbaum.

Shupert, C. L., Lindblad, I. M., & Leibowitz, H. W. (1983). Visual testing for competitive diving: a two visual systems approach. In D. Golden (ed.), *Proceedings of the 1983 U.S. Diving Sports Science Seminar* (pp. 100–115). Indianapolis, IN: U.S. Diving.

Siff, M. C. (2000). *Supertraining*. Denver, CO: Supertraining Institute.

Simons, C. (2014). Monitoring training load and health in sub-junior and junior female gymnasts – a pilot study (Unpublished master's thesis). Vrije University, Amsterdam, The Netherlands.

Singer, R. N. (1988). Psychological testing: what value to coaches and athletes? *International Journal of Sport Psychology*, *19*, 87–106.

Singh, S., Smith, G. A., Fields, S. K., & McKenzie, L. B. (2008). Gymnastics-related injuries to children treated in emergency departments in the United States, 1990–2005. *Pediatrics*, *121*(4), e954–960. doi:10.1542/peds.2007-0767

Sipila, S., Koskinen, S. O., Taaffe, D. R., Takala, T. E., Cheng, S., Rantanen, T., et al. (2004). Determinants of lower-body muscle power in early postmenopausal women. *J Am Geriatr Soc*, *52*(6), 939–944.

Smith, D. J. (2003). A framework for understanding the training process leading to elite performance. *Sports Medicine*, *33*(15), 1103–1126.

Smith, J. (1870). United States Patent No. 108401. U. S. P. Office. Retrieved from: www.google. com/patents/US108401?printsec=claims&dq=spring+floor+gymnastics+patent#v=onepage&q& f=false

Smith, J. A. (1983). The back somersault take-off – a biomechanics study. *Carnegie Research Paper, 1*, 31–39.

Smoleuskiy, V., & Gaverdouskiy, I. (1996). *Tratado General de Gimnasia Artística Deportiva*. Barcelona: Paidotribo.

Soberlak, P.A., & Côté, J. (2003). The developmental activities of elite ice hockey players. *Journal of Applied Sport Psychology, 15*, 41–49.

Soric, M., Misigoj-Durakovic, M., & Pedisic, Z. (2008). Dietary intake and body composition of prepubescent female aesthetic athletes. *Int J Sport Nutr Exerc Metab, 18*(3), 343–354.

Spielberger, C. D. (1966). *Anxiety and Behavior*. New York, NY: Academic Press.

Sprynarova, S., & Parizkova, J. (1969). Comparison of the circulatory and respiratory functional capacity in girl gymnasts and swimmers. *J. Sports Med., 9*, 165–172.

Stamford, B. A., Weltman, A., Moffat, R., & Sady, S. (1981). Exercise recovery above and below the anaerobic threshold following maximal work. *J Appl Physiol, 51*, 840–844.

Steele, V. A., & White, J. A. (1986). Injury prediction in female gymnasts. *British Journal of Sports Medicine, 20*(1), 31–33.

Stein, N. (1998). Speed training in sport. In B. Elliott (ed.), *Training in Sport* (pp. 288–349). New York, NY: John Wiley & Sons.

Steinberg, N., Siev-Ner, I., Peleg, S., Dar, G., Masharawi, Y., Zeev, A., & Hershkovitz, I. (2011). Injury patterns in young, pre-professional dancers. *Journal of Sports Sciences, 29*(1), 47–54.

Stone, M. H., Stone, M. E., & Sands, W. A. (2007). *Principles and Practice of Resistance Training*. Champaign, IL: Human Kinetics.

Stone, M. H., Wilson, D., Rozenek, R., & Newton, H. (1984). Anaerobic capacity. *National Strength and Conditioning Association Journal, 5*(6), 63–65.

Stroescu, V., Dragan, J., Simionescu, L., & Stroescu, O. V. (2001). Hormonal and metabolic response in elite female gymnasts undergoing strenuous training and supplementation with SUPRO Brand Isolated Soy Protein. *J Sports Med Phys Fitness, 41*(1), 89–94.

Suchomel, T. J., Sands, W. A., & McNeal, J. R. (2016). Comparison of static, countermovement, and drop jumps of the upper and lower extremities in U.S. Junior National Team male gymnasts. *Science of Gymnastics Journal, 8*(1), 15–30.

Suinn, R. M. (1993). Imagery. In R. N. Singer, M. Murphey, & K. Tennant (eds), *Handbook of Research on Sport Psychology* (pp. 492–510). New York, NY: Macmillan.

Sureira, T. M., Amancio, O. S., & Braga, J. A. P. (2012). Influence of artistic gymnastics on iron nutritional status and exercise-induced hemolysis in female athletes. *International Journal of Sport Nutrition and Exercise Metabolism, 22*, 243–250.

Sward, L., Hellstrom, M., Jacobsson, B., Nyman, R., & Peterson, L. (1991). Disc degeneration and associated abnormalities of the spine in elite gymnasts. *Spine, 16*(4), 437–443.

Sward, S. B. (1985). Energy cost of competitive gymnastic events. In *Human Performance: Efficiency and Improvements in Sports, Exercise and Fitness* (pp. 48–51). Reston, VA: American Alliance for Health, Physical Education, Recreation, and Dance.

Szogy, A., & Cherebetiu, G. (1971). Capacité aérobie maximum chez les sportifs de performance. *Médecine du Sport, 45*, 224–234.

Takei, Y. (1998). Three-dimensional analysis of handspring with full turn vault: deterministic model, coaches' beliefs, and judges' scores. *Journal of Applied Biomechanics, 14*, 190–210.

Tanaka, S. (1987). The Japanese gymnastic golden era between the 1960s and 1970s. In B. Petiot, J. H. Salmela, & T. B. Hoshizaki (eds), *World Identification Systems for Gymnastic Talent* (pp. 45–57). Montreal: Sport Psyche.

Tanner, J. M. (1962). *Growth at Adolescence: With a General Consideration of the Effects of Hereditary and Environmental Factors Upon Growth and Maturation from Birth to Maturity* (2 ed.). Oxford: Blackwell.

Taylor, J. C., Whindcup, P. H., Hindmarsh, P. C., Lampe, F., Odoki, K., & Cook, D. G. (2001). Performance of a new pubertal self-assessment questionnaire: a preliminary study. *Paediatric and Perinetal Epidemiology*, *15*, 88–94.

Tenner, E. (1996). *Why Things Bite Back*. New York, NY: Random House.

Terry, M. B., Goldberg, M., Schechter, S., Houghton, L. C., White, M. L., O'Toole, K., Chung, W. K., Daly, M. B., Keegan, T. H. M., Andrulis, I. L., Bradbury, A. R., Schwartz, L., Knight, J. A., John, E. M., & Buys, S. S. (2016). Comparison of clinical, maternal, and self pubertal assessments: implications for health studies. *Pediatrics*, *138*(1), e20154571.

Tesch, P. A. (1980). Fatigue pattern in subtypes of human skeletal muscle fibers. *International Journal of Sports Medicine*, *1*(2), 79–81.

Theodoropoulou, A., Markou, K. B., Vagenakis, G. A., Bernardot, D., Leglise, M., Kourounis, G., et al. (2005). Delayed but normally progressed puberty is more pronounced in artistic compared with rhythmic elite gymnasts due to the intensity of training. *J Clin Endocrinol Metab*, *90*(11), 6022–6027.

Thomas, L., Fiard, J., Soulard, C., & Chautemps, G. (1997). *Gimnasia Deportiva: De la escuela … a las asociaciones deportivas*. Lérida: Agonos.

Tilley, D. (2013). Is the way you spot setting you up for a shoulder injury? (Part I). Retrieved from: http://hybridperspective.com/tag/spotting/

Tofler, I. R., Stryer, B. K., Micheli, L. J., & Herman, L. R. (1996). Physical and emotional problems of elite female gymnasts. *The New England Journal of Medicine*, *335*(4), 281–283.

Too, D., & Adrian, M. J. (1987). Relationship of lumbar curvature and landing surface to ground reaction forces during gymnastic landing. In J. Terauds, B. A. Gowitzke, & L. E. Holt (eds), *Biomechanics in Sports III & IV* (pp. 96–102). Del Mar, CA: Academic Publishers.

Trappe, S. W., Costill, D. L., Vukovich, M. D., Jones, J., & Melham, T. (1996). Aging among elite distance runners: a 22-yr longitudinal study. *J Appl Physiol*, *80*(1), 285–290.

Turoff, F. (1991). *Artistic Gymnastics: A Comprehensive Guide to Performance and Teaching Skills for Beginners and Advanced Beginners*. Dubuque: Brown Publishers.

Ubukata, O. (1981). Objective load measurement in gymnastic training. In A. Morecki, K. Fidelus, K. Kedzior, & A. Wit (eds), *Biomechanics VII-B* (3-B ed., pp. 392–397). Baltimore, MD: University Park Press.

Ukran, M. L., Cheburaev, V. S., & Antonov, L. K. (1970). Scientific work in the U.S.S.R. gymnastics team. *Yessis Review of Soviet Physical Education and Sports*, *5*(1), 1–6.

Uneståhl, L-E. (1975). *Hypnosis in the Seventies*. Orebro: Veja.

Vain, A. (2002). Criteria for preventing overtraining of the musculoskeletal system of gymnasts. *Biology of Sport*, *19*(4), 329–345.

Van der Eb, J. Filius, M., Rougoor, G., Van Niel, C., de Water, J., Coolen, B., & de Koning, H. (2012). Optimal velocity profiles for vault. In E. J. Bradshaw, A. Burnett, P. A. Hume (eds), *Proceedings of the 30th Conference of the International Society of Biomechanics in Sports*, Australian Catholic University, Melbourne, 2–6 July, 71–75.

van Dieen, J. H., Luger, T., & van der Eb, J. (2012). Effects of fatigue on trunk stability in elite gymnasts. *European Journal of Applied Physiology*, *112*(4), 1307–1313. doi:10.1007/s00421-011-2082-1

van Mechelen, W. (1997). Sports injury surveillance systems. *Sports Medicine*, *24*(3), 164–168.

Van Praagh, E., & Dore, E. (2002). Short-term muscle power during growth and maturation. *Sports Med*, *32*(11), 701–728.

Vandewalle, H., Peres, G., Heller, J., Panel, J., & Monod, H. (1987). Force-velocity relationship and maximal power on a cycle ergometer. Correlation with the height of a vertical jump. *Eur J Appl Physiol*, *56*, 650–656.

Vandewalle, H., Peres, G., Sourabié, O., Stouvenel, O, & Monod, H. (1989). Force-velocity relationship and maximal anaerobic power during cranking exercise in young swimmers. *Int J Sports Med*, *13*, 439–445.

Vanti, C., Gasperini, M., Morsillo, F., & Pillastrini, P. (2010) Low back pain in adolescent gymnasts. Prevalence and risk factors. *Scienza Riabilitativa*, *12*(2), 45–50.

Verkhoshansky, Y. V. (1981). Special strength training. *Soviet Sports Review*, *16*(1), 6–10.

Verkhoshansky, Y. V. (1985). *Programming and Organization of Training*. Moscow, U.S.S.R: Fizkultura i Spovt.

Verkhoshansky, Y. V. (1996). Speed training for high level athletes. *New Studies in Athletics, 11*(2–3), 39–49.

Verkhoshansky, Y. V. (1998). Organization of the training process. *New Studies in Athletics, 13*(3), 21–31.

Verkhoshansky, Y. V. (2006). *Special Strength Training: A Practical Manual for Coaches*. Moscow, Russia: Ultimate Athlete Concepts.

Viana, J., & Lebre, E. (2005). Heart rate analysis during men and women artistic gymnastics. In *5th International Conference of the AFRAGA*. Edited by M. Jemni and J. F. Robin Hammamet, Tunisia, pp. 81–83.z

Vianna, N. S., Jr (2002). The role of families and coaches in the development of aspiring expert tennis players (Unpublished master's thesis). Federal University of Minas Gerais, Brazil.

Vickers, J. (2007). *Perception, Cognition, and Decision Training. The Quiet Eye in Action*. Champaign, IL: Human Kinetics.

Vigarello, G. (1988). *Une Histoire Culturelle du Sport. Techniques d'Hier ... et d'Aujourd'Hui*. Paris: R. Laffont. Revue EPS.

Vine, S. J., & Klostermann, A. (2017). Success is in the eye of the beholder: a special issue on the quiet eye. *European Journal of Sport Science, 17*(1), 70–73.

Viitasalo, J. T. (1985a). Effect of training on force-velocity characteristics. *Biomechanics IX-A. International Series on Biomechanics*, 91–95.

Viitasalo, J. T. (1985b). Measurement of the force-velocity characteristics for sportsmen in field conditions. *Biomechanics IX-A. International Series on Biomechanics*, 96–101.

Viitasalo, J. T. (1988). Evaluation of explosive strength for young and adult athletes. *Research Quarterly for Exercise and Sport, 59*(1), 9–13.

Vivanco-Allende, A., Concha-Torre, A., Menéndez-Cuervo, S., & Rey-Galán, C. (2013). Parkour: una nueva causa de lesiones internas graves [Parkour: a new cause of serious internal injury]. *An Pediatr, 79*(6), 396–397. doi:10.1016/j.anpedi.2013.03.004

Vroomen, J., & de Gelder, B. (2000). Sound enhances visual perception: cross-modal effects of auditory organization on vision. *Journal of Experimental Psychology: Human Perception and Performance, 26*, 1583–1590.

von Lassberg, C., Beykirch, K., Campos, J. L., & Krug, J. (2012). Smooth pursuit eye movement adaptation in high level gymnasts. *Motor Control, 16*(2), 176–194.

von Laßberg, C., Rapp, W., Mohler, B., & Krug, J. (2013) Neuromuscular onset succession of high level gymnasts during dynamic leg acceleration phases on high bar. *J Electromyogr Kinesiol., 23*(5), 1124–1130. doi:10.1016/j.jelekin.2013.07.006

Vuillerme, N., Danion, F., Marin, L., Boyadjian, A., Prieur, J. M., Weise, I., & Nougier, V. (2001). The effect of expertise in gymnastics on postural control. *Neuroscience Letters, 303*(2), 83–86.

Vuillerme, N., & Nougier, V. (2004). Attentional demand for regulating postural sway: the effect of expertise in gymnastics. *Brain Research Bulletin, 63*(2), 161–165. doi:10.1016/j.brainresbull.2004.02.006

Wade, M., Campbell, A., Smith, A., Norcott, J., & O'Sullivan, P. (2012). Investigation of spinal posture signatures and ground reaction forces during landing in elite female gymnasts. *Journal of Applied Biomechanics, 28*(6), 677–686.

Wanke, E. M., Thiel, N., Groneberg, D. A., & Fischer, A. (2013). [Parkour—"art of movement" and its injury risk]. *Sportverletzung Sportschaden, 27*(3), 169–176. doi:10.1055/s-0033-1350183

Warda, K. A., Robertsb, S. A., Adamsa, J. E., Lanham-Newc, S., & Mughald, M. Z. (2007). Calcium supplementation and weight bearing physical activity – do they have a combined effect on the bone density of pre-pubertal children? *Bone, 41*(4), 496–504.

Watts, J. D. (1985). Does gymnastics damage the spine? *British Medical Journal, 290*(6486), 1990.

Weiker, G. G. (1985). Introduction and history of gymnastics. *Clinics in Sports Medicine, 4*(1), 3–6.

Weimann, E. (2002). Gender-related differences in elite gymnasts: the female athlete triad. *J Appl Psysiol, 92*(5), 2146–2152.

Weimann, E., Blum, W. F., Witzel, C., Schwidergall, S., & Bohels, H. J. (1999). Hypoleptinemia in female and male elite gymnasts. *European Journal of Clinical Investigation, 29*(10), 853–860.

Weimann, E., Witzel, C., Schwidergall, S., & Bohels, H. J. (2000). Peripubertal perturbations in elite gymnasts caused by sport specific training regimes and inadequate nutritional intake. *Int J Sports Med, 21.*

Weinberg, R. S., & Gould, D. (1999) *Foundations of Sport and Exercise Psychology.* Champaign, IL: Human Kinetics.

Weissensteiner, J. R. (2015). The importance of listening: engaging and incorporating the athlete's voice in theory and practice. *British Journal of Sports Medicine, 49*, 839–840.

Werner, P. H. (1994). *Teaching Children Gymnastics.* Champaign, IL: Human Kinetics.

Wertz, J., Galli, M., & Borchers, J. R. (2013). Achilles tendon rupture: risk assessment for aerial and ground athletes. *Sports Health, 5*(5), 407–409. doi:10.1177/1941738112472165

Westermann, R. W., Giblin, M., Vaske, A., Grosso, K., & Wolf, B. R. (2014). Evaluation of men's and women's gymnastics injuries: a 10-year observational study. *Sports Health: A Multidisciplinary Approach.* doi:10.1177/1941738114559705

Whitlock, S. (1989). When is spotting appropriate? *Gymnastics Safety Update, 4*(2), 5–6.

Whitlock, S. (1992). Hand spotting. *Gymnastics Safety Update, 7*(3), 3–4.

Wiedemann, R. (2016, 30 May). A full revolution. In the run-up to the Olympics, Simone Biles is transforming gymnastics. In *The New Yorker,* The sporting scene (Condè Nast). Retrieved 11 January 2017 from: www.newyorker.com/magazine/2016/05/30/simone-biles-is-the-best-gymnast-in-the-world

Wiemann, K. (1976a). Biomechanics of a dismount from the uneven bars. *Gymnast, 18*(1), 58–59.

Wiemann, K. (1976b). Mechanical effect of the forward leg snap. *International Gymnast, 18*(3), 50–51.

Wiemann, K. (1979). Theoretical reflections on exercise with "leg snap". *International Gymnast, 21*(2), 56–57.

Wilk, K. (1990). Dynamic muscle strength testing. In A. L. R. (ed.), *Muscle Strength Testing* (pp. 123–150). New York, NY: Churchill Livingstone.

Williams, A. M., Davids, K., & Williams, J. G. (1999). *Visual Perception and Action in Sport.* London: E & FN SPON.

Williams, J. M. (1986). Psychological characteristics of peak performance. In J. M. Williams (ed.), *Applied Sport Psychology: Personal Growth to Peak Performance* (pp. 123–132). Palo Alto, CA: Mayfield.

Willmore, J., & Costill, D. (1999). *Physiology of Sport and Exercise* (2 ed.). Champaign, IL: Human Kinetics.

Wilmore, J. H., & Costill, D. L. (2005). *Physiology of Sport and Exercise* (3 ed.). Champaign, IL: Human Kinetics.

Wilson, J. M., & Flanagan, E. P. (2008). The role of elastic energy in activities with high force and power requirements: a brief review. *J Strength Cond Res, 22*(5), 1705–1715.

Wine, J. D. (1971). Test anxiety and direction of attention. *Psychological Bulletin, 76*, 92–104. doi.10.1037/h0031332

Winter, E. M. (2005). Jumping: power or impulse. *Medicine and Science in Sports and Exercise, 37*, 523.

Winter, E. M., & Fowler, N. (2009). Exercise defined and quantified according to the Systeme International d'Unites. *J Sports Sci, 27*(5), 447–460.

Woodson, W. E., Tillman, B., & Tillman, P. (1992). *Human Factors Design Handbook.* New York, NY: McGraw-Hill.

Wulf, G., Chiviacowsky, S., & Drews, R. (2015). External focus and autonomy support: two important factors in motor learning have additive benefits. *Hum Mov Sci, 40*, 176–184. doi:10.1016/j.humov.2014.11.015

Wulf, G., Chiviacowsky, S., & Lewthwaite, R. (2010). Normative feedback effects on the learning of a timing task. *Research Quarterly for Exercise and Sport, 81*, 425–431.

Wulf, G., Horger, M., & Shea, C. H. (1999). Benefits of blocked over serial feedback on complex motor skill learning. *J Mot Behav, 31*(1), 95–103. doi:10.1080/00222899909601895

Wulf, G., & Lewthwaite, R. (2009). Conceptions of ability affect motor learning. *J Mot Behav, 41*(5), 461–467. doi:10.3200/35-08-083

Wulf, G., McConnel, N., Gartner, M., & Schwarz, A. (2002). Enhancing the learning of sport skills through external-focus feedback. *J Mot Behav, 34*(2), 171–182. doi:10.1080/00222890209601939

Wulf, G., & Shea, C. H. (2002). Principles derived from the study of simple skills do not generalize to complex skill learning. *Psychon Bull Rev, 9*(2), 185–211.

Wulf, G., Shea, C., & Lewthwaite, R. (2010). Motor skill learning and performance: a review of influential factors. *Medical Education, 44*(1), 75–84. doi:10.1111/j.1365-2923.2009.03421.x

Wulf, G., Shea, C. H., & Matschiner, S. (1998). Frequent feedback enhances complex motor skill learning. *J Mot Behav, 30*(2), 180–192. doi:10.1080/00222899809601335

Wulf, G., & Toole, T. (1999). Physical assistance devices in complex motor skill learning: benefits of a self-controlled practice schedule. *Research Quarterly for Exercise and Sport, 70*(3), 265–272. doi:10.1080/02701367.1999.10608045

Wüstemann, S., & Milbradt, J. (2008). Seitigkeit von Längsachsendrehungen. Vorstellung beim Kadertrainerseminar. Berlin.

Wyon, M. A., Koutedakis, Y., Wolman, R., Nevill, A. M., & Allen, N. (2014). The influence of winter vitamin D supplementation on muscle function and injury occurrence in elite ballet dancers: a controlled study. *Journal of Science and Medicine in Sport, 17*, 8–12.

Xin, P., & Li, W.-s. (2000). Kinematic characteristics of the Steinemannstemme move on the uneven bar: a case study. In Y. Hong & D. P. Johns (eds), *Proceedings of XVIII International Symposium on Biomechanics in Sports* (I ed., pp. 367–369). Hong Kong, China: The Chinese University of Hong Kong, International Society for Biomechanics in Sports.

Yan, X., Papadimitriou, I., Lidor, R., & Eynon, N. (2016). Nature versus nurture in determining athletic ability. *Medicine and Sport Science, 61*, 15–28. doi:10.1159/000445238

Yang, G., Rothrauff, B. B., & Tuan, R. S. (2013). Tendon and ligament regeneration and repair: clinical relevance and developmental paradigm. *Birth Defects Research. Part C, Embryo Today: Reviews, 99*(3), 203–222. doi:10.1002/bdrc.21041

Yeadon, M. R., & Trewartha, G. (2003). Control strategy for a hand balance. *Motor Control, 7*, 411–430.

Yeadon, M. R. (1993a). The biomechanics of twisting somersaults part I: rigid body motions. *Journal of Sports Sciences, 11*, 187–198.

Yeadon, M. R. (1993b). The biomechanics of twisting somersaults part II: contact twist. *Journal of Sports Sciences, 11*, 199–208.

Yeadon, M. R. (1993c). The biomechanics of twisting somersaults part III: aerial twist. *Journal of Sports Sciences, 11*, 209–218.

Yeadon, M. R. (1993d). The biomechanics of twisting somersaults part IV: partitioning performances using the tilt angle. *Journal of Sports Sciences, 11*, 219–225.

Yeadon, M. R., & Brewin, M. A. (2003). Optimised performance of the backward longswing on rings. *Journal of Biomechanics, 36*, 545–552.

Yeadon, M. R., King, M. A., & Hiley, M. J. (2005). Computer simulation of gymnastics skills. Paper presented at the 5th International Conference of the Association Française pour la Recherche en Activités Gymniques et Acrobatiques (AFRAGA).

Yeadon, M. R., & Knight, J. P. (2012). A virtual environment for learning to view during aerial movements. *Comput Methods Biomech Biomed Engin, 15*(9), 919–924. doi:10.1080/10255842.2011.566563

Yeadon, M. R., & Mikulcik, E. C. (1996). The control of non twisting somersaults using configuration changes. *Journal of Biomechanics, 29*, 1341–1348.

Yeowell, H. N., & Steinmann, B. (1993). Ehlers-Danlos Syndrome, Kyphoscoliotic Form. In R. A. Pagon, M. P. Adam, H. H. Ardinger, T. D. Bird, C. R. Dolan, C. T. Fong, R. J. H. Smith, & K. Stephens (eds), *GeneReviews(R)*. Seattle, WA: University of Washington, Seattle.

Yerkes, R. M., & Dodson, J. D. (1908). The relation of strength of stimulus to rapidity of habit formation. *Journal of Comparative Neurology of Psychology, 18*, 459–482.

Yoshida, T., Udo, M., Chida, M., Ichioka, M., Makiguchi, K., & Yamaguchi, T. (1990). Specificity of physiological adaptation to endurance training in distance runners and competitive walkers. *European Journal of Applied Physiology and Occupational Physiology, 61*(3–4), 197–201.

Young, B. W., & Salmela, J. H. (2002). Perceptions of training and deliberate practice of middle distance runners. *International Journal of Sport Psychology, 33*(2), 167–181.

Young, W. B., Prior, J. F., & Wilson, G. J. (1995). Effects of instructions on characteristics on countermovement and drop jump performance. *Journal of Strength and Conditionning Research, 9*(4), 232–236.

Young, W., Wilson, G., & Byrne, C. (1999a). Relationship between strength qualities and performance in standing and run-up vertical jumps. *J Sports Med Phys Fitness, 39*(4), 285–293.

Young, W. B., Wilson, G. J., & Byrne, C. (1999b). A comparison of drop jump training methods: effects on leg extensor strength qualities and jumping performance. *Int J Sports Med, 20*(5), 295–303.

Yu, B., Lin, C. F., & Garrett, W. E. (2006). Lower extremity biomechanics during the landing of a stop-jump task. *Clin Biomech, 21*(3), 297–305.

Zaggelidis, S., Martinidis, K., Zaggelidis, G., & Mitropoulou, T. (2005). Nutritional supplements use in elite gymnasts. *Physical Training. Electronic Journals of Martial Arts and Sciences.* Retrieved from: http://ejmas.com

Zaichkowsky, L., & Takenaka, K. (1993). Optimizing arousal levels. In R. N. Singer, M. Murphey, & L. K. Tennant (eds), *Handbook of Research on Sport Psychology* (pp. 511–527). New York, NY: Macmillan.

Zaitseva, I. P., Skalny, A. A., Tinkov, A. A., Berezkina, E. S., Grabeklis, A. R., Nikonorov, A. A., & Skalny, A. V. (2015). Blood essential trace elements and vitamins in students with different physical activity. *Pakistan Journal of Nutrition, 14*(10), 721–726.

Zar, J. H. (1984). The Latin square experimental design-ultiway factorial analysis of variance. In *Biostatistical Analysis* (2 ed., p. 248). Englewoods Cliff: Prentice Hall.

Zernicke, R. F., & Loitz, B. J. (1992). Exercise-related adaptations in connective tissue. In P. V. Komi (ed.), *Strength and Power in Sport* (pp. 77–95). Oxford, UK: Blackwell Scientific Publications.

# INDEX